Units of Measurement

Unit	Definition
Ampere (A)	The unit of electric current; equal to 1 coulomb moving past a given point in 1 second
Coulomb (C)	The quantity of electric charge that flows in 1 second under 1 ampere of current; the quantity of charge carried by 6.25×10^{18} electrons
Decibel (dB)	A logarithmic unit for expressing the difference between two sound, power, voltage, or current levels; equal to 10 log of the power ratio or 20 log of the voltage or current ratio; equal to 0.1 bel
Farad (F)	The amount of capacitance that occurs when 1 volt applied across the plates of a capacitor produces a charge of 1 coulomb
Henry (H)	The amount of inductance that will produce 1 volt across an inductor when the current through it changes at the rate of 1 ampere per second
Hertz (Hz)	The SI unit of frequency; equal to 1 cycle per second
Joule (J)	The SI unit of work; equal to 0.7367 foot-pounds
Ohm (Ω)	The electrical resistance that allows 1 ampere to flow when 1 volt is applied
Siemen (S)	The unit of conductance; the reciprocal of 1 ohm
Tesla (T)	The unit of magnetic flux density; equal to 1 weber per square meter
Volt (V)	A unit that indicates the potential difference between two charged bodies if 1 joule of work is required to transfer 1 coulomb of charge from one point to another
Watt (W)	The power dissipated when 1 ampere flows under 1 volt of applied emf; equal to 1 joule per second
Weber (Wb)	The SI unit of magnetic flux; a rate of change of webers per second in a single-turn coil that induces an emf of 1 volt per second

Principles of dc and ac Circuits

Second Edition

George J. Angerbauer

College of San Mateo

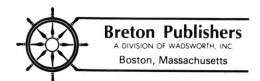

Breton Publishers
A DIVISION OF WADSWORTH, INC.
Boston, Massachusetts

Breton Publishers
A Division of Wadsworth, Inc.

Library of Congress Cataloging in Publication Data

Angerbauer, George J., 1918–
    Principles of DC and AC circuits.

    Includes index.
    1. Electric circuits.   I. Title.   II. Title: Principles of D.C. and A.C. circuits.
TK454.A53 1985   621.319′1   84–11115
ISBN 0–534–04203–1

Printed in the United States of America
2 3 4 5 6 7 8 9–89 88 87 86 85

ISBN 0-534-04203-1

Principles of dc and ac Circuits was sponsored by George J. Horesta. It was prepared for production by Jean T. Peck. Carol Beal served as copy editor. Ellie Connolly served as art coordinator and interior designer. The cover designer was Trisha Hanlon; cover photo © Eric Roth/THE PICTURE CUBE. The book was set in Century Expanded by A & B Typesetters, Inc.; printing and binding was by Halliday Lithograph.

In memory of my father and mother

Contents

vi
Contents

x
Contents

Preface

Principles of dc and ac Circuits has been written for first-year students studying electronics technology. It is applicable to colleges, technical and vocational institutes, academies, and individuals seeking to acquire a fundamental knowledge of dc and ac principles. Based on many years of teaching experience, it minimizes nonessentials and concentrates on fundamental concepts. Each chapter has been written so that the material progresses from established principles to advanced concepts. It is comprehensive yet attempts to maintain student interest throughout.

Since technicians are required to analyze circuits in order to determine proper operation, the presentation has been designed to stimulate interest and develop an analytical approach on the part of the student toward troubleshooting and problem solving. Many worked examples – many of which are new in this Second Edition – are provided throughout each chapter to illustrate the concepts presented.

A comprehensive summary follows each chapter. Frequent review of the summaries will provide students with an excellent learning aid. Following the summary in each chapter, a progress test is included so that students may determine if they have mastered the subject matter. Numerical exercise problems – many of which have also been added to this Second Edition – are included at the end of each chapter to provide ample problem-solving experience. They emphasize, for the most part, practical issues that the electronics technician will encounter. Answers to odd-numbered problems are found at the back of the text. Answers and selected solutions to all problems are contained in the Instructor's Manual, which is available from the publisher to those instructors who have adopted the text for use in class.

Elementary concepts of Kirchhoff's laws are introduced in the presentation before studying Ohm's law. This innovative approach has proved very helpful to the student in conceptualizing current and voltage distribution in electric/electronics circuits. A more rigorous application of Kirchhoff's laws is found in Chapter 8, "Network Analysis of dc Circuits."

The text material has been arranged to provide a comprehensive foundation for continued studies in electronics. For example, the use of Thévenin's, Norton's, and the superposition theorems provides a convenient and practical method of analyzing active circuits (transistors, linear integrated circuits, etc.) under varying load conditions.

The International System of Units (SI) has been used throughout the text. As mathematics has been used as a tool to explain the relationship between quantities, a working knowledge of algebra and trigonometry is a prerequisite. Some knowledge of logarithms will also help the student in understanding the information on filters. Cumbersome derivations and superfluous formulas have been eliminated so that the reader does not become confused and lose sight of the real objectives.

Content/coverage changes in the Second Edition include a new chapter on power supplies (Chapter 22) and a totally revised and expanded chapter on instruments (Chapter 10) that emphasizes the use of the oscilloscope. Complex dc/ac circuits, voltage dividers, and coupling and by-pass circuits have also received revised coverage, and some basic transistor circuits have been introduced to provide practical illustrations of current and voltage in circuits.

A comprehensive laboratory manual has been prepared to complement the text and is now available with the Second Edition. Based on a number of years of use and refinement, it modularizes the text *Principles of dc and ac Circuits* into many experiments that reinforce and verify the information and concepts studied. An updated version of the ancillary Study Guide is also available.

For the many timely suggestions and reviews of various parts of the manuscript, I wish to thank the following individuals: Dr. Robert I. Eversoll, Western Kentucky University, KY; William H. Mowbray, Community College of Rhode Island, RI; Jack R. Braun, Burlington County College, NJ; Albert J. Marcarelli, Connecticut School of Electronics, CT; Kent Tunnell, Northeastern Oklahoma A&M, OK; Melvin F. Dostal, Marshalltown Community College, IA; Robert T. Campbell and William L. Burke, Jr., North Shore Community College, MA; Clayton Peary and David Stone, Central Maine Vocational Technical Institute, ME; Lloyd Harrich, College of San Mateo, CA; J.R. Long, Virginia Highlands Community College, VA; and John Lotz, Kalamazoo Valley Community College, MI.

I also wish to give special thanks in this regard to my colleagues Mr. Don Beaty, Mr. George Bramlett, Mr. Tullio Bertini, Mr. Vern Skovgaard, and Mr. Roy Holmgren. Finally, I wish to express appreciation to my wife, Gerry, and my daughter, Jeri, for their tireless efforts in typing the manuscript.

⌐Introduction

Welcome to the World of Electronics

Actually, you are already living in a world that is greatly influenced by electronics and need no official welcome. Few people in the world today can escape the impact of electricity and electronics in thier lives. Where can you go and not find transistorized radios, TV receivers, cassette players, cordless telephones, hand-held calculators, or home and office computers? Consider how electronics has "reduced" the size of the world, in terms of communication, in just the past fifty years! With the aid of transistorized radios and satellites, the news media can put us in instant touch with almost any part of the world. On our TV screens, we are able to see the peoples of virtually every nation and learn about their countries and customs. Now, even the moon and the planets are within our reach.

The operation of the countless electric and electronic devices we enjoy every day is based on fundamental laws, some of which were discovered over one hundred years ago and are still integral to equipment design today. A study and appreciation of the underlying principles of these laws is the purpose of this text.

We all use electricity and electronics essentially as a tool to expand our awareness of and experience with the physical world around us. Few fields of endeavor have escaped the impact of these disciplines: engineering, science, business, manufacturing, automotive servicing, farming, entertainment, education, communications, medicine, the military, and others. The end is not yet in sight. The ability of the computer to solve complex problems in a few seconds, for example, has extended our ability to develop new products, and there will be many more important discoveries and new uses for electric and electronic devices in the years to come. This evolution will no doubt be based on many of the principles explained in this text.

What Is an Electronics Technician?

Your studies should lead you to an interesting and rewarding career, whether as an electronics or engineering technician. An electronics technician is a person who has developed computational skills, analytical abilities, and measurement techniques to work with electronic equipment. Generally, he or she is assigned either to an engineering group to assist in the development of prototype equipment, where work will involve the assembly, testing, and calibration of these prototypes; or to a testing, measuring, and calibration facility, where the manufactured equipment is prepared for final customer delivery. To qualify for these positions, you will study the circuit actions of resistors, capacitors, and inductors, separately and in combinations, when subjected to direct and alternating currents.

The semiprofessional (engineering) technician typically has had two or more years of post–high school education in a community college, technical institute, or extension division of a university. Some skilled industrial technicians have achieved their positions without formal college degrees, but with considerable self-study and discipline.

Electronics Technicians in Industry

Electronics technicians are employed in every kind of developmental and manufacturing facility—from companies with only a few employees to large factories employing thousands of individuals.

Technicians perform a variety of work within this broad range of commercial enterprise. The measurement of voltage and current on electronic circuits is one of the jobs performed. This measurement can be accomplished with an instrument similar to the one in Figure A, called a *digital multimeter*. This sophisticated measuring instrument finds wide-

Figure A

Digital multimeter (Courtesy of John Fluke Manufacturing Company)

spread use in industry. Its use is based on the digital multimeter's ability to measure voltage resistance and current over a wide range of values.

Students preparing to become electronics technicians must learn how to perform numerous calculations accurately. While the engineer is responsible for the circuit or system design, the technician should be completely familiar with normal circuit performance and able to analyze or troubleshoot malfunctions. Scientific pocket calculators, like the one shown in Figure B, are accurate timesavers that allow both student and technician to concentrate on circuit parameters rather than step-by-step mathematical computations. Electronics and electrical engineers, laboratory technicians, and university and industrial researchers often use a pocket-sized calculator, like the type shown in Figure C, that has full programming capability. The unit shown features a series of prerecorded mathematical programs and allows the individual to "write" a special program, if needed, to facilitate the solution of recurring problems.

Figure B

Scientific pocket calculator (Courtesy of Hewlett-Packard Company)

Figure C

Fully programmable pocket calculator (Courtesy of Hewlett-Packard Company)

Your Future in Electronics

Qualifying for a career in electronics can provide great rewards, both in terms of financial success and personal satisfaction. Excitement and security are included too – the field is continually growing, with new products being developed or new markets appearing for devices previously used only in a very specialized sense. The future and potential growth of the electronics industry means that your future as an electronics technician is promising.

In many cases, technicians are promoted to supervisory positions, where they act as service, quality control, or quality assurance engineers. Major firms often send their technicians to foreign countries to check out manufacturing methods and ensure that finished products comply with American standards and design requirements, which means that you can conceivably spend April in Paris with all expenses paid. You can also use your career as a technician as a stepping stone to more education and thus equip yourself for further research, development, and prototype design. There are, in short, many opportunities available for the person who is willing to increase his or her knowledge and, as the cliché goes, apply diligently to the job at hand. Keeping your eyes open for extra opportunities also won't hurt a bit!

Principles
of dc
and ac
Circuits

Second Edition

1

Current and Voltage

Objectives

After studying this chapter, you will be able to:

Describe the nature of electricity.

Discuss the force that makes current move from one point to another around a closed circuit of metallic conductors.

Introduction

A study of direct and alternating currents involves a study of electricity. While we are hard-pressed for an accurate definition of what electricity really is, we know its essential characteristics and how to use and control it in countless applications. We know how to generate it by mechanical, chemical, thermal, frictional, photoelectrical, and piezoelectric methods.

Basic to understanding electrical apparatus and electronic equipment is a knowledge of how electricity behaves under various conditions. We will begin by examining the elementary structure of the atom and learning about its electrical charges.

Applications

All components in electrical and electronic equipment have electrical potential or voltage applied to them that forces electricity to flow through them. This flow of electricity causes electric motors to rotate, electric lamps to light up, radio and television receivers to operate, electronic calculators to perform their operations, and numerous other electric and electronic devices to function. Without electricity, our style of life would be much more primitive than the one to which we have become accustomed.

1–1
Atoms and Their Structure

> **Matter**   Any substance that has weight (mass) and occupies space.

    An understanding of the positive and negative charges of electricity begins with a study of the electrical structure of matter. **Matter** is any substance that has weight (mass) and occupies space. It is found in any one of three states: *solid, liquid, gaseous.*

Elements

All matter consists of one or more elements (of which there are slightly more than one hundred). Typical examples of metallic elements are copper, aluminum, iron, and gold. Several gaseous elements are oxygen, helium, and nitrogen. Other elements, such as sulfur, silicon, and germanium, are neither metallic nor gaseous. Silicon and germanium are used extensively in transistors, diodes, integrated circuits, and other solid-state devices.

    A substance made up of two or more different kinds of elements is called a *compound.* An example is water (H_2O), which is made up of two atoms of hydrogen and one of oxygen. Now, if a drop of water could be continually divided until we have the smallest particle possible and still have water, the particle would be called a molecule. Only with the aid of the most powerful electron microscopes have scientists been able to distinguish the outline of some of the largest molecules present in compounds.

Atoms

The atoms of a particular element all have the same average mass but differ from the average mass of the atoms of all other elements. Therefore, *the number of different atoms is equal to the number of elements.* Just as thousands of words may be formed by combining different letters of the alphabet, thousands of different materials can be made by combining the various atoms.

Nucleus   The center of an atom, containing proton(s) and neutron(s).

Proton   A positively charged particle with considerable mass.

Neutron   A subatomic particle that is neither positively nor negatively charged and that has a mass approximately equal to that of a proton.

The center of an atom, called the **nucleus**, contains several particles. Of most significance to our study of electric circuits are the proton and the neutron. **Protons** are positively charged particles that have considerable mass, relatively speaking. Each proton weighs about 1835 times more than an electron associated with the atom. **Neutrons** are neither positively nor negatively charged; they are neutral. Consequently, the net charge of the nucleus is positive because of the presence of protons. Even though the presence of a neutron does not contribute to the flow of electricity, it adds substantially to the weight of an atom, since its mass is essentially equal to the mass of the proton.

Electron   A negatively charged subatomic particle with much less mass than a proton but with a charge equal and opposite to the charge of a proton.

Orbiting the nucleus (like planets revolving around a sun) are tiny negative particles called **electrons**. Electrons travel at high velocities and have charges that are equal but opposite to the charge of a proton.

You may wonder why the electron doesn't fly out of its orbit. It doesn't because the attraction between the electron and the proton is exactly equal to the centrifugal force acting upon it. Figure 1–1 shows

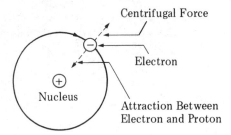

Figure 1-1

Balancing forces within the atom

this balance of forces in a hydrogen atom, which has one proton and one electron.

That electrons and protons have equal but opposite charges does not imply that they are the same size or of equal mass. Scientists have been able to calculate the weight of an electron, and they found that it weighs about 1/1835 as much as a proton. However, the proton is smaller than the electron and has a greater density or mass. Because of the very small mass of an electron, it can be easily and rapidly moved around in an electric or electronic circuit.

Atomic Structure

The simplest of all atoms is hydrogen, which consists of one proton and one orbiting electron. The atoms of all other elements contain more than one proton and one electron and are therefore more complex. Take an atom of carbon, for example, as represented in Figure 1-2. Notice that the carbon atom has six electrons and six protons.

Energy bands The arrangement of electrons into definite shells around the nucleus of an atom.

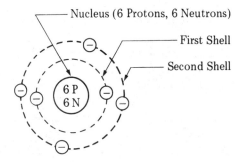

Figure 1-2

Electric charges in a carbon atom

A definite arrangement exists between the number of electrons orbiting the nucleus and the *number of shells* (often referred to as **energy bands**) a given atom may have. The first shell can have a maximum of 2 electrons. The second may have any number up to 8. The third band may have a maximum of 18. The maximum number of electrons that any shell can have is $2n^2$, where n is the number of the shell, counting from the nucleus. Shells are also designated by letters, as shown in Figure 1–3, which represents a silicon atom (used extensively in transistors and other solid-state devices) with 4 valence electrons.

Valence   The number of electrons in the atom's outermost shell, which determines the atom's ability to gain or lose an electron, which, in turn, determines the electrical and chemical properties of the atom.

Free electrons   Valence electrons that are not strongly attached to the nucleus and are relatively free to move from atom to atom.

The number of electrons in the outermost shell determines the **valence** of the atom. These electrons exist in the valence band and are called *valence electrons*. The valence of an atom determines its ability to gain or lose an electron, which, in turn, determines the electrical and chemical properties of the atom. An atom lacking only one or two electrons for a complete shell can easily gain electrons to complete the shell. If, however, the outer shell is less than half full, the atom can rather

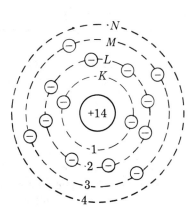

Figure 1–3

Shell designation (silicon atom)

easily lose valence electrons. For example, copper has only one valence electron, and it is not strongly attached to the nucleus. Such electrons are called **free electrons**, and they are relatively free to move from atom to atom.

I–2
Static Electricity

Charged Bodies

If a neutral atom *acquires* a free electron from some other atom or object, it will become *negatively charged*. If it should *lose* an electron, it will become *positively charged*. This process can be demonstrated by rubbing a glass rod with a silk cloth. Some of the loosely bound electrons in the glass rod will be transferred to the silk; once the transfer takes place, the glass becomes positively charged. The silk, having acquired some electrons, becomes negatively charged.

Static Charges

If a glass rod is charged and brought near small bits of paper, they will jump up to the rod. Thin pieces of many kinds of plastic material can also be easily charged.

> **Dielectric field** or **electrostatic field**   The field of force surrounding any charged body.

> **Static electricity**   Opposite static charges that occur when friction between two bodies causes electrons to transfer from one body to the other.

A field of force called a **dielectric** or **electrostatic field** surrounds any charged body (see Figure 1–4). The extent of this field is a function of the strength of the electric charge, which, in turn, depends on how many electrons have either been added to or subtracted from the

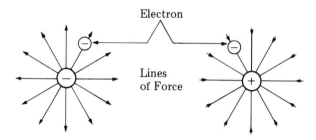

Figure 1-4

Dielectric field surrounding negatively and positively charged bodies

bodies. Electricity in this case is **static electricity**, because the electrons do not move between the charged bodies.

Should an electron be placed within the negative dielectric field (Figure 1-4), the lines of force surrounding the charged body would *repel* the electron. This repelling force is stronger near the charged body and diminishes with the square of the distance. If an electron were placed within the electrostatic field around a positively charged body, the field would *attract* the electron, and the electron would be drawn toward the positive charge.

Two small objects that are oppositely charged and brought close to each other will be attracted to each other, as shown in Figure 1-5a. If the objects are similarly charged, they will repel each other, as shown in Figure 1-5b. These phenomena indicate a fundamental law regarding electrostatic charges: *Unlike charges attract, and like charges repel.* Electrostatic fields are important to the operation of such electronic devices as capacitors, field effect transistors, and vacuum tubes.

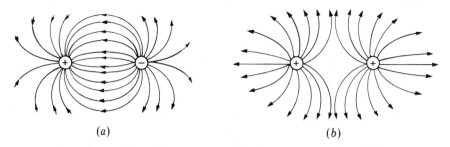

(a)   (b)

Figure 1-5

Lines of force: (a) Between unlike charges. (b) Between like charges.

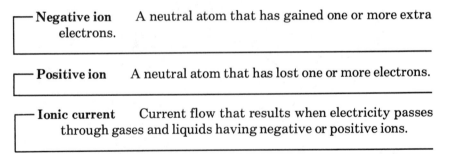

Negative ion　A neutral atom that has gained one or more extra electrons.

Positive ion　A neutral atom that has lost one or more electrons.

Ionic current　Current flow that results when electricity passes through gases and liquids having negative or positive ions.

When a neutral atom gains one or more extra electrons, it becomes a **negative ion**. If it loses one or more electrons, it acquires a net positive charge and is called a **positive ion**. *Ions*, then, are *atoms that have gained or lost electrons*. The presence of negative or positive ions allows electricity to pass through certain gases and liquids when a voltage is applied. The resulting current flow is called **ionic current**.

1-3
The Coulomb

Dynamic electricity　Electrons in motion; electrons flowing from a negative to a positive charge until the charges are equalized.

As you learned in Section 1-2, a charge between two bodies without a conducting path between them produces static electricity. By contrast, if a copper wire were placed between two differently charged bodies, electrons would flow from the negative to the positive charge until they were equalized. Electrons in motion constitute **dynamic electricity**, which characterizes most electric and electronic circuits.

The quantity of electric charge possessed by a single electron is too small for practical purposes. A practical unit is needed to represent the charge carried by many billions of electrons encountered in most electric circuits. The coulomb is the basic unit of quantity of electric charge. Scientists have determined that a *coulomb represents the quantity of electric charge carried by* 6.25×10^{18} *electrons*. (Refer to the Appendix for a brief explanation of the use of powers of 10.)

Letter symbols are used to represent variable terms in algebraic equations, and abbreviations (or unit symbols) are used for units of measurement. The letter symbol Q represents the quantity of electric charge, and C is the unit symbol for the coulomb.

The coulomb is named after Charles A. Coulomb (1736–1806), a Frenchman who demonstrated that, in both electrical and gravitational attraction, force diminishes with the square of the distance between two bodies.

Example 1–1

A neutral object receives a charge of 18.75×10^{18} electrons. What is its charge in coulombs?

Solution

$$-Q = \frac{18.75 \times 10^{18}}{6.25 \times 10^{18}} = 3 \text{ C}$$

The minus sign implies the charge is negative.

Example 1–2

A neutral dielectric material loses 3.125×10^{18} electrons. What is its charge in coulombs?

Solution

This quantity is one-half coulomb. Therefore, the dielectric will have $Q = 0.5 \text{ C}$.

Example 1–3

A dielectric with a plus charge of 2 C has 6.25×10^{18} electrons added. What is its new charge?

Solution

The minus charge of 1 C added to the plus 2 C leaves plus 1 C on the dielectric.

1–4
Electromotive Force

We have seen how we can produce a charge by friction, as when we rub a glass rod with a silk cloth. The energy expended in rubbing the glass rod with the silk cloth causes some of the loosely bound electrons in the glass rod to transfer to the silk. Thus, the glass rod acquires a positive charge. The charge between two bodies has the ability to do work, since any charged body exerts a force on another body having a greater or lesser charge.

Work

Work is defined in terms of force applied and the distance through which it acts. In the English system, the practical unit of work is the foot-pound (ft-lb). The equation for work is as follows:

$$W = Fs$$

where
$$W = \text{work}$$
$$F = \text{force}$$
$$s = \text{distance}$$

If a person pushes a piano with a force of 50 ft-lb and it moves 10 ft along the floor, then 500 ft-lb of work has been expended. If a 60 lb weight is lifted from the floor to a platform 4 ft high, the work done in lifting the weight is 60 lb $\times$ 4 ft = 240 ft-lb. In electricity, work is done, for example, when an electromagnetic force moves an electron beam across a TV screen.

Potential   The ability to do work caused by the charge between two bodies.

Potential difference   The difference between the charges on two bodies; the excess or lack of electrons at one point compared with the excess or lack of electrons at some other point.

In electricity, we refer to the ability to do work as **potential**. The difference between the charges on the two bodies is called **potential difference**.

Two things must be kept in mind regarding this potential difference. First, it is a force that may or may not cause electrons to flow between two differently charged bodies. When they have an electric path between them, electrons will flow from the negative to the positive charge until they are equalized. When they have no electric path, the potential difference still exists. An example of a charged body with no electric path is a battery (see Figure 1–6a) that has nothing connected

(a)   (b)

Figure 1-6

Battery: (a) Typical batteries. (b) Symbol for battery.

between its terminals. (The symbol of a battery is shown in Figure 1-6b.) Second, a potential difference may not exist at one point only. The difference in the charge of one point is noted only by comparison with the charge of some other point. *Potential difference is the excess or lack of electrons at one point compared with the excess or lack at some other point.*

Volt

> **Electromotive force (emf)**   The force, or pressure, resulting from a potential difference.

The force, or pressure, resulting from a potential difference is called the **electromotive force**, abbreviated **emf**. A special unit has been derived to express potential difference. The SI unit for potential difference is the volt. The unit symbol for volt is V. The letter symbol for potential difference is E or V. Potential difference, or voltage, is measured by an instrument called a voltmeter.

The distinction between E and V will be explained in subsequent paragraphs. The volt is named in honor of Alessandro Volta (1745–1827), who is most famous for inventing the first battery.

Specifically, a *volt is defined as the potential difference between two charged bodies* if one joule (J) of work is required to transfer one coulomb (C) of charge from one point to another in an electric circuit.

Joule

The joule is the practical unit for measuring work in the metric system. One joule (J) equals 0.7376 foot-pound. In electrical terms, *one joule equals one watt per second*, which is a small unit. Electric power is usually measured in watts per hour (W/h) or in kilowatts per hour (kW/h) when larger amounts of power are involved.

The relationship between volts and joules may be expressed mathematically as follows:

$$E \text{ (or } V) = \frac{W \text{ (joules)}}{Q \text{ (coulombs)}} \qquad \text{1–1}$$

where E (or V) is the potential difference in volts, W is the energy in joules, and Q is the quantity of electric charge in coulombs.

Energy

Energy may be thought of as anything that can be converted into work. When we say that an object has energy, we mean that it is capable of exerting a force on another object in order to do work on it. Conversely, if we do work on some object, we have added to it an amount of energy equal to the work done. The units of energy are the same as those for work; namely, the joule and the foot-pound.

There are two kinds of energy: kinetic and potential. **Kinetic energy** is possessed by a body by virtue of its motion. **Potential energy** is possessed by a system by virtue of its position or condition. Examples of kinetic energy are a moving car, bullet, or flywheel. Examples of potential energy are a compressed spring or a lifted object.

The capacity to do work is also called energy. A battery has electric energy since it can do work if it is connected to a small lamp (as in a flashlight) or a portable AM/FM radio.

Example 1–4

What is the voltage between two points if 6 J is required to cause 2 C to flow?

Solution

$$E \text{ (or } V) = \frac{6}{2} = 3 \text{ V}$$

When we referred to emf's, we used the term *potential differ-ence* (PD). There are two kinds of potential difference. A potential *rise* is provided by a battery or other voltage source, and a potential *fall* occurs in the electric conductors and the load (lamps, motors) that are con-nected to it. Numerically, these two potential differences are equal, since the potential fall *must* equal the potential rise. The same unit of measurement can apply to both. These potentials are often called volt-ages, but the use of that term leaves doubt about whether the potential is a rise or fall.

A source of voltage, such as a battery or a generator, produces an emf (rise in potential) that forces electrons (current) around a circuit. The letter symbol for emf is E. The drop in potential that occurs as elec-trons flow through the external circuit is usually called *voltage drop*. The letter symbol for voltage drop is V.

Figure 1–7 illustrates the difference between emf and voltage drop. With the switch (SW) open, the voltmeter across the battery V_1 indicates the potential generated by the battery as a result of its inter-nal chemical action. The voltmeter across the lamp V_2 registers zero since no current is flowing through the lamp and no voltage drop is cre-ated. Now, if the switch is closed, the battery emf forces current through the lamp, creating a voltage drop equal to the battery poten-tial. Therefore, V_2 reads the same as V_1. A voltage drop exists across the lamp only when electrons are passing through it, as the electrons return the potential difference in the circuit to zero. In this text, we will use the symbol E to represent a supply potential, or emf, and the symbol V to represent a voltage drop.

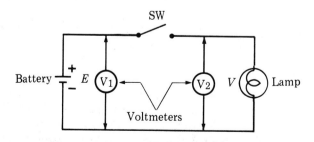

Figure 1–7

Simple circuit illustrating the difference between emf and voltage drop

1–5
Electric Current

Movement of Electrical Charges

—Electric current   A flow of electrons through a conductor or
other device.

The word *current* implies running or flowing; an **electric current** is a flow of electrons. To define an electric current, we must note two things:

1. A certain quantity of electrical charge is involved.
2. It is moved in a certain period of time.

Therefore, in measuring an electric current, we are determining the *quantity of charge that flows in a definite interval of time*, or the *rate of flow* of electric charge. The symbol for electric current is I, from the French word *intensité*, or intensity of electron flow.

No electric current flows through a wire if no voltage is across it. For example, in a piece of copper wire, an extremely large number of free electrons move aimlessly from atom to atom. Because this random movement results in no uniform movement of electrons from one end of the wire to the other, there is no current flow.

If both ends of a copper wire (or any wire) are connected to the same terminal of a battery, they are at the same potential. No current flows because there is no potential difference, or voltage, across the wire. Current will flow through a wire only if one end is connected to a point of positive charge and one end is connected to a point of negative charge.

Current, then, is a drifting or movement of electrons in one general direction. We can refer to current as *a movement of electrical charges*.

Two schools of thought exist regarding the direction of current. Some people hold that current flows from the negative terminal of the voltage source, around the circuit or current path, to the positive terminal (*electron flow*). Others are more comfortable in assuming that the current is in the opposite direction (*conventional current flow*). In either case, the calculated current, the voltage drops across circuit elements, and their polarities (all to be discussed in detail in Chapter 3) are the same. This text will use the idea of electron flow.

Ampere

The practical unit for measuring electric current is the ampere, abbreviated A. The ampere is named in honor of André Ampère (1775–1836), who laid the foundation for electrodynamics with his discovery of the dynamics of electrical current flow. The 1881 Electrical Congress in Paris agreed to measure volume of current flow in amperes after him. *One ampere is equal to one coulomb moving past a given point in one second.* The ampere may also be defined as a unit of electric current that will deposit 1.118 milligrams (mg) of silver per second in a standard type of electrolytic solution. The instrument used to measure current (or amperage) is an ammeter. The use of this instrument will be explored in Chapter 10.

Because the current I is the amount of electrical charge Q per unit time t, we have the following equation:

$$I = \frac{Q \text{ (coulombs)}}{t \text{ (seconds)}} \qquad \text{1–2}$$

where I is current in amperes, Q is the charge transferred in coulombs, and t is the time in seconds (s) during which a constant or average current of I amperes is moving past a given point. This formula shows that *ampere* is just another term for *coulombs per second.*

Example 1–5

A photographic strobe lamp will fire when the capacitor (a device for storing electrical charges) to which it is connected charges to $0.5Q$ in 2 s. What will the current be at the moment of capacitor discharge?

Solution

$$I = \frac{Q}{t} = \frac{0.5}{2} = 0.25 \text{ A}$$

Ammeters

An ammeter has two terminals, one marked positive (+) and the other negative (−). The symbol for an ammeter is shown in Figure 1–8.

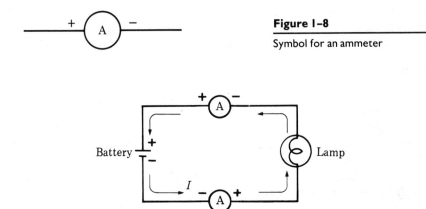

Figure 1-8

Symbol for an ammeter

Figure 1–9

Simple series circuit showing two possible ammeter connections

> ─ **Series circuit**   A circuit in which resistances or other compo-
>   nents are connected end to end so that only one path exists
>   through which current can pass.

Because the ammeter is designed to measure the current through a wire or circuit (to be discussed later in this chapter), it must be connected *in series* with it. That is, all the current through the circuit must flow through the meter. Therefore, to measure current, we must break, or open, the circuit and insert the ammeter. *The positive terminal of the meter must be connected to the positive side of the break, and the negative terminal to the negative, or minus, side.* If the connections should be accidentally reversed, the meter will try to indicate current flow in the reverse direction and may be damaged. Two possible ammeter connections are shown in Figure 1-9, which illustrates a simple **series circuit** containing a lamp bulb and its symbol on a schematic diagram.

1–6
Resistance

> ─ **Conductors**   Materials with many free electrons that facilitate
>   the passage of electric current.

Resistance   The opposition a material or device offers to electric current.

Insulators   Materials that have no free electrons and thus prevent the passage of electric current.

A good electrical **conductor** has many free electrons. Nevertheless, some opposition is encountered as an electric current passes through it. If the current passing through the conductor is excessive, it gets warm or even hot. This heat indicates that there is some opposition, or **resistance**, to current. The less resistance a conductor has, the better it can conduct an electric current. **Insulators**, on the other hand, have very few free electrons and will not allow current to pass through them. Electrical wires have insulation around them to prevent making electrical contact (shorts) with other wires.

Some conductors are better than others. Silver is the best conductor, but it is expensive and is only used in specialized applications. Copper is not quite as good as silver but is much less expensive and is very extensively used. Homes, buildings, automobiles, and airplanes are all wired with copper wires and cables containing a number of separate, insulated conductors. Gold is not as good a conductor as copper but has certain qualities that require its use in specialized applications. Chapter 2 has more details on conductors and insulators.

Resistance, whose symbol is R, is a property of conductors and other circuit elements that, depending on their dimensions, material, and temperature, determine the amount of current produced by a given difference of electric potential. The unit of resistance is the ohm, named for the German physicist Georg Simon Ohm (1787–1854). The ohm is symbolized by the Greek letter omega (Ω). Some typical resistors are shown in Figure 1-10a, and the schematic symbol is shown in Figure 1-10b.

One ohm is defined as the electrical resistance that allows exactly one ampere to flow when exactly one volt is applied. Expressed mathematically, the equation for resistance is as follows:

$$R \text{ (ohms)} = \frac{E \text{ (volts)}}{I \text{ (amperes)}} \qquad \textbf{1–3}$$

The following examples give an idea of the relative resistances of several different items. A piece of copper wire used in electronic

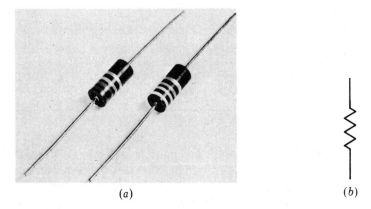

(a)   (b)

Figure 1–10

Resistor: (a) Typical carbon composition resistors. (b) Schematic symbol for resistance.

equipment has about 0.01 Ω/ft. A 100 W lamp has 144 Ω when connected to a 120 V power line. The resistance element in a small electric heater has about 9 Ω resistance.

1-7
Closed and Open Circuits

Closed circuit A continuous path from the voltage source through interconnecting wiring and device(s) and back to the source.

For operation, an electric lamp or other device must be connected to a source of voltage. The connections involved must provide a *continuous* path from the voltage source, through interconnecting wiring, the device, and back to the source. Such a path is called a closed, or series, circuit. An example of a simple series circuit is shown in Figure 1-11. This circuit is a **closed circuit** because there is a *complete* path for the electrons to flow through. Turning on a light switch in your home is an example of closing a circuit to make the light come on. Nothing happens in an electric circuit until it is closed.

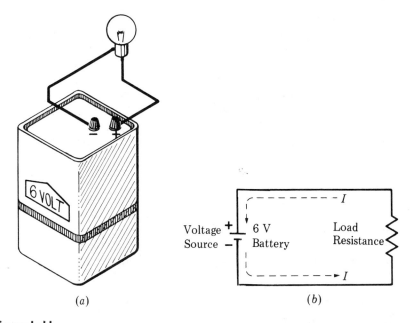

(a)   (b)

Figure 1–11

Simple closed circuit: (a) Pictorial view showing battery and lamp. (b) Schematic diagram.

Open circuit   A circuit that has an incomplete path so that no current can flow.

An **open circuit** is one that has an *incomplete* path so that no current can flow. Consider the circuit in Figure 1–12, which is the basic circuit shown in Figure 1–11 with a switch added. The circuit is open because the switch provides a break, preventing current from flowing. Switches are used to control circuits, that is, to open or close them.

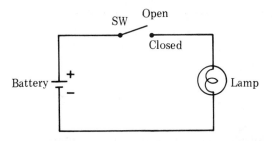

Figure 1–12

Simple example of an open circuit

Purpose of Voltage Source

The voltage source must have the capability of providing a potential difference and a supply of electric current around the circuit. To be of practical value, the supply should have the capacity to provide the current for relatively long periods of time. If you refer to Figure 1–11, you will see that the supply voltage appears *across the load*; the voltage does not flow through the load. Current will continue through the load as long as there is a potential difference.

Direction of Current Flow

In Figure 1–11, we have assumed a battery as the voltage source. Chemical action within the battery produces an emf. Electrons (current) leave the negative terminal, are forced through the load, and return to the positive terminal. The dotted arrows in Figure 1–11*b* show the direction of current.

I–8
Sources of Direct Current

Direct current (dc)   Electric current in which the electron flow around the circuit is always in the same direction.

Alternating current (ac)   Electric current that periodically changes its direction, usually many times per second.

The examples given thus far have been with **direct current (dc)**: The electron flow around the circuit is always in the same direction. The strength or amplitude of the current may change, though, under the following conditions:

1. A larger or smaller potential difference is applied.
2. The circuit's total resistance changes.

Many applications require the use of **alternating currents (ac)**, which will be explained in Chapter 15. Alternating current, which periodically changes its direction 50 to 60 times per second, is the electric current supplied to our homes and industries. We will study direct currents first because they are easier to understand.

Some of the major sources of electricity are as follows:

1. Static electricity: Friction between two moving bodies can cause electrons to transfer from one to the other, leaving opposite static charges on them.

2. Electromagnetism: There is a close relationship between magnetism and electricity. A generator, such as is used in an automobile, is an example of an electromagnetic device. It produces voltage by means of a coil of wire (called an armature) rotating within a magnetic field. (Refer to Chapter 15 for more information on this subject.)

3. Chemical action: Chemical action produces opposite charges on two dissimilar materials, which then become the positive and negative terminals. Common examples of chemical action devices are dry and wet cells and batteries.

4. Photoelectricity: Certain elements are photosensitive: They have the ability to release electrons when light strikes them, or they will change their electrical resistance when exposed to varying light intensities. Examples of photoelectric devices are photoelectric cells, TV camera tubes, and phototransistors.

5. Thermal emission: Certain materials, when heated to a sufficiently high temperature, can boil off electrons from their surfaces. An example of a device using thermal emission is the vacuum tube such as a TV tube. The heated surface is called a cathode, and the element receiving the boiled-off electrons is called the anode.

Summary

Molecules are made up of atoms, which are the smallest part of an element that can exist and still retain its identity.

The atom is made up of subatomic particles, the most important of which are the electrons and protons.

Electrons are negatively charged particles that orbit a positively charged nucleus.

Free electrons are not tightly bound to the nuclei of their atoms and can be forced around a circuit in response to an applied voltage.

A coulomb (C) consists of 6.25×10^{18} electrons.

A coulomb moving past a given point in a circuit in 1 s constitutes 1 A of current.

The field surrounding a charged body is called the electrostatic or dielectric field of force.

The law of charges states that unlike charges attract and like charges repel.

A potential difference exists between two bodies having different charges.

A potential difference is required to force electrons around a circuit.

The letter symbol representing a voltage source, or emf, is E. The letter symbol representing voltage drop is V.

The letter symbol representing electric current is I, and the letter symbol for resistance is R.

Resistance is measured in ohms, symbolized by the Greek letter Ω (omega). Schematically, it is represented by a zigzag line.

Current is measured in amperes (A).

Current is the transfer of electric charge.

An electric charge is measured in coulombs (C).

The number of amperes is the number of coulombs of charge being transferred per second.

Potential difference or electromotive force (emf) is measured in volts (V).

Resistance in a circuit opposes current flow.

The load in a circuit is the device where the power or energy is consumed.

Direct current (dc) implies the current flows in one direction only.

An alternating current (ac) periodically reverses direction.

An electrical circuit has three parts: (*a*) a source of potential to produce current flow; (b) connecting wires to complete the electrical path from the voltage source to the device to be operated and back to the source; and (c) a device to be operated, called the load. In Figure 1-12, the lamp is the load

┌──Progress Test

The bracketed number after each question indicates the section of this chapter where the answer can be found.

1. From an electronics standpoint, the smallest particles of which all matter is composed are: (a) compounds. (b) elements. (c) electrons and protons. (d) molecules.   [1–1]

2. An electron: (a) has a mass greater than a proton. (b) is smaller than a proton. (c) has a positive charge. (d) weighs approximately 1/1835 as much as a proton.   [1–1]

3. The maximum number of electrons that can exist in the third energy band or level of an atom is: (a) 8. (b) 4. (c) 18. (d) 32.   [1–1]

4. A coulomb is: (a) equal to 6.25×10^{18} electrons. (b) equal to 2.1×10^9 electrons. (c) a term used to measure current flow. (d) equal to 6.25×10^{28} electrons.   [1–3]

5. A body becomes negatively charged by: (a) rubbing it with another body of the same material. (b) losing a quantity of electrons. (c) acquiring an excess of electrons. (d) connecting it to a body that is positively charged.   [1–2]

6. The electrostatic field surrounding a charged body: (a) varies in strength directly as the distance. (b) varies inversely as the square of the distance. (c) is uniform at all points from the body. (d) may be represented by parallel lines of force radiating outward from the body.   [1–2]

7. If the rate of flow of current is 6 A, how many seconds are required to allow 27 C to pass a given point in the circuit? (a) 0.222 (b) 162 (c) 27 (d) 4.5   [1–5]

8. The unit of resistance is the: (a) ohm. (b) ampere. (c) watt. (d) volt.   [1–6]

9. Current flow is measured in: (a) watts. (b) volts. (c) amperes. (d) coulombs.   [1–5]

10. Potential difference is measured in: (a) ergs. (b) electrostatic units. (c) coulombs. (d) volts. [1–4]

11. Good conductors of electricity have: (a) relatively low resistance. (b) relatively few free electrons. (c) relatively high resistance. (d) many free protons. [1–6]

12. Which of the following statements apply to a closed circuit? (a) It has no potential difference impressed across it. (b) It has a discontinuous path for current flow. (c) It has a complete path for current flow. (d) It has no resistance. [1–7]

13. An open circuit has: (a) no applied voltage. (b) no resistance. (c) an incomplete path for current. (d) maximum current flow for the amount of applied emf. [1–7]

14. Which of the following statements is true? (a) The current in a circuit forces voltage through the resistance. (b) The voltage impressed across a load forces current through it. (c) The current through a circuit depends on the load resistance, not the voltage. (d) The current is stationary and the voltage flows. [1–7]

Problems

Answers to odd-numbered problems are at the back of the book.

1. How many coulombs are represented by a charge of 12.5×10^{18} electrons?

2. What is the charge in coulombs of a body having 8.6×10^{16} electrons?

3. A charged body has lost 1.87×10^{16} electrons. What is its polarity and charge in coulombs?

4. What is the voltage of a cell that expends 0.225 J in moving 0.15 C?

5. If a 2.2 V battery supplied 7.7 J of energy, what is the charge in coulombs it is moving?

6. How much current is flowing in a circuit if 8.7 C has moved past a given point in 3.2 s?

7. If the rate of flow is 0.25 A, how many seconds are required for 2 C to pass a given point in the circuit?

2

Resistive Elements

Objectives

After studying this chapter, you will be able to:

Select the proper resistor for a specific application.

Read the ohmic value of color-coded resistors.

Determine whether wire-wound or carbon resistors are required for certain applications.

Recognize the characteristics of nonlinear resistors.

Introduction

Resistors are the most common elements used in electronic circuits. Considerable research is still going on in an effort to further increase reliability and to reduce the size needed to incorporate a given amount of power. New techniques of materials analysis and automated production have led to improved characteristics and high reliability to meet the demands of the computer, aerospace, and other modern industries.

Applications

Resistors are used in most electric appliances and devices, such as toasters, electric irons, and hair dryers. Radios, TVs, and stereo equipment also include many resistors among their electric components.

There are various types of resistors to suit the requirements of different electronic circuits. Carbon composition resistors are the least expensive and most common. However, in high-quality radios, amplifiers, and tape recorders, metal film resistors are likely to be used. These resistors do not produce nearly as much electrical noise as carbon composition resistors.

Potentiometers are special types of resistors that are used as volume and tone controls in the home entertainment field. They are also widely used in nearly every other kind of electronic equipment to control the level of an electrical signal going into amplifiers or other kinds of circuits.

Some electronic components, such as diodes, transistors, and other specialized devices, have nonlinear resistance characteristics. These devices will be discussed in this chapter.

2–1
Types of Resistors

Resistor   An element used to reduce supply voltages to some desired value or to limit current.

A **resistor** is a small component with two leads. A wide variety of resistors is used in the electronics industry today.

Carbon Composition

The carbon composition resistor is the basic mass-produced resistor of the electronics industry. Billions are used each year. They have low failure rates when properly used. Power ratings available are 1/8, 1/4, 1/2, 1, and 2 W in voltage ratings of 250, 350, and 500 V. Standard tolerances are 5% and 10%. Figure 2–1 shows these resistors in the wattage ratings mentioned.

The resistive element is a combination of carbon particles and a binding resin; the proportions are varied to provide the desired resistance. Attached to the ends of the resistive element are metal caps with axial leads of tinned copper wire for soldering the resistor into a circuit.

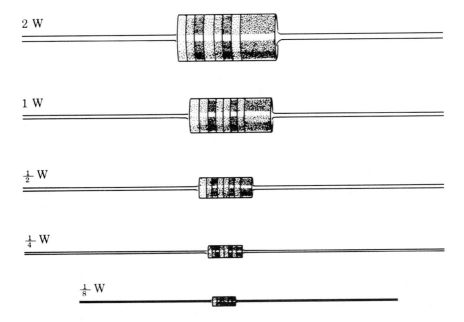

Figure 2–1

Carbon composition resistors arranged in their standard wattage ratings (ruler is for size comparison)

The unit is enclosed in a plastic case to prevent moisture and other harmful elements from entering. Figure 2–2 shows the construction.

Carbon composition resistors have a tendency to produce electric *noise* as a result of the current passing from one carbon particle to another. For example, noise can appear as hiss in a loudspeaker connected to a hi-fi system. This noise can overcome very weak signals.

Carbon composition resistors are primarily suitable where performance requirements are not demanding and where the lowest possible cost can be achieved. They are extensively used in entertainment electronics, though better resistors such as metal glaze are often used in critical circuits. Overload, moisture, and temperature extremes are typical causes of failure.

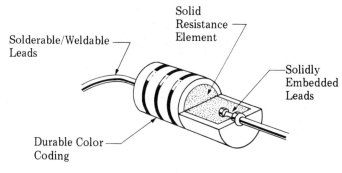

Solid
Resistance
Element

Solderable/Weldable
Leads

Solidly
Embedded
Leads

Durable Color
Coding

Figure 2–2

Internal construction of typical carbon composition resistor

Deposited Carbon

Deposited-carbon resistors are made by placing ceramic rods in a methane-filled flask and heating it until, by a gas-cracking process, a carbon film is deposited on the cores. A helix-grinding process forms the resistive path.

Compared with carbon composition, deposited carbon offers a major improvement in lower current noise and particularly in closer tolerance. In the lower-resistance ranges, 1% tolerance is standard, and in the higher ranges, 2% is typical. These resistors are being replaced by better resistors such as metal film and metal glaze.

High-Voltage Ink Film

For various applications in cathode-ray circuits, in physics, in radar, and in medical electronics, resistors capable of withstanding high voltages are required. In such resistors, a special resistive ink film is laid down in a helical band on a ceramic base (see Figure 2–3). The wide space required between bands is much greater than could be accomplished by abrasive cutting of a resistive path. Resistances available in different high-voltage (HV) types range from 1 kΩ to 100,000 MΩ, and voltages range up to 100 kV.

Metal Film

Metal film resistors are made by the deposition of vaporized metal in vacuum on a ceramic-core rod. The resistive path is helix-ground in the same way as for the deposited-carbon resistor. The metal film resist-

Figure 2–3

Typical ink film resistor

ance range is not as wide as the carbon range, going from about 10 Ω to 10 MΩ.

Where metal film excels is in tolerance and temperature coefficient (TC). For example, 1% tolerance is standard, and 0.1% is available. The TC ranges from 500 to a remarkable 25 parts per million per degree Celsius (ppm/°C). Metal film resistors have the greatest reliability of any type.

Metal film is definitely the superior resistor for numerous high-grade applications, such as in low-level stages of certain instruments. However, the cost is 20 to 40 times the cost of carbon composition.

At the low end of metal film performance, another resistor type, the metal glaze, is comparable, and in a great many applications, it is a better choice than metal film.

Metal Glaze

The newest of the resistor types is metal glaze, a metal-glass mixture that is applied as a thick film to a ceramic substrate and fired to form a film. The amount of metal in the mixture determines the resistance. With helix grinding, the resistance ranges from as little as 1 Ω to as high as many megohms. Power ratings of 1/8, 1/4, 1/2, 1, and 2 W are standard, with higher powers available.

At the low end of the performance range, metal glaze offers a TC of 500 ppm/°C, and 5% tolerance. At the high end, the TC is 100 ppm/°C, and standard tolerance is 1% or 2%. Thus, metal glaze covers the vast middle ground of electronics applications.

Broadly speaking, the metal glaze category encompasses another resistor consisting of a tin oxide film on a glass substrate. Metal glaze is as close as any resistor can come to being a universal resistor.

Wire-Wound

Wire-wound resistors are substantially different from all other types in that no film or resistive coating is used. A drawn wire with precisely controlled characteristics is wound around a ceramic-core form. Different wire alloys are used to provide different resistance ranges. Wire-wound resistors are characterized by highest stability, highest power rating, and special TC or resistor fuse functions. And only a wire-wound can withstand up to 250 W.

However, certain drawbacks to wire-wound resistors place them out of the running for most nonspecialized requirements. In low-power, low-cost, or high-density, limited-space applications, other types are usually preferred. Even with relatively noninductive windings (see Chapter 13 for information on inductance), wire-wounds are inclined to introduce an undesirable reactive element in high-frequency circuits.

Figure 2–4 shows the construction details of a modern power wire-wound resistor. The completed unit is coated with an insulating material such as baked enamel. Power wire-wound resistors are not color-coded. Their resistance values, wattages, and type numbers are stamped on with a special high-temperature paint that does not burn off.

Cermet (Ceramic Metal)

The cermet resistors are made by firing certain metals blended with ceramics on a ceramic substrate. The type of mix and its thickness determine the amount of resistance. Cermet resistors have precise resistance values and great stability under extreme temperatures. They are often produced as small rectangles with leads to attach to printed circuit (PC) boards. As with other resistor types, they are available with several different resistance values all in one package. Figure 2–5 shows a dual in-line package (DIP), which is commonly used with PC boards.

2–2
Resistor Color Code

Carbon resistors are color-coded – that is, they have several color bands painted around the body near one end – to identify their ohmic values. Other types of resistors are not color-coded; instead, they have their

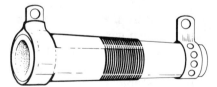

Figure 2–4

Construction details of typical power wire-wound resistor, with only part of resistance element shown

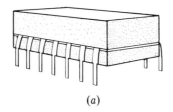

(a)

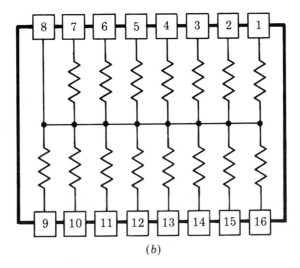

(b)

Figure 2–5

Dual in-line package (DIP): (a) Precision resistor network. (b) Typical resistance network.

ohmic values and, sometimes, identifying part numbers printed on them. The code has been established by the Electronics Industries Association (EIA) and appears in Table 2-1.

Using the Color Code

The use of the code can be understood by referring to the resistor shown in Figure 2-6. The first two bands of the resistor represent the *first* and *second significant numbers* and are red and violet, which, from Table 2-1, represent 2 and 7, respectively. The *multiplier band* is orange, which indicates that three zeros must be added. Therefore, the resistor's ohmic value is 27,000 Ω, frequently written as 27 kΩ (k = 1000).

    The *tolerance band* is silver, which represents ±10% (see Table 2-1). Consequently, any value between 24,300 and 29,700 Ω is within the tolerance range and is satisfactory.

Table 2–1   Carbon Resistor Color Code

	First Band	Second Band	No. of Zeros or Multiply by	Tolerance
Black	0	0	0	
Brown	1	1	1	
Red	2	2	2	
Orange	3	3	3	
Yellow	4	4	4	
Green	5	5	5	
Blue	6	6	6	
Violet	7	7	7	
Gray	8	8	8	
White	9	9	9	
Gold	–	–	0.1	5%
Silver	–	–	0.01	10%
No color	–	–	–	20%

Example 2–1

Determine the ohmic value of a carbon resistor with color bands of green, blue, red, and gold.

Solution

Green (5), blue (6), and red (00) gives 5600 Ω, and gold indicates ±5%.

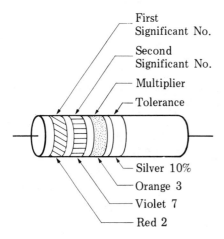

First Significant No.

Second Significant No.

Multiplier

Tolerance

Silver 10%

Orange 3

Violet 7

Red 2

Figure 2–6

Color-coded carbon composition resistor, with axial leads

Resistors Less Than 10 Ω

The third color band for resistors under 10 Ω will be either gold or silver, which indicates a multiplying value of 0.1 or 0.01, respectively.

Example 2–2 ─────────────────────────────

What is the ohmic value of a resistor whose color code is blue, gray, gold, and silver?

Solution

Blue (6), gray (8), and gold ($\times$ 0.1) gives 6.8 Ω, and silver indicates ±10%.

Example 2–3 ─────────────────────────────

What is the ohmic value of a resistor coded orange, white, silver, and gold?

Solution

Orange (3), white (9), and silver ($\times$ 0.01) gives 0.39 Ω, and gold indicates ±5%.

For more exacting requirements, resistors are available in 1% and 2% tolerances. These resistors have brown or red fourth bands, respectively, and are usually hot-molded, metal film, or composition types.

For equipment requiring high reliability (low failure rates), resistors having a fifth color band are used. These resistors are manufactured to military specification (MILSPEC), which indicates the reliability level, expressed as a percentage, per 1000 h operation. Table 2–2 shows the color coding for this band and the associated reliability levels.

Table 2–2   Reliability Levels of Fifth Band

Color	Level (per 1000 h)
Brown	M = 1.0%
Red	P = 0.1%
Orange	R = 0.01%
Yellow	S = 0.001%

Example 2–4

What is the failure rate that can be expected per 1000 h operation for a resistor with a brown fifth band?

Solution

Convert 1% to a decimal fraction to find the failure rate:

$$\frac{1}{100} = 0.01$$

Example 2–5

A group of 100 MILSPEC resistors are all coded blue, gray, red, brown, and yellow. What is their tolerance? What is the failure rate that can be expected per 1000 h operation for this group?

Solution

A brown fourth band indicates ±1% tolerance. A yellow fifth band indicates 0.001%. Convert that percentage to a decimal fraction to find the failure rate:

$$\frac{0.001}{100} = 0.00001 \times 100 = 0.001$$

For more information on MILSPEC and industrial-grade resistors, and for an explanation of their part numbers, refer to the Appendix.

2–3
Standard Fixed-Resistance Values

Resistance values have been standardized by the EIA. These values are shown in Table 2–3 in the three columns marked ±20%, ±10%, and ±5%. Successive values cover all intermediate values within the tolerances of that column. Resistors beyond the range of 10 to 100 Ω are multiplied by an appropriate power of 10 (0.1, 10, 100, and so on). Several examples will be given to indicate how to apply the information shown in Table 2–3.

Table 2–3 Standard Resistance Values

± 20%	± 10%	± 5%
10	10	10
		11
	12	12
		13
15	15	15
		16
	18	18
		20
22	22	22
		24
	27	27
		30
33	33	33
		36
	39	39
		43
47	47	47
		51
	56	56
		62
68	68	68
		75
	82	82
		91
100	100	100

Example 2–6

A 330 Ω ± 10% resistor may have an actual value anywhere in the range of 297 to 363 Ω. A 270 Ω ± 10% resistor may be as much as 297 Ω, and a 390 Ω ± 10% resistor may be as low as 351 Ω.

Example 2–7

A 150,000 Ω ± 20% resistor can vary between 120,000 and 180,000 Ω. A 220,000 Ω resistor in the same column can be as low as 176,000 Ω.

Example 2–8 ─────────────────────────────────

An 82 Ω ± 5% resistor may have a high value of 86.1 Ω. A 91 Ω ± 5% resistor may have a low value of 86.45 Ω.

2–4
Resistor Power Ratings

The physical size of a resistor is not determined by its resistance but by how much *power*, or heat, it can dissipate. In electric circuits, the unit of power is the watt (W), named in honor of James Watt (1736–1819), inventor of the steam engine. *One watt is the power dissipated when one ampere flows under one volt of applied emf.* Small resistors can dissipate or radiate less heat than larger ones because of their reduced surface area. If the temperature exceeds a specified value, the resistor may be damaged or may not return to its original resistance when allowed to cool.

Because electrical operating conditions are so varied, resistors are made in a number of *power ratings*, ranging from 0.1 W to hundreds of watts. Figure 2–1 shows several resistors in small wattage ratings. Some wire-wound resistors of higher power ratings are shown in Figure 2–7.

The power rating applies to a resistor mounted in free space with adequate ventilation. In practice, these conditions are generally not met. Resistors are usually mounted under a chassis or on printed circuit boards, where the ventilation is restricted. It is good engineering

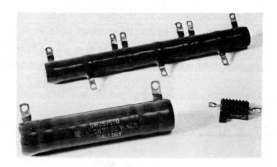

Figure 2–7

Several types of power wire-wound resistors

practice, therefore, to at least double the calculated wattage rating to provide a margin of safety.

When resistors are confined to restricted areas where the temperature significantly rises above ambient, the power rating must be derated, because the resistor cannot dissipate its rated power. A typical *derating curve* is shown in Figure 2–8.

Example 2–9

A resistor rated at 0.5 W is to be operated in an ambient temperature of 100°C. What is the maximum power the resistor can dissipate under these conditions?

Solution

Refer to Figure 2–8. Locate 100°C on the X axis (horizontal axis). Project upward (at 90° to the X axis) until you intersect the sloping part of the curve. Project horizontally to the left from this intercept to the Y axis (vertical axis). Read approximately 60%. Therefore:

$$0.5 \text{ W} \times 0.6 = 0.3 \text{ W}$$

Note: The 60% figure must be converted to a decimal fraction by dividing by 100.

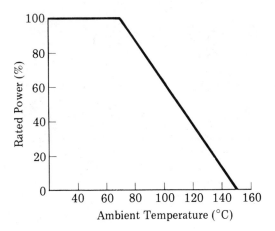

Figure 2–8

Typical power-derating curve

2–5
Variable Resistors

> Rheostat   A two-terminal variable resistor that consists of a re-
> sistance element, wound around a circular, insulated form, and
> an axially rotating contact to vary the resistance.

A **rheostat**, shown in Figure 2–9, is a form of variable resistor. The resistance element is wound around a circular, insulated form. Mounted axially is a rotating contact that allows the resistance to be varied between the terminals.

The schematic symbol for a rheostat is shown in Figure 2–10. Note that it is a two-terminal device; that is, it has two electrical connections.

Rheostats are used to *control the circuit current by varying the amount of series resistance.* Figure 2–11 illustrates the use of a rheostat

Figure 2–9

Rheostat used in power applications

Figure 2-10

Schematic symbol for rheostat

to vary the intensity of small incandescent lamps, as in an automobile instrument panel. The speed of very small electric motors can be controlled in the same manner. In Figure 2-11, the subscript B in E_B means battery. Therefore, E_B represents the battery voltage.

> **Potentiometer**   A variable resistor having three electrical connections, the center one being the movable contact, or wiper.

Potentiometers are a very common type of variable resistor. They are widely used in the home entertainment field and in industrial applications. Figure 2-12 shows some typical potentiometers (frequently abbreviated "pot") and the schematic symbol. A larger, older type of potentiometer is shown at the left in Figure 2-12*a*; it gives a good idea of the basic construction of a potentiometer.

Notice that a potentiometer has *three* electrical connections, the center being the movable contact. The terminals have been arbitrarily numbered 1, 2, and 3. Assume that the total resistance between terminals 1 and 3 is 1000 Ω and that the movable contact is exactly in the center. Then, the resistance between terminals 1 and 2 is 500 Ω, and between terminals 2 and 3, it is 500 Ω. If the movable contact is positioned closer to terminal 1, the resistance between 1 and 2 will be less than 500 Ω, and between 2 and 3, it will be more than 500 Ω. Correctly positioning the movable contact allows the resistance between terminals 1 and 2 (or 2 and 3) to be varied between 0 and 1000 Ω.

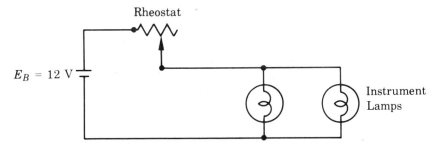

Figure 2-11

Rheostat to control circuit resistance for varying brightness of instrument lamps

(a)

(b)

Figure 2-12

Potentiometers: (a) Typical examples. (b) Schematic symbol.

In most electrical and electronics applications, a dc or ac voltage is impressed across the outside terminals of the potentiometer (terminals 1 and 3). As mentioned previously, proper positioning of the movable contact permits any value of voltage between maximum and zero to be obtained between terminals 2 and 3 (or 1 and 2). Hence, the potentiometer serves as a simple device for selecting any value of voltage between zero and maximum. Potentiometers are used as volume and tone controls in radio, hi-fi, and TV receivers and also as contrast and brightness controls in the latter. An application of a potentiometer to control the level of signal into an audio power amplifier is shown in Figure 2-13. The audio input signal could be from any source, such as a microphone amplifier or a tuner.

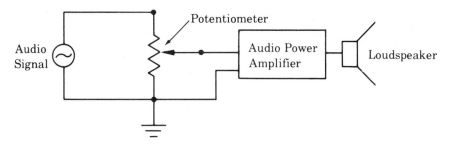

Figure 2-13

Potentiometer as a volume control

In many applications, it is convenient to attach an off-on switch on the back of the potentiometer to control power to a circuit. Rotating the shaft clockwise from its most counterclockwise position causes the switch to close. Reversing the direction opens the switch. Volume controls on radios and TV sets frequently use this arrangement.

Potentiometers are frequently ganged together, as shown in Figure 2–14. Their wipers are individually controlled by concentric shafts; the outermost shaft controls the pot closest to the threaded bushing.

Printed circuit boards require special kinds of very small potentiometers known as *trimmer pots*. Several types are shown in Figure 2–15. Sometimes, potentiometers are connected as rheostats. Figure 2–16 illustrates two methods of rheostat connection.

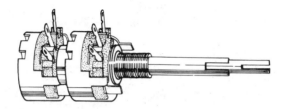

Figure 2–14

Dual potentiometer with concentric shafts

Figure 2–15

Three types of trimmer potentiometers

Figure 2–16

Two methods for connecting a potentiometer as a rheostat

2–6
Conductance

Conductance (G)   The reciprocal of resistance; the ease with which current flows through a circuit.

Sometimes, it is more convenient to express the ease with which current flows through a circuit rather than the amount of resistance encountered. The ease with which current flows is referred to as **conductance**, designated by G. It is the reciprocal of resistance. Thus, we see that any device or circuit has conductance as well as resistance. The unit of conductance is the siemens (S). Previously, the unit of conductance was the mho, which is ohm spelled backward, and was written as ℧.

Since conductance is the reciprocal of resistance, it is expressed mathematically as follows:

$$G \text{ (siemens)} = \frac{1}{R \text{ (ohms)}}$$

If the conductance of a device is known, its resistance can be found by the following formula:

$$R = \frac{1}{G}$$

Example 2–10 ───────────────────────

A certain piece of copper wire has a resistance of 0.2 Ω. What is its conductance?

Solution

$$G = \frac{1}{R} = \frac{1}{0.2} = 5 \text{ S}$$

Example 2–11 ───────────────────────

What is the conductance of a transistor with an internal resistance of 5000 Ω?

Solution

$$G = \frac{1}{5000} = \frac{1}{5 \times 10^3}$$

$$= 0.0002 \text{ S} = 0.2 \text{ mS}$$

2–7
Nonlinear Resistors

In most circuits, we can assume that the resistance is constant with relation to current and voltage. This *linear relation* can be graphically shown as in Figure 2–17. For example, if 3 V is applied to a certain resistor and 1 A flows, then 6 V applied to the same resistor will cause double the current, or 2 A, to flow. Chapter 4 deals more fully with this subject.

> **Nonlinear resistance**   The property exhibited by devices whose resistance does not change uniformly with changes of voltage or current.

Many different elements have characteristic curves (*I–V* plots) that are not straight lines, and they are, therefore, not linear elements. Such elements are called *nonlinear* since their resistances are **nonlinear**

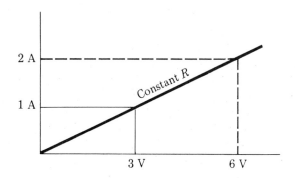

Figure 2–17

Plot of linear relation between current and voltage

resistances. Typical nonlinear elements are filaments of incandescent lamps, diodes, thermistors, and varistors (a special resistor made of Carborundum crystals held together by a binder). The curve in Figure 2-18a shows how the current through a varistor increases very rapidly when the voltage across the device increases beyond a certain amount (about 100 V in this case). A corresponding rapid decrease in resistance also occurs. Varistors are used to provide overvoltage protection in certain circuits.

A thermistor is another special resistor. It is made of metallic oxides in a suitable binder and has a large negative coefficient of resistance (that is, resistance goes down with an increase in temperature). Figure 2-18b illustrates this resistance-temperature relationship. The curve in Figure 2-18c shows how the resistance changes with voltage

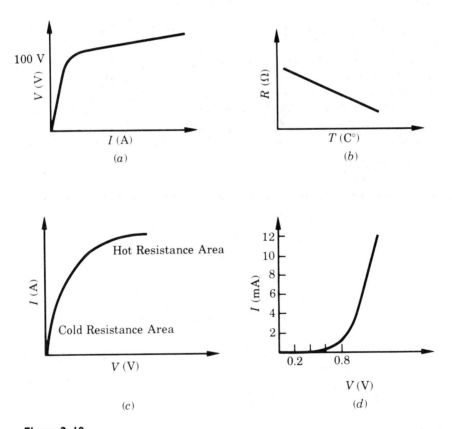

Figure 2-18

Characteristic curves of different nonlinear devices: (a) Varistor. (b) Thermistor. (c) Incandescent lamp. (d) Diode.

for an incandescent lamp. Figure 2–18*d* shows the *I–V* characteristic of a typical silicon diode.

The important thing to remember about any nonlinear resistance is that the resistance calculated for one value of *V* and *I* cannot be used when *V* and *I* change. But by making a number of changes in voltage and current, we can develop a resistance curve like the curves in Figures 2–18*a*, 2–18*c*, and 2–18*d*.

2–8
Resistor Troubles

Overheating, the principal reason for resistor failure, is usually caused by excessive current flow, which causes the resistor to open. If the resistor is in series with other components, as is usually the case, then the circuit becomes inoperative.

Carbon Composition

Carbon composition resistors are more apt to be damaged by heat than other resistors are. If the heat is insufficient to cause burnout, it will most likely cause the resistance to change to some value other than the coded value. Often, the color bands become so burned that it is impossible to read them.

Wire-Wound

Some types of wire-wound resistors have adjustable sliders that can be moved up or down an exposed section of the resistance wire. This procedure permits you to set the resistor to a particular ohmic value within the limits of the total resistance. But if the wire is fine, it may be easily broken by careless handling. *Caution*: Never attempt to adjust this kind of resistor with the voltage applied! Such an adjustment could be lethal if high voltages are involved.

The wattage rating of adjustable resistors applies only when *all* the resistance is in the circuit. The slider divides the wattage rating in direct proportion to the divided resistance. Therefore, avoid overloading any section by limiting the current to a wattage that can be safely handled by the section of resistance involved.

Potentiometers

The constant turning of a potentiometer shaft eventually causes some wear of the resistive material. Then, the wiper contact will not make a good electrical connection, and a noisy control will develop. This situation causes the scratchy noise you hear when turning the volume or tone control on a receiver. Some cleaners are available that can be sprayed on the resistive element and wiper to provide temporary relief, but the best solution is to replace the control.

2–9
Checking Resistance Values

It is frequently necessary to check the value of a resistor. Values can be checked with special resistance bridges or an instrument called an *ohmmeter* (see Chapter 10). Ohmmeters are not quite as accurate as a bridge, but they are faster and consequently the more popular instrument. It is important to select a proper scale on the ohmmeter when you are measuring resistance. For example, a one million Ω resistor should be checked on a high-resistance scale. If a low-resistance scale were used, the resistor would appear open.

If you are checking a resistor in a circuit, first, turn off the power, and then, temporarily disconnect one end of the resistor to eliminate erroneous readings resulting from possible parallel paths.

┌──Summary

Wire-wound resistors are used mostly for high-power requirements.

The physical size of a resistor determines its power-handling capability.

Power wire-wound resistors are not color-coded.

Because most wire-wound resistors act somewhat like inductors, their use is limited to dc and lower frequencies.

Carbon composition resistors are the most frequently used resistors.

Composition resistors create more electrical noise than other types.

Composition resistors are color-coded to show their ohmic values.

The fourth color band may be silver, gold, red, or brown, representing 10%, 5%, 2%, and 1% tolerances, respectively.

When fifth color bands are used, the reliability level is indicated.

When no fourth band is employed, the tolerance is 20%.

For resistance values between 1 and 10 Ω, the multiplier band is gold.

For resistance values less than 1 Ω, the multiplier band is silver.

Resistance values have been standardized by the EIA, although precision resistors can be made to any value.

The power rating of a resistor must be derated when the device is operating at temperatures significantly higher than ambient.

The wattage rating of adjustable resistors applies only when all of the resistance is in the circuit.

Composition resistors frequently change value when they are overheated.

Rheostats are variable resistors having two terminals.

Potentiometers are three-terminal devices that are used as simple voltage dividers. Typical applications are volume and tone controls in receivers and amplifiers.

Trimmer pots are used to provide high resolution (the ability to set the resistance to an exact value). They are widely used in industrial applications.

Conductance is the reciprocal of resistance and is expressed as $G = 1/R$. The unit of conductance is the siemens.

A linear resistor has a resistance that is constant with relation to current and voltage.

Any device with a resistance that changes with varying currents or voltages has nonlinear characteristics. Examples are incandescent lamps, semiconductor diodes, and transistors.

Overheating is the principal cause of resistor failure.

Potentiometers tend to become noisy after much use.

Resistance values are most commonly checked with an instrument called an ohmmeter.

┌──Progress Test

The bracketed number after each question indicates the section of this chapter where the answer can be found.

1. Composition resistors are the most popular because they are: (a) more accurate. (b) the least expensive. (c) smaller. (d) color-coded. [2–1]

2. Composition resistors can be damaged by: (a) excessive current. (b) not derating them when they are operating in a confined space. (c) excessive voltage. (d) all of the above. [2–8]

3. A characteristic of wire-wound resistors is: (a) color coding. (b) fragility. (c) reliability. (d) low power rating. [2–1]

4. A resistor having 6800 Ω resistance may be expressed as: (a) 0.68 kΩ. (b) 6.8 kΩ. (c) 0.068 MΩ. (d) 6.8 MΩ. [2–2]

5. A resistor having an orange fifth band has: (a) ±3% tolerance. (b) a value between 1 and 99 kΩ. (c) a 0.01% reliability. (d) a 0.001% reliability. [2–2]

6. A resistor having no fourth band has a tolerance of: (a) 5%. (b) 10%. (c) 15%. (d) 20%. [2–2]

7. What is the color code of a 1000 Ω ± 5% resistor? (a) brown, black, black, black (b) brown, black, red, gold (c) brown, red, black, silver (d) brown, orange, black, gold [2–2]

8. Rheostats are: (a) resistors with several adjustable taps. (b) three-terminal resistive devices. (c) used like potentiometers. (d) two-terminal adjustable resistors. [2–5]

9. A potentiometer: (a) can be connected as a rheostat. (b) has three connecting terminals. (c) can be used as a voltage divider. (d) is all of the above. [2–5]

10. Conductance is: (a) synonymous with resistance. (b) the reciprocal of R. (c) measured in ohms. (d) a measure of a circuit's ability to resist current. [2–6]

11. A nonlinear resistance: (a) is characteristic of semiconductors. (b) varies directly with E. (c) is not affected by temperature. (d) varies directly with I. [2–7]

12. Resistor failures are generally caused by: (a) excessive current flow. (b) failure to derate under high ambient temperatures. (c) resistance increasing with temperature. (d) a and b. [2–8]

Problems

Answers to odd-numbered problems are at the back of the book.

1.   What is the ohmic value of a resistor having a color code of orange, white, red, and silver?

2.   What is the resistance of a composition resistor that is color-coded brown, gray, green, and gold?

3.   Indicate the value of a resistor with the following color code: yellow, violet, gold, and silver.

4.   If a resistor has color bands of green, blue, and silver, what is its resistance?

5.   A circuit has 400 Ω resistance. What is its conductance?

6.   What are the possible lower and upper resistance values of a resistor coded red, red, red, and silver?

7.   What is the resistance of a circuit with a conductance of 0.015 S?

8.   A certain transistor has an output resistance of 5000 Ω. What is its conductance?

9.   A certain vacuum tube has an output conductance of 2200 μS (microsiemens). What is its resistance?

10.   Schematically show how a potentiometer can be connected as a rheostat.

11.   Complete the following chart.

	Resistor Colors				Rated Value (Ω)	Tolerance (%)
	Band 1	Band 2	Band 3	Band 4		
(a)	Red	Red	Orange	Gold	_____	_____
(b)	Yellow	Violet	Brown	Silver	_____	_____
(c)	Brown	Black	Brown	None	_____	_____
(d)	Green	Blue	Black	Gold	_____	_____
(e)	Brown	Green	Black	Red	_____	_____
(f)	Orange	White	Green	Silver	_____	_____
(g)	Blue	Gray	Red	Gold	_____	_____
(h)	Red	Violet	Red	Gold	_____	_____

12.   Calculate the upper and lower limit of the resistors described in Problem 11.

13. What is the rated value of the following color-coded resistors? (a) yellow, violet, gold, red (b) red, red, gold, gold (c) brown, green, silver, gold (d) brown, black, gold, brown.

14. Furnish the missing resistor code colors in the following chart.

| | Resistor Colors | | | Rated Value | |
	Band 1	Band 2	Band 3	Band 4	(Ω)	Tolerance
(a)	_____	_____	_____	_____	47,000	5%
(b)	_____	_____	_____	_____	2,200,000	20%
(c)	_____	_____	_____	_____	33	10%
(d)	_____	_____	_____	_____	270	1%
(e)	_____	_____	_____	_____	4.7	5%
(f)	_____	_____	_____	_____	39,000	10%
(g)	_____	_____	_____	_____	100,000	5%
(h)	_____	_____	_____	_____	680	20%
(i)	_____	_____	_____	_____	10,000	5%
(j)	_____	_____	_____	_____	100	10%

15. Calculate the conductance (in micro- or milli- units) of the resistors described in Problem 14.

16. A group of 1000 MILSPEC resistors all have red fifth bands. What is the maximum number of resistors that would be allowed to fail after 1000 h operation?

17. A certain piece of satellite electronic equipment contains 500 resistors having yellow fifth bands. What is the maximum number of resistors that could be allowed to fail after 43,800 h operation?

3

Kirchhoff's Laws

Objectives

After studying this chapter, you will be able to:

Recognize the difference between series and parallel circuits.

Explain how voltages are distributed around a series circuit.

Explain how current divides in a parallel circuit.

Introduction

In the early days of electricity, observations were made and laws established regarding electricity's behavior. Since then, these laws have had widespread application and are the basis for the study of electric circuits. Among the first laws formulated for electricity were two developed by Gustav Kirchhoff (1824–1887). One deals with the *distribution of current* at different junctions in a circuit. The other has to do with the *distribution of voltages* around the circuit. As you become familiar with these laws, you will be able to analyze what occurs in simple and complex circuits.

Applications

An introduction to Kirchhoff's laws is presented at this time because these laws provide insight into the behavior of current flow and voltage distribution in electronic circuits. Ohm's law and the general characteristics of series, parallel, and series-parallel circuits are more quickly understood once you have a basic knowledge of these two laws. A more detailed application of Kirchhoff's laws will be given in Chapter 8.

3–1
Series Circuit Defined

The definition of a series circuit was given in Section 1–5. You have already learned that, when two or more elements are connected end to end in an electric circuit, they are said to be in series and form what is called a series circuit. A simple series circuit is shown in Figure 3–1, where two resistors are connected to a 6 V battery. Electrons leaving the negative terminal of the battery flow through resistor 1, then resistor 2, and back to the positive terminal of the battery. There is only one path or circuit for the current. The current through each resistor is the same, because it cannot either gain or lose any electrons around the circuit. Therefore, we may conclude that, *if two or more elements are series-connected, the current is the same through each.*

Pictorial diagrams, such as in Figure 3–1, are infrequently used in the electrical and electronics industries. It is more customary to show the circuit in *schematic* form, as in Figure 3–2.

Most series circuits consist of more than just two resistors and a single voltage source. For example, some may have one or more resistors in series with a coil or inductor, as shown in Figure 3–3a. Inductors are described in Chapter 13. Others may include one or more resistors in series with a transistor, as in Figure 3–3b. In many cases, there is more than one voltage source. Some circuits may have two sources of direct current or one direct and one alternating current (ac) source. This latter source of emf is discussed in Chapter 15.

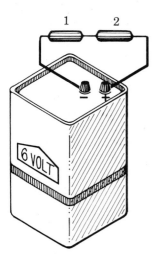

Figure 3–1

Pictorial view of simple circuit showing two power resistors series-connected to a 6 V battery

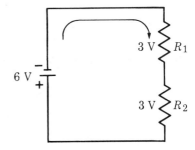

Figure 3-2

Schematic diagram of pictorial circuit
shown in Figure 3-1

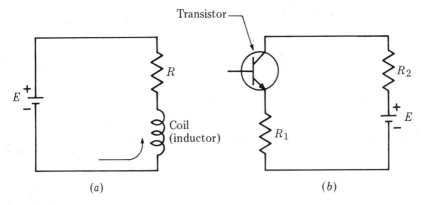

Figure 3-3

Simple examples of series circuits: (a) Resistor and coil in series and connected across voltage source E. (b) Two resistors, R_1 and R_2, connected in series with a transistor and the
combination connected to voltage source E.

3-2
Kirchhoff's Current Law
for a Series Circuit

Assume two resistors, R_1 and R_2, are connected as shown in Figure 3-4.
If they are connected to a battery, an electric current will be established.
This current will move in the direction of the arrows. The current
through R_1 can be designated as I_1 and the current through R_2 as I_2.
Current I_1 entering junction x must have the same value, or *magnitude*,

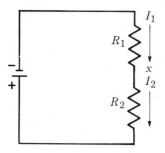

Figure 3–4

Series-connected resistors

as I_2 leaving the junction, because electrons are neither added nor lost as they pass through R_1, junction x, and R_2. Algebraically, we have the following formulas:

$$I_1 = I_2 \qquad \text{3-1}$$
$$I_1 - I_2 = 0 \qquad \text{3-2}$$

Equation 3-2 represents Kirchhoff's current law (abbreviated KCL) in its simplest form. In words, it can be stated as follows: *The algebraic sum of the current entering and leaving any junction is zero.* With such a simple configuration as shown in Figure 3-4, the conclusion seems immediately apparent. Kirchhoff's current law has greater application as a circuit's complexity increases.

Example 3–1 ────────────────────────

In Figure 3-5, what is the junction current at x?

Solution

The current through R_1 and R_2 is I_1 and I_2, respectively. From Equation 3-2, we obtain the following:

$$I_1 - I_2 = 0$$
Substitute $\qquad 1.5 \text{ A} - I_2 = 0$
Then $\qquad\qquad 1.5 \text{ A} = I_2$

Therefore, the junction current at x is 1.5 A. This current is the same current that enters and leaves junction y.

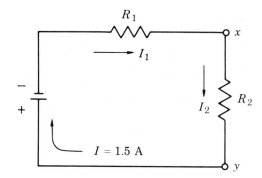

Figure 3-5

Simple series circuit

Example 3-2

In the circuit of Figure 3-6, $I_1 = 0.35$ A. What is the value of the current entering junction y?

Solution

Since this circuit is a series circuit, the current entering junction x is the same as that leaving the junction, and $I_1 = I_2$. Therefore, the current entering junction y is 0.35 A.

From this example, we see that the total current I_T in the circuit is equal to I_1, which is equal to I_2:

$$I_T = I_1 = I_2 \qquad\qquad \textbf{3-3}$$

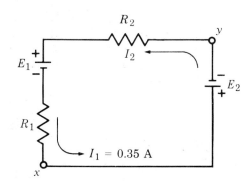

Figure 3-6

Series circuit containing two dc voltage sources with series-aiding connection

3-3
Kirchhoff's Voltage Law for a Series Circuit

As current is forced through an electric circuit by the application of a voltage, it encounters resistance. Some of the voltage of the supply source will appear across each resistance, forcing current through it. If a circuit has two resistors of equal value, then one-half the supply will be across each resistor, as shown in Figure 3-7. If the resistors are increased or decreased in ohmic value by the same amount, the voltage drops will be the same: *The voltage across resistors in a series circuit are proportional to the individual resistors.*

A simple equation that shows the relationship of the voltage drops (V_{R_1} and V_{R_2}) and the supply voltage (E_S) is as follows:

$$E_S - (V_{R_1} + V_{R_2}) = 0 \qquad \textbf{3-4}$$

This equation represents Kirchhoff's voltage law (KVL): *The algebraic sum of the voltage drops around any closed loop is zero.* Or, *the algebraic sum of all the voltage drops in a series circuit must equal the supply voltage.* Rewriting Equation 3-4, we have the following equation:

$$E_S = V_{R_1} + V_{R_2} + \cdots + V_{R_n} \qquad \textbf{3-5}$$

where R_n = any number of resistors in series.

Note: In this text, we will use E_S to designate supply potentials and V to indicate voltage drops.

For the circuit of Figure 3-6, close inspection reveals that E_1 is in series with E_2 (through R_2). In this connection, the positive terminal

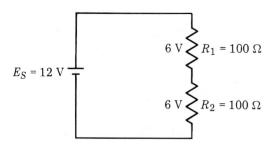

Figure 3-7

Series circuit with two resistors showing voltage drops

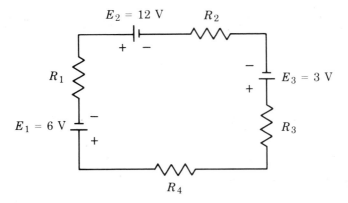

Figure 3–8

More complex series circuit with three voltage sources

of E_1 connects to the negative terminal of E_2 (via R_2). Hence, the two voltage sources are additive and are said to be *series-aiding* (S–A) sources. Conversely, we could say that the negative terminal of E_1 connects to the positive terminal of E_2 (via R_1). Again, the voltage sources are series aiding. Therefore, the total voltage E_T acting in the circuit is as follows:

$$E_T = E_1 + E_2$$

A more complex series circuit is shown in Figure 3–8. Starting with the negative terminal of E_1, we note that it connects to the positive terminal of E_2, when we trace clockwise around the circuit, via R_1. So E_1 is S–A with E_2. However, the negative terminal of E_2 connects, via R_2, to the negative terminal of E_3. So E_3 is a *series-opposing* (S–O) source; it opposes $E_1 + E_2$. Consequently, the total effective or net voltage E_T acting on the circuit is as follows:

$$E_T = E_1 + E_2 - E_3 \qquad\qquad \textbf{3–6}$$

This equation is a variation of Equation 3–5. Substituting the voltages indicated in Figure 3–8 into Equation 3–6 gives the following result:

$$E_T = 6\text{ V} + 12\text{ V} - 3\text{ V} = 15\text{ V}$$

Therefore, the three voltage sources provide only a net 15 V to force current around the circuit.

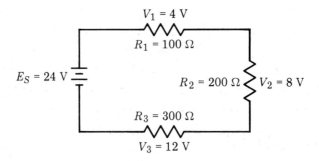

Figure 3–9

Series circuit for verifying KVL

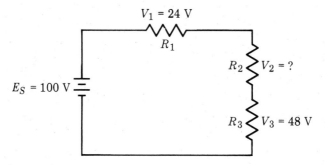

Figure 3–10

Series circuit illustrating KVL

Example 3–3

Check the validity of Kirchhoff's voltage law by using the circuit shown in Figure 3–9.

Solution

$$E_S = V_{R_1} + V_{R_2} + V_{R_3}$$
$$24\text{ V} = 4\text{ V} + 8\text{ V} + 12\text{ V}$$

Example 3–4

Find the voltage drop across R_2 in the circuit shown in Figure 3–10.

Solution

Use Equation 3–5.

$$100 \text{ V} = 24 \text{ V} + V_2 + 48 \text{ V}$$

Solve for V_2
$$100 \text{ V} - (24 \text{ V} + 48 \text{ V}) = V_2$$
$$V_2 = 28 \text{ V}$$

3–4
Parallel Circuit Defined

> **Parallel circuit**   A circuit in which two or more components are connected to the same pair of terminals, such as a battery, so that the current divides between them (sometimes referred to as a *shunt connection*).

If a circuit is so arranged that there is more than one current path connected to a common voltage source, it is called a **parallel circuit**. Parallel circuits, therefore, contain two or more resistances that are not connected in series. An example of a basic parallel circuit is shown in Figure 3–11.

That there are two separate current paths in the circuit of Figure 3–11 can be verified as follows. Current I_T leaves the negative terminal of the voltage supply E_S. At junction x, the current divides, I_1 going through R_1 and I_2 going through R_2. These two currents combine at junction y and return to E_S. Thus, two current paths have been established, one through each resistor.

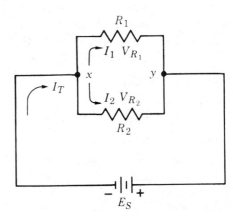

Figure 3–11

Simple parallel circuit

3–5
Kirchhoff's Current Law for a Parallel Circuit

From the definition and discussion in Section 3–4, we see that the current entering a junction is the same as the current leaving a junction. In the circuit of Figure 3–11, therefore, the following equation holds:

$$I_T - (I_1 + I_2) = 0 \qquad \textbf{3–7}$$

Equation 3–7 may be rewritten in a general form as follows:

$$I_T = I_1 + I_2 + I_3 + \cdots + I_n \qquad \textbf{3–8}$$

where n = any number of current paths. This equation says that *the total current in a parallel circuit is equal to the sum of the individual branch currents.* Equation 3–8 is Kirchhoff's current law for parallel circuits.

In Figure 3–11, junction x is also called a *node*. A node is simply a common connection for two or more components. Therefore, I_T enters node x and divides into I_1 and I_2.

To illustrate Kirchhoff's current law, let's consider the junction shown in Figure 3–12. The 3 A entering the junction divides into $I_1 = 1$ A and $I_2 = 2$ A. We substitute these values into Equation 3–7:

$$3\,\text{A} - (1\,\text{A} + 2\,\text{A}) = 0$$

Example 3–5 ——————————————————————————————

In Figure 3-13, current $I_1 = 1$ A is flowing toward junction x and $I_3 = 5$ A is flowing away from x. What is the value of I_2?

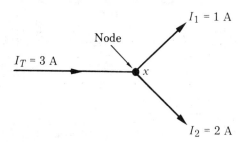

Figure 3–12

Current entering junction x equals the current leaving x

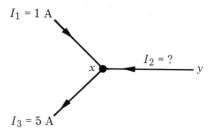

Figure 3–13

Example of Kirchhoff's current law

Solution

The current leaving the junction is subtracted from the currents entering the junction (see the arrows). Therefore:

$$I_1 + I_2 - I_3 = 0$$

Substitute $1\,\text{A} + I_2 - 5\,\text{A} = 0$

$$I_2 = 5\,\text{A} - 1\,\text{A} = 4\,\text{A}$$

3–6
Kirchhoff's Voltage Law
for a Parallel Circuit

In a parallel circuit, *the same voltage is present across each resistor or component and is equal to the supply*. Hence, in Figure 3–11, V_{R_1} and V_{R_2} equal E_S. Kirchhoff's voltage law for parallel circuits can be expressed in equation form as follows:

$$E_S = V_{R_1} = V_{R_2}$$ **3–9**

If voltage measurements were taken across each resistor in Figure 3–11, we could verify Equation 3–9.

In the following chapter, you will learn how to calculate the voltage drop across a resistor from the current flowing through it. Once the voltage drop has been calculated for one resistor, you will find it to be the same for each resistor in the parallel combination.

3–7
Series-Parallel Circuit Defined

> **Series-parallel circuit**   A circuit that contains both series- and parallel-connected elements, where the total current flows through the series elements and divides when going through the parallel combinations.

Most electronic circuits have some elements that are in parallel with each other, which, in turn, are in series with other circuit elements. Such a combination constitutes a **series-parallel circuit**. In these circuits, the total current flows through the series elements and divides when going through the parallel combinations.

An example of a simple series-parallel circuit is shown in Figure 3–14. Resistor R_1 is the series element, and resistors R_2 and R_3 are the parallel elements. The total current I_T flows through R_1 and *divides at junction x* into currents I_2 and I_3. These two currents *recombine at junction y* and flow to the positive terminal of the voltage source. A simple equation can be used to express the current in a series-parallel circuit:

$$I_T = I_1 = I_2 + I_3 \qquad\qquad \textbf{3–10}$$

The current I_2 through R_2 produces a voltage drop, which we can designate as V_{R_2}. Similarly, I_3 through R_3 produces a voltage drop,

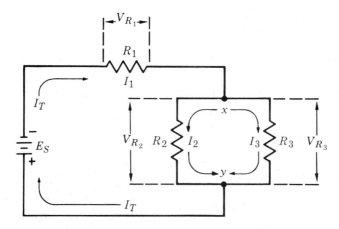

Figure 3–14

Simple series-parallel circuit

which can be designated as V_{R_3}. As you have previously learned, the voltage is the same across each element in a parallel combination. Therefore, $V_{R_2} = V_{R_3}$. Kirchhoff's voltage law tells us that the algebraic sum of all the voltages around a circuit is zero. In this circuit, we have the following equation:

$$E_S - (V_{R_1} + V_{R_2}) = 0 \qquad\qquad \textbf{3-11}$$

We could use V_{R_3} in place of V_{R_2} in this equation if we so desired, but we *cannot use both of them.* Equation 3–11 may be rewritten in the following familiar form:

$$E = V_1 + V_2 \quad \text{or} \quad E = V_1 + V_3 \qquad\qquad \textbf{3-12}$$

Note: For convenience, V_{R_1} can be written as V_1 and V_{R_2} as V_2, and so forth.

3–8
Using Kirchhoff's Laws to Solve Series-Parallel Circuit Problems

Several examples that illustrate the use of KVL and KCL in simple series-parallel electrical circuits are given next.

Solving for Voltage Distribution

Example 3–6 ————————————————————————

In the circuit shown in Figure 3–15, find the voltages across R_2 and R_3 from the information given.

Solution

Because R_1 and R_2 are in parallel, their voltage drops will be the same. Therefore:

$$V_{R_1} = V_{R_2} = 4\,\text{V}$$

Use Equation 3–5 $\qquad E_S = V_{R_1} + V_{R_3}$

Substitute $\qquad\qquad 10\,\text{V} = 4\,\text{V} + V_{R_3}$

$$10\,\text{V} - 4\,\text{V} = V_{R_3}$$

Therefore $\qquad\qquad\quad V_{R_3} = 6\,\text{V}$

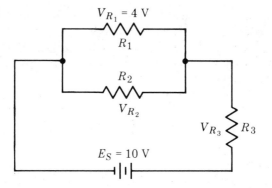

Figure 3–15

Series-parallel circuit for Example 3–6

Example 3–7

From the information given in Figure 3–16, determine the voltage drops across R_1 and R_2.

Solution

$$V_{R_2} = V_{R_3} = 37 \text{ V}$$
$$V_{R_1} = E_S - (V_{R_2} + V_{R_4}) = 100 \text{ V} - (37 \text{ V} + 41 \text{ V}) = 22 \text{ V}$$

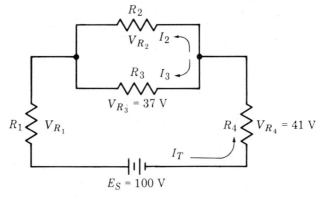

Figure 3–16

Series-parallel circuit for Example 3–7

Solving for Current Distribution

Example 3–8

Refer to Figure 3–16. The current through R_3 is 0.5 A and through R_1 is 2 A. What is the current through R_2?

Solution

Let $I_1 = I_T$. Then (KCL):

$$I_T = I_2 + I_3$$

Substitute

$$2\text{ A} = I_2 + 0.5\text{ A}$$

$$1.5\text{ A} = I_2$$

Example 3–9

In Figure 3–17, what is I_3 if $I_T = 6$ A, $I_2 = 1$ A, and $I_4 = 3$ A?

Solution

$$I_T - (I_2 + I_3 + I_4) = 0$$

$$6\text{ A} - (1\text{ A} + I_3 + 3\text{ A}) = 0$$

$$6\text{ A} - (4\text{ A} + I_3) = 0$$

$$2\text{ A} - I_3 = 0$$

$$I_3 = 2\text{ A}$$

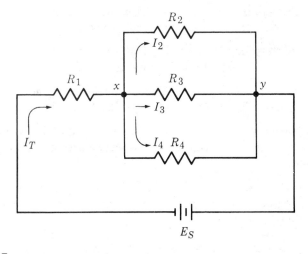

Figure 3–17

Series-parallel circuit for Example 3–9

⌐—Summary

When two or more elements are connected end to end, they are in series.

In a series electric circuit, there is only one path for the current to flow through.

Kirchhoff's current law (KCL) says that the algebraic sum of the currents arriving at and leaving any junction point is zero. Expressed mathematically:

$$I_1 - I_2 = 0$$

Kirchhoff's voltage law (KVL) says that the algebraic sum of all the voltage drops in a series circuit must equal the supply voltage. Expressed mathematically:

$$E_S - (V_{R_1} + V_{R_2} + \cdots + V_{R_n}) = 0$$

Kirchhoff's voltage law may also be written as follows:

$$E_S = V_{R_1} + V_{R_2} + \cdots + V_{R_n}$$

A parallel circuit has more than one path for current when the circuit elements are connected to a common voltage source.

A node is a common connection for two or more components.

A parallel circuit may also be defined as one containing two or more resistances that are not connected in series.

In a simple parallel circuit, the same voltage is present across each resistor or component and is equal to the supply voltage.

A series-parallel circuit has some elements that are in series, which, in turn, are connected to others that are parallel-connected.

Two series voltage sources with the positive terminal of one connecting to the negative terminal of the other are series-aiding (S–A) sources.

Two series voltage sources with the positive terminal of one connecting to the positive terminal of the other are series-opposing (S–O) sources.

Progress Test

The bracketed number after each question indicates the section of this chapter where the answer can be found.

    1.   Express, mathematically, KCL for junction x in Figure 3–4. [3–2]

    2.   The electrical quantity that is common to several series-connected elements is _____. [3–1]

    3.   Kirchhoff's current law states that the algebraic sum of the currents entering a junction must equal the algebraic sum of the currents leaving the junction (true or false). [3–2]

    4.   Refer to Figure 3–4. If the supply voltage is 36 V and $V_{R_2} = 13$ V, what is the voltage drop across R_1? [3–3]

    5.   What will be the ratio of voltage drops across three series-connected resistors if each one is double the value of the other? [3–3]

    6.   The electrical quantity that is common to elements in a parallel circuit is _____. [3–4 and 3–6]

    7.   The total current in a parallel-resistive circuit is the sum of the currents in the several branches (true or false). [3–5]

    8.   In a series-parallel circuit, the sum of the voltage drops across all resistors is equal to the supply (true or false). [3–7]

    9.   Refer to Figure 3–14. If the current through R_2 is 0.75 A and $I_1 = 1.25$ A, what is the magnitude of I_3? [3–8]

    10.   When two voltage sources are series-connected so that the positive terminal of one connects to the negative terminal of the other, they are said to be _____. [3–3]

Problems

Answers to odd-numbered problems are at the back of the book.

    1.   In the circuit of Figure 3–18, determine the voltage across resistor R_2.

    2.   Determine the voltage drop across resistor R_2 in the circuit of Figure 3–19.

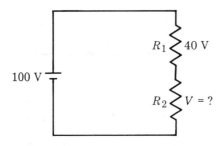

R_1 40 V

100 V

R_2 V = ?

Figure 3–18

Schematic diagram for Problem I

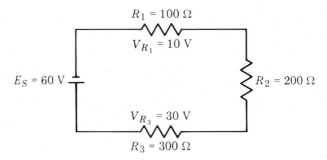

$R_1 = 100\ \Omega$

$V_{R_1} = 10\ V$

$E_S = 60\ V$

$R_2 = 200\ \Omega$

$V_{R_3} = 30\ V$

$R_3 = 300\ \Omega$

Figure 3–19

Schematic diagram for Problem 2

3. Determine the value of V_{R_3} in the circuit of Figure 3–20.

4. A certain series circuit is made up of three resistors connected across a dc voltage source. (a) If $R_3 = 3 \times R_2$ and $R_1 = 0.5 \times R_2$, what are V_{R_3} and V_{R_1} if $V_{R_2} = 6$ V? (b) What is the supply voltage?

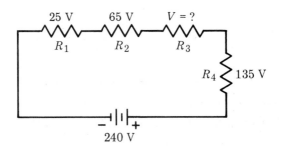

25 V   65 V   V = ?

R_1 R_2 R_3

R_4 135 V

240 V

Figure 3–20

Schematic diagram for Problem 3

5. Refer to Figure 3-11. If $I_1 = 3$ A and $I_T = 5$ A, what is the value of I_2?

6. Calculate the magnitude of I_2 for the junction shown in Figure 3-21.

7. In the circuit of Figure 3-14, if the current through R_1 is 0.5 A and the current through R_3 is 0.15 A, what is I_2?

8. For the circuit of Figure 3-22, calculate: (a) V_{R_1} (b) V_{R_3}.

9. Refer to Figure 3-23. Suppose $V_{R_2} = 13.5$ V, $V_{R_1} = 9.75$ V, and $E_S = 32$ V. Find: (a) V_{R_3} (b) V_{R_4}.

10. Refer to Figure 3-24. (a) What is the value of I? (b) What is the direction of the current through I?

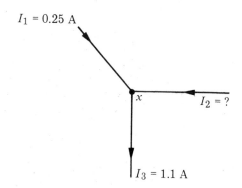

Figure 3-21

Diagram for Problem 6

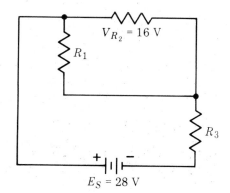

Figure 3-22

Schematic diagram for Problem 8

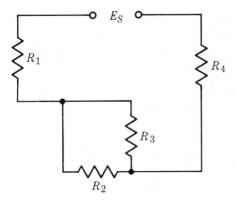

Figure 3–23

Schematic diagram for Problem 9

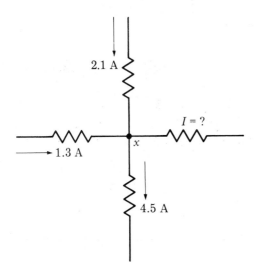

Figure 3–24

Diagram for Problem 10

11. In the series-parallel circuit shown in Figure 3–17, suppose $I_3 = 0.065$ A, $I_4 = 0.125$ A, and the current leaving junction y is 0.4 A. (a) What is the value of I_T? (b) What is the value of I_2?

12. In the circuit of Figure 3–25, use KVL to find the value of V_{R_3} from the data given.

13. In the circuit of Figure 3–26, use KVL to find the value of V_{R_2} from the data given.

14. In the circuit of Figure 3–27, use KVL to find the value of V_{R_1} from the data given.

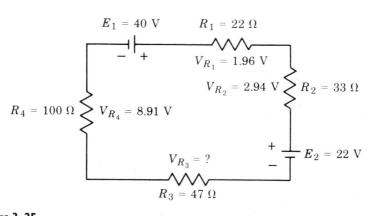

Figure 3–25

Schematic diagram for Problem 12

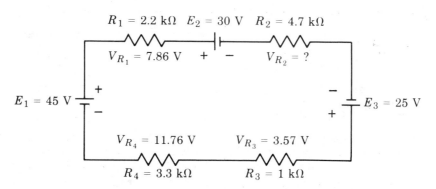

Figure 3–26

Schematic diagram for Problem 13

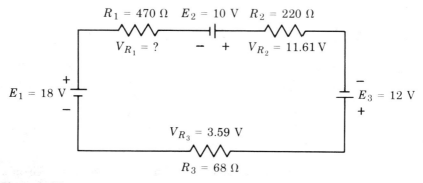

Figure 3–27

Schematic diagram for Problem 14

4

Ohm's Law

Objectives

After studying this chapter, you will be able to:

Calculate the current in a simple series circuit with a given applied voltage.

Explain the effect that resistance has on current flow and voltage drops in an electric circuit.

Calculate the amount of series resistance necessary to drop a voltage to its desired value.

Introduction

In the previous chapters, you have learned about the behavior of electric currents in and out of junctions and the relationship of voltage drops to resistance. You will now study the relationships between voltage, resistance, and current, which will be explained by using Ohm's law, the most fundamental of all electrical laws.

Applications

With an understanding of Ohm's law, you will be able to easily calculate the current in a simple series circuit with a given applied voltage. Also, you will understand the effect of resistance in a circuit. If the available voltage to a circuit, from a power supply or battery, is greater than required, you will know how to calculate the amount of series resistance necessary to drop the voltage to the desired value. With slight variation, Ohm's law can be extended for use with your future studies of alternating current circuits.

4–1
Ohm's Law Defined

Ohm's law states that *the current in an electric circuit is proportional to the applied voltage and inversely proportional to its resistance.* As the voltage increases in a circuit (resistance remaining constant), the current increases by the same amount. Hence, *if the voltage is doubled, the current will double.* Also, the amount of current in a circuit is inversely proportional to its resistance when the voltage remains unchanged. Stated another way, if the resistance in a circuit increases, the amount of current will decrease. For example, if the resistance is increased three times, the current will be reduced to one-third of its former value (voltage remaining constant).

It is convenient to express Ohm's law by the following simple equation:

$$I \text{ (amperes)} = \frac{E \text{ (volts)}}{R \text{ (ohms)}} \qquad \textbf{4–1}$$

By simple algebra, Equation 4–1 can be restated in terms of resistance or voltage as follows:

$$R = \frac{E}{I} \qquad \textbf{4–2}$$

$$E = IR \qquad \textbf{4–3}$$

Here is another way of expressing Ohm's law: An electrical pressure of one volt across a resistance of one ohm will cause one ampere of current to flow.

4–2
Solving for Current

A simple electric circuit is shown in pictorial form in Figure 4–1 so that you can see the physical relationship of the several components. Generally speaking, schematic diagrams rather than pictorial drawings are used in electronics work. The diagram shown in Figure 4–2 represents schematically the pictorial drawing in Figure 4–1.

Observe the polarity of the ammeter connections in Figure 4–2. Note that the positive terminal of the ammeter connects to the positive

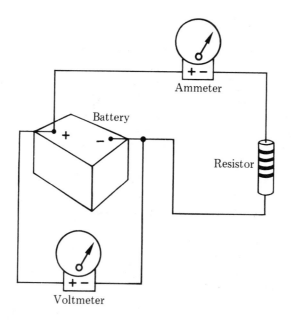

Figure 4–1

Pictorial diagram of simple electric circuit

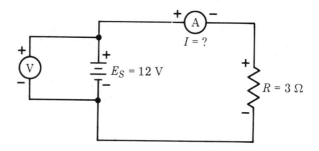

Figure 4–2

Schematic diagram of series circuit

terminal of the battery, while the negative terminal connects to the resistor. Notice also that the ammeter is *connected* in series with the resistor so that all the current in the circuit must pass through it. Because ammeters have very low resistance, they do not materially increase the circuit resistance. If an ammeter were accidentally connected across (in parallel with) the battery or resistor, a very large current would momentarily flow and would be likely to damage the meter.

Connected across the battery is a voltmeter to read the battery voltage. Because voltmeters are normally very high resistance instruments, they do not draw any significant amount of current from the battery. Observe the polarity of the voltmeter connections. The positive lead connects to the positive terminal of the battery, while the negative lead connects to the battery's negative terminal. A very important rule to remember is that *voltmeters are always connected in parallel with a voltage source or load, while ammeters are always connected in series with the circuit or load.*

Here are some examples to illustrate how Ohm's law can be used to solve for the current in a series circuit.

Example 4–1

Determine the current in the simple series circuit shown in Figure 4–2 from the information given.

Solution

Use the Ohm's law formula for finding current:

$$I = \frac{E}{R}$$

Substitute known values in the formula:

$$I = \frac{12 \text{ V}}{3 \text{ }\Omega} = 4 \text{ A}$$

Therefore, 12 V connected across a resistance of 3 Ω results in a current of 4 A through the resistor. In this case, the ammeter will read 4 A.

Example 4–2

Calculate the current in the circuit shown in Figure 4–3 with the values indicated.

Solution

$$I = \frac{E}{R} = \frac{6 \text{ V}}{100 \text{ }\Omega} = 0.06 \text{ A}$$

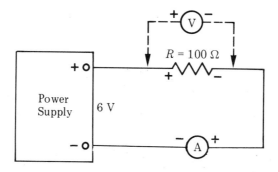

Figure 4–3

Series circuit

Ammeter Connections

In the circuit of Figure 4–3, the ammeter is shown connected to the negative terminal of the power supply rather than the positive terminal. The ammeter can be connected *any place in the circuit* as long as it is in *series* with the load. Regardless of whether the ammeter is connected as shown in Figure 4–2 (positive side of meter connected to positive terminal of battery) or as shown in Figure 4–3 (negative side of ammeter connected to negative terminal of power supply), the current passes *through the meter* in the same direction.

Voltmeter Connections

If a voltmeter were to be used to read the voltage drop across the resistor in Figure 4–3, it would be connected as shown by the dotted lines. Observe the polarity of the voltage drop across the resistor and the polarity of the voltmeter connection. The positive side of the voltmeter *must* connect to the positive side of the resistor.

4–3
Calculating Resistance

The resistance of an electric circuit can be easily determined by using the Ohm's law formula previously given and solving for resistance, as follows:

$$R = \frac{E}{I} \qquad \textbf{4-4}$$

This formula tells us that the resistance in a circuit is inversely proportional to the amount of current. If the current is small, the circuit resistance must be large, assuming that the voltage remains constant. The examples that follow illustrate the use of this formula.

Example 4–3

Refer to Figure 4–4. Determine the ohmic value of the load resistor R_L from the data given.

Solution

Use Equation 4–2 and substitute known values:

$$R_L = \frac{E}{I} = \frac{10 \text{ V}}{2 \text{ A}} = 5 \text{ } \Omega$$

This circuit would be considered a relatively low resistance circuit because 2 A is flowing with only 10 V applied.

Example 4–4

Calculate the value of the resistor R_L in the circuit of Figure 4–5 from the information given.

Solution

$$R_L = \frac{16 \text{ V}}{0.05 \text{ A}} = 320 \text{ } \Omega$$

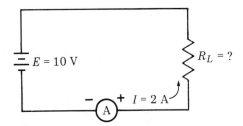

Figure 4–4

Determining resistance in a simple circuit

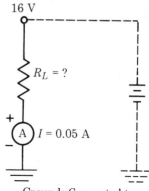

16 V

$R_L = ?$

$I = 0.05$ A

Figure 4-5

Resistance calculation in a simple circuit

Grounds Connected to Common Chassis

The resistor and ammeter seem to form an incomplete circuit in Figure 4-5. However, you must assume that, when a voltage is shown connected to one end of the arrangement and ground at the other, some form of voltage source is connected across them. The dotted lines and battery represent this source in Figure 4-5. But the voltage source may be a battery, power supply, or generator. Regardless of the source, it is evident that the circuit is indeed complete.

When no *polarity sign* (that is, + or −) is shown in front of the voltage value, such as 16 V in Figure 4-5, it is *always assumed to be a positive voltage*. The ground connection would therefore be negative. In some circuits (certain types of transistors, for example), reversed polarities are used. In such cases, a minus sign (−) is placed in front of the voltage.

Example 4-5

Calculate the resistance of R_L in the circuit of Figure 4-6 from the data given.

Solution

$$R_L = \frac{9 \text{ V}}{0.006 \text{ A}} = 1500 \ \Omega$$

Note: A negative supply voltage is assumed in Example 4-5. Ground would be positive, as evidenced by the reversed order of the long and short lines representing the battery connections in Figure 4-6. Again, notice how the ammeter is connected in series in the circuit and

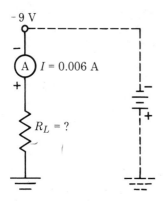

- 9 V

A $I = 0.006$ A

R_L = ?

Figure 4–6

Calculating R_L from E and I

notice the polarity. A circuit must be opened (disconnected wires) so that the current meter can be installed.

4–4
Determining Voltage

If the resistance and current of a circuit are known, it is an easy matter to calculate the amount of applied voltage. We use the Ohm's law formula and solve for voltage:

$$E = IR \qquad\qquad \textbf{4–5}$$

From this formula, we can see that *voltage is the product of current and resistance*. The voltage drop across a resistance, or circuit, will vary directly with either the current or resistance. For example, if the current through a resistor is doubled, the voltage drop (*IR* drop) will double. Or if the current can be maintained at a given value but the resistance is doubled, then the voltage drop will double. The following two examples show how to calculate voltage, or *IR*, drop.

Example 4–6

Determine the value of the supply voltage in the circuit of Figure 4–7 from the information given.

Solution

$$E = IR = 2\,\text{A} \times 50\,\Omega = 100\,\text{V}$$

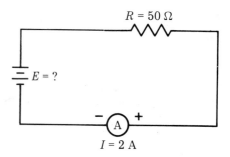

Figure 4–7

Calculating *E* when *R* and *I* are known

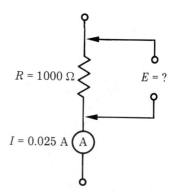

Figure 4–8

Calculating *IR* drop

Hence, from Example 4–6, we see that it requires a 100 V supply to force 2 A of current through a 50 Ω resistor. We can say that the *IR* drop across the resistor is 100 V, the same as the supply. *No IR drop occurs across the ammeter, since its resistance is assumed to be zero for all practical purposes.*

Example 4–7

From the data indicated, calculate the *IR* drop across the resistance shown in the drawing in Figure 4–8.

Solution

The voltage drop will be equal to the product of *I* and *R*:

$$E = IR = 0.025 \text{ A} \times 1000 \text{ Ω} = 25 \text{ V}$$

4–5
Prefixes

The usual units of amperes, volts, and ohms are often too large or too small for most electronics applications. In these cases, it is more convenient to use new units of measurement, which are formed by placing a special word, or *prefix*, in front of the unit. Each prefix has a definite meaning or value. Five of the most common prefixes are listed next.

Prefix	Abbreviation	Value
Milli	m	1/1000 or 0.001
Kilo	k	1000
Micro	μ	1/1,000,000 or 0.000001
Mega	M	1,000,000
Nano	n	1/1,000,000,000

For example, the 0.025 A of current flowing in Figure 4–8 could be written as 25 mA. The 0.006 A of current in Figure 4–6 could be written as 6 mA. A complete list of prefixes, adopted by the International Committee on Weights and Measures, is shown in Table 4–1.

Table 4–1 Prefixes

Multiples and Submultiples	Prefix	Symbol
10^{12}	Tera	T
10^{9}	Giga	G
10^{6}	Mega	M
10^{3}	Kilo	k
10^{2}	Hecto	h
10^{1}	Deka	da
10^{-1}	Deci	d
10^{-2}	Centi	c
10^{-3}	Milli	m
10^{-6}	Micro	μ
10^{-9}	Nano	n
10^{-12}	Pico	p
10^{-15}	Femto	f
10^{-18}	Atto	a

Here are several examples that illustrate how one electrical unit may be converted to another.

Example 4–8

Convert 2.7 V to millivolts (mV).

Solution

There are 1000 or 10^3 mV in each volt. Therefore:

$$2.7 \text{ V} \times 1000 = 2700 \text{ mV}$$

Example 4–9

Convert 0.68 A to milliamperes (mA).

Solution

$$0.68 \text{ A} \times 1000 = 680 \text{ mA}$$

Example 4–10

Convert 0.05 s to microseconds (μs). *Note:* The letter μ is the Greek letter *mu*.

Solution

There are 1,000,000 or 10^6 μs in each second. Therefore:

$$0.05 \text{ s} \times 1{,}000{,}000 = 50{,}000 \text{ } \mu\text{s}$$

Example 4–11

Convert 75,000 nanoseconds (ns) to milliseconds (ms).

Solution

Because there are 1,000,000 ns to each millisecond (the difference between 10^{-9} and 10^{-3}), we must divide 75,000 ns by 1,000,000. Therefore:

$$75{,}000 \text{ ns} \div 1{,}000{,}000 = 0.075 \text{ ms}$$

Example 4–12

Convert 82 kΩ to megohms.

Solution

There are 1000 or 10^3 kΩ to a megohm. Therefore, we must divide 82 kΩ by 1000 to obtain megohms.

$$82 \text{ kΩ} \div 1000 = 0.082 \text{ MΩ}$$

4–6
Graphical Representations of Ohm's Law

In Section 4–1, you learned, from the formula $I = E/R$, that the current in a circuit is directly proportional to the applied voltage and inversely proportional to the resistance. If the voltage is doubled, the current will increase two times. If the voltage is increased three times, the current will be three times greater, assuming the resistance remains constant. This *linear relationship* is shown by the top diagonal straight line in Figure 4–9, which is a graph of the equation $I = E/R$ for a resistance of 20 Ω. Note that the voltage is plotted on the horizontal axis and the current on the vertical axis.

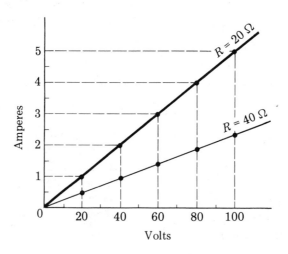

Figure 4–9

Linear relationship between current and voltage in a constant-resistance circuit

If we had assumed a load resistance of 40 rather than 20 Ω, the diagonal line marked $R = 40$ Ω would be the result. If a resistance less than 20 Ω had been used, the resulting line would have been steeper than the one for the 20 Ω load. The curves in Figure 4–9 show the direct proportionality between voltage and current for different values of load resistances.

Now, consider the effect of a changing resistance in a series circuit where the voltage is held constant. Study the simple circuit shown in Figure 4–10. If we assume $R = 20$ Ω, then 6 A of current will flow. We plot this value on the graph shown in Figure 4–11, and we increase the resistance to 40 Ω. The resulting current will be 3 A (plotted on the same graph). By increasing the resistance in increments of 20 Ω, we obtain currents of 2, 1.5, 1.2, and 1 A, as plotted. The loci of these points form the curve shown in Figure 4–11. This curve forms a graph of the equation $I = E/R$ and shows the *inverse relationship* between current and resistance for a constant voltage.

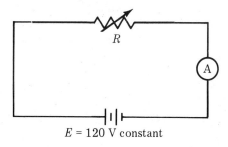

Figure 4–10

Series circuit with a changing resistance and a constant voltage

$E = 120$ V constant

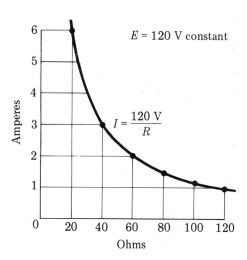

$E = 120$ V constant

$$I = \frac{120 \text{ V}}{R}$$

Amperes

Ohms

Figure 4–11

Graph of *I* versus *R* for constant voltage in a series circuit

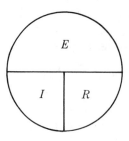

Figure 4-12

Memory aid for learning Ohm's law

4-7
Memory Aid

Ohm's law can be easily remembered by using the simple memory aid shown in Figure 4-12. By covering any one of the letters, you obtain the arrangement of the other two in the right-hand side of the formula for determining the value of the covered letter. For example, placing a finger over the I gives E/R, indicating that $I = E/R$. Covering the letter E leaves IR, indicating that E is the product of IR. Similarly, if R is covered, E/I remains, signifying that R equals E divided by I.

⌐Summary

Ohm's law states that current is proportional to applied voltage.

The amount of current in a series circuit is inversely proportional to the resistance.

The three forms of Ohm's law are as follows:

$$I = \frac{E}{R} \quad R = \frac{E}{I} \quad E = IR$$

The voltage drop across a resistor is the product of current times resistance (IR).

Electric and electronic circuits are represented by schematic diagrams.

Ammeters are always connected in series with the circuit elements.

Ammeters are very low resistance instruments.

Voltmeters are always connected across (in parallel with) the component for which the voltage is to be measured.

Voltmeters are normally very high resistance instruments.

Ammeter and voltmeter polarities must be observed when these instruments are used in circuits.

Prefixes are used to convert standard units of current, voltage, and resistance to frequently used smaller and larger units.

┌──Progress Test

The bracketed number after each question indicates the section of this chapter where the answer can be found.

1. Ohm's law states that the current through a resistor is: (*a*) proportional to its resistance and inversely proportional to the voltage across it. (*b*) inversely proportional to the voltage across it. (*c*) equal to the voltage across it squared, divided by the resistance. (*d*) inversely proportional to its resistance. [4–1]

2. The voltage drop across a resistor equals the: (*a*) product of the current squared and resistance. (*b*) square of the voltage across it divided by the resistance. (*c*) product of the current and resistance. (*d*) current divided by the resistance. [4–1]

3. The total resistance of a circuit may be calculated by: (*a*) dividing the impressed emf by the total current. (*b*) multiplying the circuit voltage by the total current. (*c*) multiplying the total voltage by the current. (*d*) squaring the total current and dividing by the applied voltage. [4–3]

4. A certain transistor has a resistance of 10,000 Ω connected in its collector circuit. If the voltage across the resistor is 9.4 V, the current through it will be: (*a*) 0.94 A. (*b*) 0.094 A. (*c*) 0.0094 A. (*d*) 0.00094 A. [4–2]

5. Which of the following is characteristic of a series circuit containing several resistors? (*a*) Current increases as the total resistance increases. (*b*) Current through each resistance varies with the resistors. (*c*) The *IR* drops are proportional to the resistance. (*d*) The voltage is the same across all resistors. [4–4]

6. If the voltage impressed across a series circuit is increased

three times and the resistance is doubled, how many times greater (or less) will the resultant current be compared with the original value? (a) one-half (b) one- and one-half (c) two (d) three   [4–4]

7.   The voltage required to force a current of 1.2 A through a resistance of 60 Ω is: (a) 66 V. (b) 72 V. (c) 80 V. (d) 92 V.   [4–4]

8.   A certain load resistor has a 6.8 V drop across it. If the current through the resistor is 0.0025 A, what is its resistance? (a) 4450 Ω (b) 3240 Ω (c) 2720 Ω (d) 270 Ω   [4–3]

9.   The load resistance of a certain transistor is 3900 Ω. If the voltage across the load is 8 V, the current flowing through it is: (a) 0.0021 A. (b) 0.0042 A. (c) 0.016 A. (d) 0.021 A.   [4–2]

10.   Converting 430 V to millivolts gives: (a) 4,300,000 mV. (b) 430,000 mV. (c) 43,000 mV. (d) 4300 mV.   [4–5]

11.   Converting 135 μA to milliamperes gives: (a) 0.0135 mA. (b) 0.135 mA. (c) 1.35 mA. (d) 13.5 mA.   [4–5]

12.   A resistor is marked 680,000 Ω. Its resistance in megohms is: (a) 680. (b) 68. (c) 6.8. (d) 0.68.   [4–5]

Problems

Answers to odd-numbered problems are at the back of the book.

1.   How much current will flow through a resistance of 20 Ω if connected across a potential of 120 V?

2.   A small soldering iron has 0.4 A of current through its heating element when connected to a 120 V outlet. What is the heater resistance?

3.   A doorbell requires 0.25 A of current to make it operate. If the resistance of the coils is 24 Ω, what must the applied voltage be to ring the bell?

4.   The transmitting tube type 801 requires a filament current of 1.25 A and a filament voltage of 7.5 V. Calculate the hot resistance of the filament.

5.   An electric toaster draws 10.5 A from the power line when the voltage across the heater element is 117 V. What is the hot resistance of the toaster?

6. A certain resistor in a television receiver is 2200 Ω. Determine the current through the resistor if the impressed potential is 45 V.

7. A transistor has an 8.5 V drop across its collector resistance of 680 Ω. Calculate the current through the resistor.

8. An incandescent lamp is connected to 120 V and uses 0.52 A current. What is the resistance of the lamp?

9. An LED (light-emitting diode) has an internal resistance of 40 Ω. How much current will flow when the LED is connected to 5 V?

10. A small portable tape player requires 280 mA of current when connected to 4.5 V dc. What is the equivalent internal resistance of the tape player?

11. A small 3 in. black-and-white portable TV has an internal resistance of 21.2 Ω. The battery pack provides 12 V dc (8 D cells in series). How much current flows in the TV when it is operating?

12. An electric coffee maker, when connected to 120 V, draws 9.6 A when perking. What is the resistance of the nichrome wire heating element?

13. The coffee maker in Problem 12 automatically switches to a second heating element when the coffee is ready. This element keeps the coffee warm. The current is reduced to 350 mA. What is the resistance of the warming coil?

14. A camera uses a 1.25 V battery to power its light meter; 60 μA flows in the light meter circuit. What is the resistance of the circuit?

15. The motor of an automobile heater fan has a resistance of 12.8 Ω, and when used, it draws 937.5 mA. What is the applied voltage?

16. If the automobile fan motor in Problem 15 were connected in a car with a 6 V system, how much current would flow?

17. A portable, battery-operated electric drill works with a 6 V battery. The internal resistance of the motor is 15 Ω. How much current does the motor use?

The following problems involve the conversion of metric prefixes for volts, amperes, and ohms. Fill in the blank spaces.

18. 2.5 V = (a) _____ mV = (b) _____ μV

19. 0.025 A = (a) _____ μA = (b) _____ mA

20. 2.2 kΩ = (a) _____ Ω = (b) _____ MΩ

21. 400 μA = (a) _____ A = (b) _____ mA

22. 48 V = (a) _____ mV = (b) _____ μV

23. $525 \, \mu V = (a)$ _____ mV = (b) _____ V

24. $220 \, k\Omega = (a)$ _____ MΩ = (b) _____ Ω

25. $470 \, \Omega = (a)$ _____ kΩ = (b) _____ MΩ

26. $23 \, V = (a)$ _____ mV = (b) _____ μV

27. $215 \, mA = (a)$ _____ A = (b) _____ μA

28. $0.468 \, A = (a)$ _____ mA = (b) _____ μA

29. $5125 \, \mu A = (a)$ _____ mA = (b) _____ A

30. $16.7 \, ms = (a)$ _____ s = (b) _____ μs

31. $33 \, k\Omega = (a)$ _____ MΩ = (b) _____ Ω

32. $2500 \, \mu s = (a)$ _____ ms = (b) _____ ns

33. $68 \, S = (a)$ _____ μS = (b) _____ mS

34. $31 \, kV = (a)$ _____ V = (b) _____ mV

5

Series Circuits

Objectives

After studying this chapter, you will be able to:

Calculate electric power and determine how the total power is dissipated in a series circuit.

Examine a series circuit and determine the relative polarities of voltages with respect to a reference point.

Explore the conditions under which maximum power and efficiency are obtained in a series circuit.

Describe the effects opens and shorts in electronic components have on series circuits.

Introduction

In the explanations and examples that have been given thus far regarding basic electrical principles, you have learned how to use very simple series circuits. You will now consider a series circuit that contains several resistive elements and learn how the total resistance of such a circuit can be calculated. The relationship between resistors of different values and their respective voltage drops and polarities will be analyzed.

Applications

The simplicity of the series circuit and its frequent use does not require that we look far to find applications. The switching of a lamp on and off is perhaps the most familiar application. The switch is *in series* with the lamp and alternately *closes* or *opens* the circuit and thereby controls the current. The old-fashioned series-connected Christmas tree lights are another example. If any one lamp burned out, the entire string went out. And as another example, some years ago, many ac–dc radios and television sets had radio tube heaters connected in series. If any one tube burned out, the circuit became open and none of the tubes would light up.

All electronic equipment – simple and complex – includes many series circuits. These circuits may include load resistors in series with the output of a transistor or resistor networks to make a transistor perform in a certain operating range.

5–1
Definition of Series Circuit

To restate what has been mentioned in prior chapters, we can define a *series circuit* as one that has all related components connected end to end so that there exists but one complete path for the current. The schematic diagram shown in Figure 5–1 is an example of a series circuit. The dc generator (note symbol G in the diagram) provides a po-

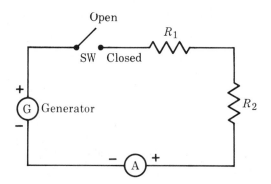

Figure 5–1

Simple series circuit

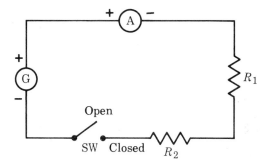

Figure 5–2

Variation of the circuit shown in Figure 5–1

tential that forces current around the circuit *provided the switch (SW) is closed*. The ammeter will indicate the value of the current.

The position of the various devices in a series circuit is generally not important. For example, the several elements of the series circuit shown in Figure 5–1 can be rearranged as indicated in Figure 5–2 without affecting the operation. It doesn't matter where the ammeter is located, because the same current is in every part of the circuit. Likewise, the switch can be located wherever it is convenient, because its effect is the same regardless of its placement in the circuit.

Series circuits often have more than one voltage source. The schematic in Figure 5–3 is typical of such a circuit. Notice that the battery is connected in a series-aiding manner with regard to the generator. Thus, the battery potential will assist the generator potential in forcing

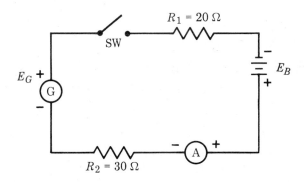

Figure 5–3

Series circuit with two voltage sources

current around the circuit, assuming that the switch is closed. If the battery polarity were reversed, then the total voltage acting around the circuit would be the difference between the two potentials. The same holds true if the battery polarity remained as shown but the generator polarity were reversed.

5–2
Calculation of Total Resistance

Because the various devices used in a series circuit are connected end to end, their several resistances are added to determine the total resistance R_T. Expressed mathematically, we have the following formula:

$$R_T = R_1 + R_2 + R_3 + \cdots + R_n \qquad \textbf{5-1}$$

Refer to the circuit in Figure 5–3. Total circuit resistance is as follows:

$$R_T = R_1 + R_2 = 20\ \Omega + 30\ \Omega = 50\ \Omega$$

Example 5–1 ──────────────────────────────────

Determine the total resistance of three series-connected resistors with values of 120, 470, and 820 Ω.

Solution

$$R_T = 120\ \Omega + 470\ \Omega + 820\ \Omega = 1410\ \Omega$$

Example 5–2 ──────────────────────────────────

A series string of resistors is made up of the following values: $R_1 = 2.7$ kΩ, $R_2 = 820$ Ω, $R_3 = 6.8$ kΩ, and $R_4 = 390$ Ω. What is the total resistance?

Solution

Where some resistance values are expressed in ohms and others in kilohms, we should convert all resistors to either one form or the other. In this case, we will express all values in kilohms.

$$R_T = 2.7\ \text{k}\Omega + 0.82\ \text{k}\Omega + 6.8\ \text{k}\Omega + 0.39\ \text{k}\Omega = 10.71\ \text{k}\Omega$$

5-3
Voltage Drops Proportional to Resistance

When measuring *voltage drops* in a series-resistive circuit, *if we find two or more to have the same value, we know their resistances must be equal.* Because the same current passes through each resistor, their *IR* drops will be the same. Refer to the circuit of Figure 5-4 for illustration. The total resistance of the circuit is 300 Ω; the current is calculated as follows:

$$I = \frac{60 \text{ V}}{300 \text{ }\Omega} = 0.2 \text{ A}$$

The *IR* drop across each resistor is 0.2 A × 100 Ω = 20 V, which indicates the *proportionality of voltage to resistance* in the series circuit.

For further illustration of this concept, refer to the schematic in Figure 5-5. Note that each resistor is 100 Ω larger than the one preceding it, starting with R_1 and going clockwise around the circuit. The voltage drop across R_2 should be twice that of R_1 because it is twice as large. Also, V_{R_3} should be three times V_{R_1} because R_3 is 300 Ω and R_1 is 100 Ω. We can expect that V_{R_4} will be double that of V_{R_2} because of the ratio of R_4 to R_2. Similarly, V_{R_4} will be four times V_{R_1} because R_4 is four times R_1.

The conclusions above can be verified by using Ohm's laws:

$$I_T = \frac{E}{\Sigma R} \qquad\qquad \textbf{5-2}$$

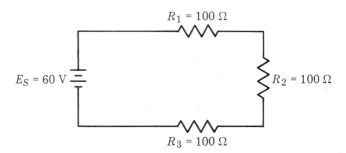

Figure 5-4

Series circuit with equal-value resistors

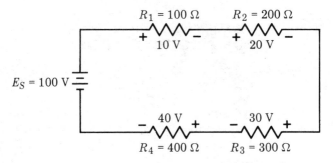

Figure 5–5

Series circuit showing proportionality of *IR* drops to resistance values

where Σ is the Greek letter *sigma*, meaning the sum of a number of quantities. Therefore, for the circuit in Figure 5–5, the current is as follows:

$$I_T = \frac{100 \text{ V}}{1000 \text{ } \Omega} = 0.1 \text{ A}$$

Then
$$V_{R_1} = 10 \text{ V} \quad V_{R_2} = 20 \text{ V} \quad V_{R_3} = 30 \text{ V} \quad V_{R_4} = 40 \text{ V}$$
$$\Sigma V = 100 \text{ V}$$

which satisfies KVL.

Example 5–3

Show that the voltage drops across the resistive circuit in Figure 5–6 are proportional to the resistances.

Solution

$$I = \frac{E}{\Sigma R} = \frac{24 \text{ V}}{1200 \text{ } \Omega} = 0.02 \text{ A}$$

Therefore
$$\begin{aligned} V_{R_1} &= 0.02 \times 150 = & 3 \text{ V} \\ V_{R_2} &= 0.02 \times 300 = & 6 \text{ V} \\ V_{R_3} &= 0.02 \times 750 = & \underline{15 \text{ V}} \\ & \Sigma V = & 24 \text{ V} \end{aligned}$$

By inspection, we see that the voltage drops are proportional to the resistances.

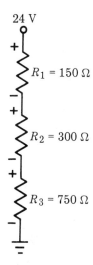

Figure 5–6

Series circuit to illustrate voltage
drops

5–4
Polarity of *IR* Drops

The potential across every resistor in a dc circuit is *polarized*: One side
is positive and the other is negative. Study the resistors in Figure 5–5
and observe the polarities of each. Note that *the end of each resistor
closest to the positive terminal of the supply is positive; the other end
is negative.* The long line of the battery symbol is positive and the short
line is negative. The resistor string in Figure 5–6 also illustrates this
rule.

Another, and equally valid, way of determining polarities of
voltage drops is to observe from which direction the current enters the
resistor. Assuming electron flow, the end of the resistor the current enters is negative (see schematics in both Figures 5–5 and 5–6). You must
know the polarities of the *IR* drops across circuit elements so that ammeters and voltmeters can be connected properly (as discussed in Chapter 4) and thus will indicate readings correctly.

5–5
Rise or Fall of Potential

In the circuit of Figure 5–7, each resistor will have identical voltage
drops because the ohmic values are equal. Therefore, the voltage drops
are as follows:

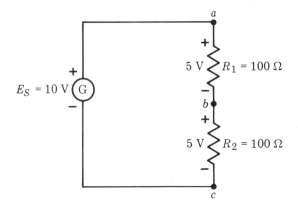

Figure 5–7

Circuit to illustrate voltage levels

$$V_{ab} = 5 \text{ V} \quad V_{bc} = 5 \text{ V}$$

Point b is common to both R_1 and R_2. Because of the polarity of the voltage drops, we can say that point a is 5 V positive with respect to point b. Therefore, the rise of potential from b to a is 5 V. With the same reasoning, point b is 5 V positive relative to point c. The voltage rise from point c to b is then 5 V. From point c to a, the total rise of potential is 10 V, which is equal to the supply voltage. This circuit satisfies Kirchhoff's voltage law since the sum of the voltage drops equals the supply potential.

Suppose we measure the IR drops across these resistors by starting at point a, the most positive point in the circuit, and going to point b. The drop is 5 V. Point b then is 5 V negative with respect to a. Continuing in the same direction, point c is 5 V below (negative) with respect to b. There is a drop of potential of 5 V in going from b to c. Now, with respect to b, we see there is a rise of potential of 5 V as we go to a. But from b to c, there is a fall or drop in potential of 5 V. Consequently, *when we refer to a rise or fall in potential, we are referring to the voltage with regard to a specific point in a circuit.* Further illustration of this concept is given in the next example.

Example 5–4

Explore the voltage rises or drops across the resistors in the circuit of Figure 5–8.

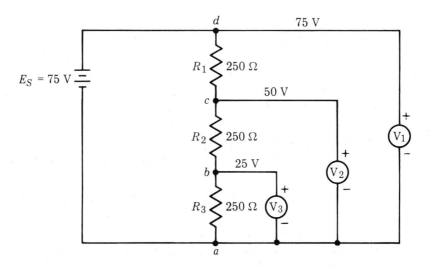

Figure 5–8

Circuit to illustrate voltage rises and drops

Solution

For simplicity, each resistor has the same ohmic value so that
all resistors will have identical voltage drops. From a to b, there
is a rise of 25 V, which would be indicated on voltmeter V_3. Now,
from b to c is an additional rise of 25 V, or a total rise of 50 V
with reference to a. Voltmeter V_2 would read 50 V. Finally, there
is another rise of 25 V as we go from c to d, making a total rise of
75 V from a to d, as indicated by V_1.

If the voltages represented in Figure 5–8 had been measured
with respect to point d, then all would have some lower or negative
value. We can say the voltages are dropping as we go down the resistive
network in a negative direction.

Example 5–5

What is the voltage from d to b in Figure 5–8?

Solution

The difference between d and b is 75 V $-$ 25 V, or 50 V. This
difference is a voltage drop because the potential at b is lower
than the potential at d.

5–6
Electric Ground

> **Ground**   A conductor providing either a common return path to the voltage source or a zero reference point, which, depending on its position, makes it possible to obtain positive and negative voltages (with respect to ground) from the same power supply.

In schematic diagrams, **ground** can signify both a *common connection* and a *zero reference point*. If it is a reference point, then only one ground appears on the schematic. If two or more grounded points are shown, they indicate a common return path to the voltage source. This common conductor may be a metal chassis in which the circuit is contained or some other form of conducting media, such as the braided wire shield around certain electric cables. This technique simplifies the wiring of most circuits by reducing the total number of conductors required.

Ground as a Common Connection

A simple circuit illustrating the use of a common ground connection appears in Figure 5–9. The negative lead of the supply is connected to ground, as is one side of the load. Thus, the circuit is completed via the ground connection, as represented by the dotted line. This practice is common with the electric systems of automobiles, aircraft, and so forth.

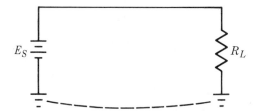

E_S R_L

Figure 5–9

Simple circuit showing chassis as ground return

Ground as a Reference Point

A circuit showing a ground as a reference point is presented in Figure 5–10. Because the resistors are of equal value, point a will be 100 V positive relative to b. Because point b is grounded, we can say that point a is 100 V above ground. Similarly, point c is 100 V negative with respect to b, or 100 V below ground. The voltage drops across R_2 and R_3 make point d 200 V negative with respect to b. Therefore, d is 200 V below ground. Now, ground could have been placed at either points c or d, as indicated by the dotted ground connections, but not simultaneously with its connection at point b. If ground were placed at c, then point a would be 200 V above ground and point d 100 V below ground. From this discussion, you can see how it is possible to obtain positive and negative voltages (with respect to ground) from the same power supply.

Example 5–6

If ground is established at point d in Figure 5–10, how high above ground is point b?

Solution

With an IR drop of 100 V across each resistor, point b will be 200 V above ground.

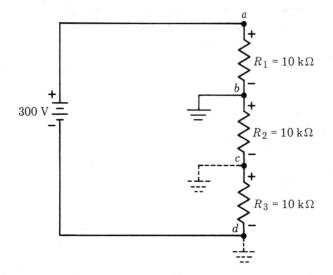

Figure 5–10

Circuit showing arbitrary ground connections

5-7
Voltage Dividers or Attenuators

The Potentiometer as a Voltage Divider

> **Voltage divider**   Two or more series-connected resistors across a
> voltage source, so fractions of the total supply can be obtained
> by connecting from one side of the source to either of the several
> resistor junctions.

In many applications, the potential available from a voltage
source is more than is needed. It becomes necessary to "drop," or scale
down, the voltage to some desired level by means of a **voltage divider.**
Perhaps the simplest voltage divider (sometimes called *attenuator*) is a
potentiometer (pot). If a voltage, either dc or ac, is connected across
a pot, an output can be obtained that is determined by the position of
the wiper.

Consider the circuit in Figure 5-11. An ac signal (voltage) E_S is
impressed across the outside terminals of the pot. The output signal V_o
is taken between the sliding contact (wiper) and one outside terminal.
By varying the position of the sliding contact, we may obtain any value
of V_o between the limits of E_S and zero.

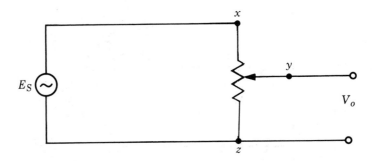

Figure 5-11

Potentiometer as voltage divider

Example 5-7

Assume the potentiometer in the circuit of Figure 5–11 is 10 kΩ and the slider is positioned so that the resistance between x and y is 4 kΩ. What is V_o if $E_S = 15$ V?

Solution

The output voltage will be determined by the ratio of R_{yz} to R_{xz} times the generator voltage. Expressed mathematically, the formula is as follows:

$$V_o = \frac{R_{yz}}{R_{xy} + R_{yz}} \times E_S$$

Since $R_{xy} = 4$ kΩ, then R_{yz} is as follows:

$$R_{yz} = 10 \text{ k}\Omega - 4 \text{ k}\Omega = 6 \text{ k}\Omega$$

Therefore $\quad V_o = \dfrac{6 \text{ k}\Omega}{4 \text{ k}\Omega + 6 \text{ k}\Omega} \times 15 \text{ V} = 9 \text{ V}$

Series Circuit as a Voltage Divider

Any combination of series resistors across a voltage source forms a voltage divider network. The voltage drop across any one resistor is a fraction of the total voltage impressed across the series combination. If no load is connected across any part of the divider network, the input voltage will be scaled down, or *attenuated*, in direct proportion to the resistors' ohmic values.

In Section 5–3, the proportionality of voltage drops to resistance was introduced. To explore this idea further, refer to the circuit of Figure 5–12, which is an example of a proportional voltage divider (or scaling network) using standard resistor values connected to an alternating current source. (*Note:* Voltage dividers can be used with ac as well as dc voltage supplies.) Let us assume that we want to find the voltage drop across R_2. Of course, we can use Ohm's law and calculate I_T. Then, $V_2 = I_T R_2$. However, we can find V_2 without calculating I_T. The voltage drop across R_2 will be determined by the ratio of R_2 to R_T times E_S. Expressed as an equation, the voltage drop is as follows:

$$V_2 = \frac{R_2}{R_T} \times E_S \qquad\qquad \textbf{5-3}$$

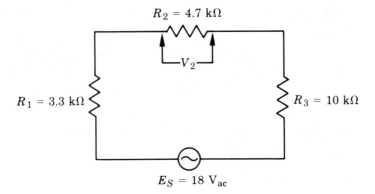

Figure 5–12

Series-resistive circuit as a voltage divider with an alternating current supply

Substituting values into this equation, we have the following calculation:

$$V_2 = \frac{4.7 \text{ k}\Omega}{18 \text{ k}\Omega} \times 18 \text{ V} = 0.261 \times 18 \text{ V} = 4.7 \text{ V}$$

This calculation leads to the general equation for any series voltage divider, sometimes referred to as a *scaling network*. The following formula is known as the *voltage divider equation*:

$$V_o = \frac{R_o}{R_T} \times E_S \qquad\qquad \textbf{5–4}$$

where R_o = resistor across which V_o is taken.

Example 5–8

Find V_o across R_1 in the circuit of Figure 5–12 if $R_1 = 500 \ \Omega$, $R_2 = 1000 \ \Omega$, $R_3 = 1500 \ \Omega$, and $E_S = 15 \text{ V}$.

Solution

$$V_o = \frac{500 \ \Omega}{500 \ \Omega + 1000 \ \Omega + 1500 \ \Omega} \times 15 \text{ V} = 2.5 \text{ V}$$

5–8
Calculation of Power

Energy and work are similar terms in that they are both independent of time. Power, however, *involves the element of time in doing work.* For example, if a person were to move an object weighing 50 lb a distance of 10 ft, the work accomplished would be 50 lb times 10 ft, or 500 ft-lb, regardless of how fast or slow the work was done. Power, however, equals *the work done divided by the time required* to accomplish it. For example, if the object were moved in 1 s, the power would equal 500 ft-lb/s.

Power is the rate of doing work. The letter symbol for power is P. Therefore, we can express power (P) as follows:

$$P = \frac{\text{work (foot-pounds)}}{\text{time (seconds)}} \qquad \textbf{5–5}$$

Example 5–9

How much power is required to move a 50 lb object 10 ft in 2 s?

Solution

$$P = \frac{500 \text{ ft-lb}}{2 \text{ s}} = 250 \text{ ft-lb/s}$$

Because the joule is the basic unit of work and the second is the basic unit of time, we can express power in joules per second. As mentioned in Chapter 2, the unit of electric power is the *watt,* named for James Watt in honor of his work in converting steam energy (heat) into mechanical energy. The symbol for the watt is W.

The following definition describes the magnitude of the watt: *One watt is the rate of doing work when one joule of work is expended in one second.* Expressed mathematically, the power is as follows:

$$P \text{ (watts)} = \frac{W \text{ (joules)}}{t \text{ (seconds)}} \qquad \textbf{5–6}$$

Power is defined as the rate of doing work. The rate at which electric energy is delivered or used is termed *electric power.* The unit of electric power is the watt, which is the power expended when one ampere flows under a pressure of one volt. Thus, we have the following equation:

$$P = EI \qquad\qquad 5\text{--}7$$

An alternative, and widely used, definition of electric power is as follows: *One watt is the power expended when one ampere of direct current flows through a resistance of one ohm.* The following examples illustrate the use of Equation 5–7.

Example 5–10

A certain mobile transceiver (transmitter-receiver) requires 12.6 V at 2.25 A for proper operation. How much power is the battery supplying?

Solution

$$P = EI = 12.6 \text{ V} \times 2.25 \text{ A} = 28.35 \text{ W}$$

Example 5–11

A certain transistor in an amplifier has 0.0008 A passing through it when 6 V is applied. What is the power dissipated by the transistor?

Solution

$$P = EI = 6 \text{ V} \times 0.0008 \text{ A} = 0.0048 \text{ W}$$

This result would be more conveniently expressed as 4.8 mW.

From Ohm's law, we know that $I = E/R$. If this value of I is substituted for I in Equation 5–7, we obtain the following:

$$P = E \times \frac{E}{R} = \frac{E^2}{R} \qquad\qquad 5\text{--}8$$

This equation indicates that *power varies directly as the square of the applied emf and inversely as the resistance.*

Example 5–12

What power is dissipated by a 2 kΩ resistor that has 235 V across it?

Solution

$$P = \frac{E^2}{R} = \frac{235^2}{2000} = 27.6 \text{ W}$$

A graph of Equation 5–8 is shown in Figure 5–13 for a 20 Ω resistor. Note that as the applied voltage increases from 40 to 80 V, for example, the power increases from 80 to 320 W. This graph illustrates that *doubling the voltage results in a quadrupling of power.*

We can calculate the power consumed in a circuit in terms of current and resistance. If, in the basic power equation of 5–7, we substitute for *E* its equivalent *IR*, we have the following:

$$P = IR \times I = I^2R \qquad\qquad \textbf{5–9}$$

This equation indicates that *the power delivered to a resistance is the product of the current squared times the resistance.*

The graph of this equation is also illustrated by the curve in Figure 5–13 for a 20 Ω resistor. For example, 2 A flowing through 20 Ω gives the following power:

$$P = 2^2 \times 20 = 80 \text{ W}$$

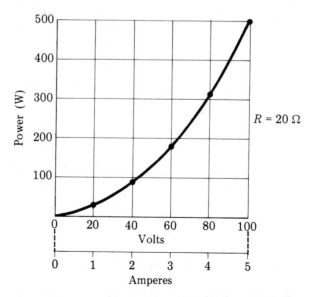

Figure 5–13

Graph of $P = E^2/R$ and $P = I^2R$ for a 20 Ω resistor

If the current is doubled, the power is as follows:

$$P = 4^2 \times 20 = 320 \text{ W}$$

This result indicates a fourfold increase of power.

Example 5–13

Calculate the power delivered to a radio antenna if the current is 12.25 A and the antenna resistance is 20 Ω.

Solution

$$P = I^2R = 12.25^2 \times 20 = 3001.25 \text{ W}$$

or approximately 3 kW.

Note that power varies as the square of either the voltage or the current and varies inversely as the circuit resistance. Also, if the voltage applied to a resistor is held constant and the resistance is halved, the power delivered to the resistor will be doubled.

Example 5–14

Calculate the increased power dissipation in a resistor if the applied voltage is held constant at 50 V and the resistance is reduced from 100 to 50 Ω.

Solution

For 100 Ω

$$P = \frac{E^2}{R} = \frac{50^2}{100} = 25 \text{ W}$$

For 50 Ω

$$P = \frac{50^2}{50} = 50 \text{ W}$$

Clearly, the power doubles when the resistance is halved, assuming a constant E.

5–9
Total Power in a Series Circuit

The total power dissipated in a series circuit is *the sum of the power supplied to each resistor*. This sum may be conveniently expressed as follows:

$$P_T = P_1 + P_2 + P_3 + \cdots + P_n \qquad \textbf{5–10}$$

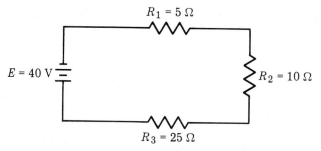

Figure 5–14

Circuit to illustrate total power, which is the sum of the power delivered to each resistor

Example 5–15

Find the total power used in the circuit of Figure 5–14.

Solution

The total resistance R_T of the circuit is 40 Ω.

$$I_T = \frac{40 \text{ V}}{40 \text{ Ω}} = 1 \text{ A}$$

Use Equation 5–9:

$$P_1 = 1^2 \times 5 = 5 \text{ W}$$
$$P_2 = 1^2 \times 10 = 10 \text{ W}$$
$$P_3 = 1^2 \times 25 = 25 \text{ W}$$

Therefore $P_T = 5 \text{ W} + 10 \text{ W} + 25 \text{ W} = 40 \text{ W}$

5–10
Internal Resistance of Voltage Sources

> **Internal resistance**   A characteristic of all sources of voltage that tends to reduce the voltage and current they can deliver to the load.

All sources of voltage are characterized by some **internal resistance**, which tends to *reduce* the voltage and current they can deliver to the load. In the case of power supplies, this internal resistance (whose symbol is R_i) is a function of the design of the power supply and, generally, does not vary greatly with changes of load. With batteries, however, we have a different situation. When batteries are fresh, or recently recharged, as with storage batteries, their internal resistance is very low, and they are consequently able to supply maximum current. As the charge of the battery is used up, its internal resistance increases rapidly because of chemical action within the battery. The net result is a reduction of terminal voltage that reduces the available current to the load.

Calculation of R_i

The internal resistance of a cell or power supply can be determined with the aid of Figure 5–15. First, the load must be disconnected so that the *unloaded terminal voltage* can be measured across terminals x–y. The load is reconnected, and then V_{xy} is measured again. The difference between the first and second readings represents the IR drop across the internal resistance of the cell or power supply.

The value of this internal resistance can now be readily calculated by dividing the difference between the unloaded and the fully loaded terminal voltages by the full current through the load. Expressed mathematically, the equation is as follows:

$$R_i = \frac{V_{NL} - V_{FL}}{I_L}$$

5–11

where
$$R_i = \text{internal resistance of source in ohms}$$
$$V_{NL} = \text{no-load terminal voltage}$$
$$V_{FL} = \text{full-load terminal voltage}$$
$$I_L = \text{full-load current}$$

The following two examples illustrate the procedure used to calculate internal resistance.

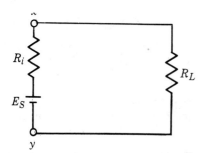

Figure 5–15

Circuit illustrating effect of R_i on terminal voltage of battery

Example 5–16

Calculate R_i of the battery in the circuit of Figure 5–15 if V_{xy} with no load is 1.5 V and V_{xy} loaded is 1.1 V when R_L is 200 Ω.

Solution

First, calculate the load current I_L:

$$I_L = \frac{1.1 \text{ V}}{200 \text{ } \Omega} = 0.0055 \text{ A}$$

Next, find V_{R_i} (the difference between V_{xy} loaded and unloaded):

$$V_{R_i} = 1.5 \text{ V} - 1.1 \text{ V} = 0.4 \text{ V}$$

$$R_i = \frac{V_{R_i}}{I_L} = \frac{0.4 \text{ V}}{0.0055 \text{ A}} = 72.73 \text{ } \Omega$$

Example 5–17

Find R_i of a power supply with a no-load voltage (V_{NL}) of 24 V and a full-load voltage (V_{FL}) of 23.98 V when a 12 Ω load is connected.

Solution

$$V_{R_i} = V_{NL} - V_{FL} = 24 \text{ V} - 23.98 \text{ V} = 0.02 \text{ V}$$

$$I_L = \frac{V_{FL}}{R_L} = \frac{23.98 \text{ V}}{12 \text{ } \Omega} = 1.998 \text{ A}$$

$$R_i = \frac{V_{R_i}}{I_L} = \frac{0.02 \text{ V}}{1.998 \text{ A}} = 0.01 \text{ } \Omega$$

Effect of R_i on Power Output

The internal resistance of a power source has an effect on the amount of power it can deliver to a load. For example, refer to the circuit shown in Figure 5–16. When the load is 0 Ω (short circuit), the current is limited only by R_i and the resistance of the connecting wires (assumed negligible in this example). The current flowing under this condition is as follows:

$$I = \frac{E}{R_i} = \frac{100 \text{ V}}{5 \text{ } \Omega} = 20 \text{ A}$$

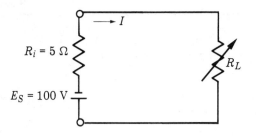

$R_i = 5\ \Omega$

$\longrightarrow I$

R_L

$E_S = 100\ V$

Figure 5–16

Effect of R_i on power output

This amount is the maximum current that can be drawn from the source. The terminal voltage is zero, which means that the 100 V is dropped across R_i.

If R_L is increased (R_i unchanged), the current drawn from the source and the IR drop across R_i will decrease. This decrease results in an increase in terminal voltage, and the terminal voltage will approach maximum as the current approaches zero.

5–11
Maximum Power Transfer Theorem ·

In order for a power supply, dc or ac generator, audio amplifier, radio transmitter, or other source to deliver maximum power to its load (for example, a loudspeaker connected to an audio amplifier), the internal resistance of the power source must be "matched" to (that is, made the same as) its load. With an audio amplifier, its output resistance must be designed to be the same as that of the loudspeaker to which it is connected. (*Note*: Although we refer to ac *resistance* here, the correct term is *impedance*, as discussed in Chapter 17.) Since loudspeakers have a nominal impedance of 8 Ω, the amplifier driving it must have an 8 Ω output in order to deliver its maximum power to the loudspeaker. A power supply will deliver maximum power when connected to a load equal to its internal resistance.

The maximum power transfer theorem can be illustrated by means of the schematic drawing in Figure 5–17. Assume that the source has an internal resistance of 10 Ω and a constant output of 20 V ac, and assume that the load is variable. If the load is varied between 0 and 80 Ω, the data tabulated in Table 5–1 results. Observe that when $R_L = 0\ \Omega$, power in the load (P_L) is 0 W. As the load resistance increases in the increments shown, P_L increases to a maximum of 10 W.

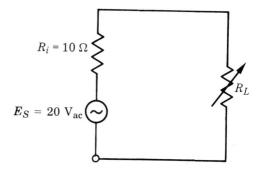

Figure 5–17

Maximum power transferred for R_L
$= R_i$

Table 5–1   Effect of Load Resistance
on Generator Output

R_L (Ω)	I_L (A)	P_L (W)	Efficiency (%)
0	2.0	0	0
2	1.67	5.56	16.7
4	1.43	8.16	28.6
6	1.25	9.37	37.5
8	1.11	9.88	44.5
10	1.0	10.0	50.0
20	0.67	8.89	66.3
40	0.40	6.4	80.0
60	0.29	4.9	84.5
80	0.22	3.95	90.0

At this point, $R_L = R_i$. Further increases of R_L result in a diminution of power in the load.

While an increase of R_L beyond the point where $R_L = R_i$ decreases the power into the load, it increases the efficiency of the circuit. The *efficiency* of power transfer (ratio of output to input power) from the source to the load *increases as the load increases*. Only where R_L is many times R_i does the efficiency approach 100%. Note that when $R_L = R_i$, the efficiency is 50%.

In amplifier stages, where voltage amplification is of paramount importance, the load will often be several times larger than the internal resistance of the transistor in order to achieve the highest efficiency. Power is of little concern in a voltage amplifier.

A graph relating load power and efficiency for the circuit of Figure 5–17 is shown in Figure 5–18. The peaking of load power at an R_L of 10 Ω is very evident.

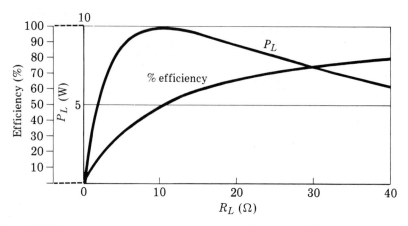

Figure 5–18

Graph of P_L and efficiency, with $R_i = 10 \, \Omega$ and variable load

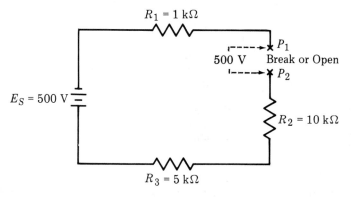

Figure 5–19

Effect of an open in a series circuit

5–12
Effect of an Open in a Series Circuit

> **Open circuit**   A break or discontinuity in the current path pro-
> ducing resistance approaching infinity.

An **open circuit** is a break or discontinuity in the current path.
The resistance across the break approaches infinity. A circuit illustrating this condition is shown in Figure 5–19. With no current, there will be

no *IR* drop across any of the three resistors. Consequently, points P_1 and P_2 will have the same potential across them as the voltage source. In this example, the voltage across the open will be 500 V, the voltage of the supply.

Many components in electronic circuits are series-connected. If any one component should burn out – or develop an open – the supply voltage would appear across the open. Troubles of this kind are easily located with a voltmeter, which reads 0 V across all components *except* the one that is open, where it reads the supply voltage.

5–13
Effect of a Short in a Series Circuit

Short circuit   An abnormal connection of very low resistance between two points of a circuit, resulting in a flow of excess (often damaging) current.

A **short circuit** is tantamount to placing a conductor of low resistance directly across the shorted element. The result is excessive current, and often overheating, of other devices that are in series. A visual inspection frequently reveals overheated components. For example, the color bands on resistors become discolored, and transformers may have tar or potting compound oozing out.

Shorts are caused by such factors as excessive voltage across a capacitor and overheating of transformer windings. Ohmmeters are generally used to locate shorts. A shorted component shows zero, or near zero, resistance. Power to the circuit must be turned off before doing any troubleshooting to locate the shorted (defective) component.

Summary

The current is the same in all parts of a series circuit.

The sum of the individual *IR* drops equals the applied voltage.

The total resistance of a series circuit is equal to the sum of the individual resistances:

$$R_T = R_1 + R_2 + R_3 + \cdots + R_n$$

The IR drops across series resistors are proportional to their resistance.

The polarity of IR drops is such that the end of the resistor closest to the negative supply terminal is negative, while the end closest to the positive terminal is positive.

Voltage rises and drops are relative to a common point in a circuit.

Ground signifies either a common connection or a zero reference point — sometimes both.

Any combination of series resistors across a voltage source forms a voltage divider network.

Potentiometers constitute the simplest kind of voltage or signal divider.

The voltage across any one resistor in a voltage divider is equal to its ohmic value divided by the total series resistance times the input voltage. For example:

$$V_{R_1} = \frac{R_1}{R_1 + R_2 + R_3} \times E_S$$

Electric power is measured in watts and can be calculated by any of the following formulas:

$$P = EI \quad P = I^2R \quad P = \frac{E^2}{R}$$

The total power dissipated in a series-resistive circuit is the sum of the power supplied to each resistor.

The internal resistance of a voltage source limits the amount of power it can deliver.

The internal resistance of a cell or battery varies with charge and age.

A source can deliver its maximum power when the load equals its internal resistance.

The efficiency of power transfer increases as the load increases with respect to R_i.

The total supply voltage will appear across an open in a circuit.

Shorted components usually result in excessive current flow, and overheating of devices that are in series is likely.

Progress Test

The bracketed number after each question indicates the section of this chapter where the answer can be found.

1.   In a series circuit: (a) all IR drops are equal. (b) the current is the same through all components. (c) the supply voltage is across each resistor. (d) the voltage drops are inversely proportional to the resistances.   [5-1]

2.   The total resistance of three resistors of 10, 15, and 30 Ω is: (a) 55 Ω. (b) 45 Ω. (c) 40 Ω. (d) 30 Ω.   [5-2]

3.   Four resistors of 1.2 kΩ, 3.9 kΩ, 820 Ω, and 5.6 kΩ are connected in series. The total resistance is: (a) 8.482 kΩ. (b) 10.782 kΩ. (c) 11.52 kΩ. (d) 18.90 kΩ.   [5-2]

4.   The total resistance of a circuit containing three resistors is 297 Ω. Two of the resistors are 56 and 150 Ω. The ohmic value of the third is: (a) 81 Ω. (b) 91 Ω. (c) 95 Ω. (d) 206 Ω.   [5-2]

5.   When two dc voltage sources are connected in a series circuit, the effective voltage will be: (a) the arithmetic sum of the two. (b) the difference between them. (c) the algebraic sum. (d) the product of the two.   [5-1]

6.   Two resistors are connected in series. Resistor $R_1 = 390\ \Omega$ and has 12 V across it. If $V_{R_2} = 25.23$ V, what is the value of R_2? (a) 302.7 Ω (b) 820 Ω (c) 2.18 kΩ (d) 4.68 kΩ [5-3]

7.   Refer to Figure 5-11. If a 1 kΩ potentiometer is used and the slider is positioned so that $R_{xy} = 195\ \Omega$, what is the value of V_o if $E_S = 12$ V? (a) 10.46 V (b) 9.66 V (c) 4.55 V (d) 2.34 V   [5-7]

8.   Suppose 1.89 V appears across a 600 Ω resistor. The power in watts is: (a) 6 W. (b) 1.13 W. (c) 3.2 mW. (d) 6 mW.   [5-8]

9.   A resistor with 0.04 A flowing through it dissipates 1.6 W. If the applied voltage is increased so that the current doubles, the power becomes: (a) 6.4 W. (b) 4.8 W. (c) 3.2 W. (d) 2.6 W.   [5-8]

10.   If the resistance of a circuit is halved but the voltage remains constant, the power: (a) halves. (b) doubles. (c) quadruples. (d) becomes one-fourth.   [5-8]

11.   Three series resistors dissipate 88 mW, 1.04 W, and 265 mW, respectively. The total dissipation is: (a) 0.457 W. (b) 0.628 W. (c) 1023 mW. (d) 1393 mW.   [5-9]

12. A battery has a no-load voltage of 1.5 V and a load voltage of 1.3 V. If the load resistance is 10 Ω, the internal resistance is: (a) 0.85 Ω. (b) 1.04 Ω. (c) 1.54 Ω. (d) 1.78 Ω. [5–10]

13. A generator whose R_i is 0.12 Ω delivers 16 A at 28 V dc to a 1.75 Ω load. What value of R_L is needed to draw maximum power from the generator? (a) 233.3 Ω (b) 1.75 Ω (c) 0.21 Ω (d) 0.12 Ω [5–11]

⌐Problems

Answers to odd-numbered problems are at the back of the book.

1. The circuit of Figure 5–3 has two voltage sources, E_G and E_B. If $E_G = 26.5$ V and $E_B = 12.6$ V, what current will flow?

2. What is the voltage drop across R_2 in the circuit of Problem 1?

3. What current would flow in the circuit of Problem 1 if the polarity of E_B were reversed?

4. Three resistors $R_1 = 56$ Ω, $R_2 = 82$ Ω, and $R_3 = 120$ Ω are connected in series across a power supply. A voltmeter connected across R_1 reads 10.5 V. What is the supply voltage?

5. Three series resistors are connected across a 75 V source. Calculate the ohmic value of R_2 if $V_{R_1} = 15$ V, $V_{R_3} = 35$ V, and $I_T = 50$ mA.

6. A 2.2 kΩ load resistor in a transistor output circuit has 3.3 mA flowing through it. Calculate the power dissipated in the load.

7. A certain resistor in an attenuator network has 90 V across it. What wattage will be dissipated if its resistance is 6.8 kΩ?

8. A 270 Ω emitter resistor in a transistor circuit has 0.01 A through it. What is its power dissipation?

9. A certain 10 Ω resistor dissipates 5W. What is the emf across it?

10. What must the voltage be in the circuit of Problem 9 if the power is reduced to one-half?

11. A 200 Ω relay is to be operated from a 12 V dc source. If the relay requires 25 mA for proper operation, how much resistance must be added to the circuit?

12. A 1 kΩ potentiometer has 2.4 V impressed across it (see Figure 5–20). If $R_{bc} = 350$ Ω, what is V_o?

13. In the series circuit of Figure 5–12, suppose $R_1 = 3.3$ kΩ, $R_2 = 2.2$ kΩ, and $R_3 = 1.8$ kΩ. Calculate V_{R_3} if $E_S = 9$ V.

14. In the circuit of Figure 5–12, suppose $R_1 = 390$ kΩ, $R_2 = 270$ kΩ, and $R_3 = 410$ kΩ. Calculate V_{R_1} if $E_S = 85$ mV.

15. A 40 V power supply is connected across three series resistors with values of $R_1 = 30$ Ω, $R_2 = 50$ Ω, and $R_3 = 80$ Ω. Calculate the power expended in R_2.

16. What is the total power expended in the circuit of Problem 15?

17. A transmitter's antenna current meter reads 18.5 A. If the antenna resistance is 24 Ω, what power is radiated?

18. Two resistors, $R_1 = 10$ kΩ and $R_2 = 5$ kΩ, are series-connected across the output of a power supply. If $E_{R_1} = 300$ V, what is the power expended by each resistor? What is P_T?

19. An unloaded power supply delivers 235 V dc. When a 1.5 kΩ load is connected, the output drops 15 V. What is the internal resistance of the supply?

20. A storage battery has a fully charged voltage of 12.6 V. When the power input circuit of a mobile transmitter is connected across its terminals, the voltage drops to 12.15 V. What is the battery's internal resistance if the transmitter draws 40 A?

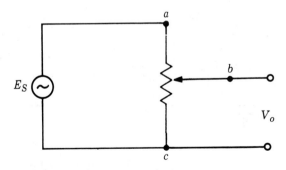

Figure 5–20

Circuit for Problem 12

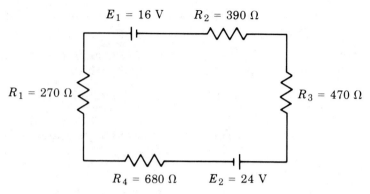

Figure 5–21

Circuit for Problem 21

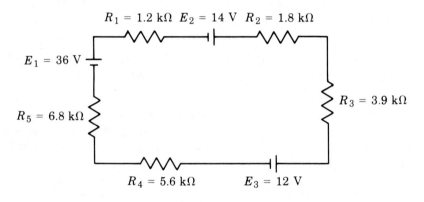

Figure 5–22

Circuit and data table for Problem 22

21. For the circuit of Figure 5-21, calculate the following parameters from the information given: (a) E_T (b) R_T (c) I_T (d) V_1 (e) V_2 (f) V_3 (g) V_4.

22. For the circuit shown in Figure 5-22, calculate the information required for the data table in the figure from the circuit parameters given.

23. From the information given in the circuit of Figure 5-23, calculate all of the circuit parameters called for in the accompanying data table.

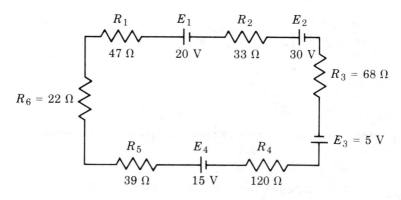

Data Table

$E_T =$ ——— $R_T =$ ——— $I_T =$ ——— $P_T =$ ———

V_1	V_2	V_3	V_4	V_5	V_6	P_1	P_2	P_3	P_4	P_5	P_6

Figure 5–23

Circuit and data table for Problem 23

6

Parallel Circuits

Objectives

After studying this chapter, you will be able to:

Calculate the equivalent resistance of several parallel resistors.

Explore the fact that the total power consumed in a parallel circuit is the sum of the power used in each branch.

Describe the effects of an open or shorted component in an electric circuit.

Introduction

Many components of electric circuits are connected so that they are in parallel with each other – that is, directly across the same source of voltage. An arrangement of this type, as opposed to a series circuit, provides more than one path through which current can flow. The more paths, or *branches*, that are added in parallel, the less opposition there is to current from the supply.

Applications

When components are connected in parallel, we can disconnect one or more of them without affecting the performance of the others. The simplest example is a very common situation: several electrical appliances plugged into the same electric outlet. If one should be unplugged or disconnected, the performance of the others is not impaired.

Another system that is familiar is the electric system of an automobile. The several circuits, such as headlights, radio, and turn signals, are all connected across—or in parallel with—the alternator-battery supply. When they are turned on, individually or in concert, they all share the same power source.

The central processing unit, or "brain," of a computer comprises many parallel circuits, all working together to perform complicated tasks crucial to today's technology. Each central processing unit, or CPU, contains hundreds of separate circuits connected across a common power source. Logic gates, for example, of which there may be thousands in a single CPU, act like switches that are either turned on or turned off. Turning off one or several logic gates does not affect the operations of those logic gates that are turned on. As the computer speeds through its operations, these parallel circuits switch on and off millions of times per second, as determined by the demands of whichever programs are running through the CPU. Stated simply, each CPU works on the basis of thousands of simple on-and-off, parallel-connected electrical signals. The more logic gates there are in a CPU, the more complicated are the tasks the individual computer is capable of performing.

In subsequent chapters on alternating currents, we will find that certain circuit components and devices are actually in parallel with each other, although they may not have the appearance of being so connected.

6–1
Definition of Parallel Circuit

Branch   Each current path in a parallel circuit.

A *parallel circuit* can be defined as a circuit that provides more than one current path. Each path is called a **branch**. An example of a parallel circuit is shown in Figure 6–1. One branch consists of R_1, and the other consists of R_2.

Voltage across Each Branch

One characteristic of a parallel circuit is that the *voltage is the same across each branch*. Since R_1 is in parallel with R_2 in Figure 6–1, each resistor has the same potential across it as the supply voltage. This fact may be expressed as follows:

$$E_S = V_{R_1} = V_{R_2}$$

6–1

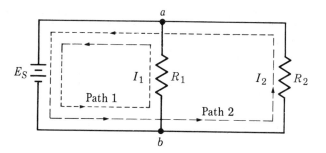

Figure 6–1

Simple parallel circuit containing two branches

Each Branch Current Equal to *E/R*

The amount of current through each branch depends on the ratio of the impressed voltage to the branch resistance. Therefore, *the lower the branch resistance, the larger the current is.* The branch currents can be the same only if the branch resistances are equal.

6–2
Total Current Equal to Sum of Branch Currents

A second characteristic of a parallel circuit is that *the total current flowing from the source is the sum of the individual branch currents.* This fact can be expressed mathematically as follows:

$$I_T = I_1 + I_2 + \cdots + I_n \qquad \text{6–2}$$

This equation is in agreement with KCL, which states that the current entering a junction, as at point *b*, must equal the current leaving the junction.

For another example of the division of currents in a parallel circuit, refer to Figure 6–2. The total current I_T divides when reaching junction *a*. A part of this total current, I_1, flows through R_1. The remainder flows on to R_2 and R_3. At junction *b*, the current divides again, with I_2 flowing through R_2 and I_3 continuing on to R_3.

After flowing through the resistors, the currents recombine.

Current I_3 joins I_2 at junction c. Currents $I_2 + I_3$ combine with I_1 at junction d. The total current I_T, made up of $I_1 + I_2 + I_3$, now flows back into the positive terminal of the supply.

Example 6–1

In the circuit of Figure 6–2, assume $I_3 = 2$ A, $I_1 = 1.5$ A, and R_2 is 0.5 times the value of R_3. Calculate I_T.

Solution

The current in each branch is *inversely proportional* to its resistance. Therefore, if R_2 is one-half the value of R_3, it will have twice the current flow. Consequently, using Equation 6–2, we calculate I_T as follows:

$$I_T = I_1 + I_2 + I_3 = 1.5 \text{ A} + (2 \times 2 \text{ A}) + 2 \text{ A} = 7.5 \text{ A}$$

Example 6–2

In the parallel circuit of Figure 6–2, assume $I_2 = 68$ mA, $I_3 = 45$ mA, and $I_T = 145$ mA. Determine the value of I_1.

Solution

$$I_T = I_1 + I_2 + I_3$$

Solve for I_1:

$$I_1 = I_T - (I_2 + I_3) = 145 \text{ mA} - (68 \text{ mA} + 45 \text{ mA}) = 32 \text{ mA}$$

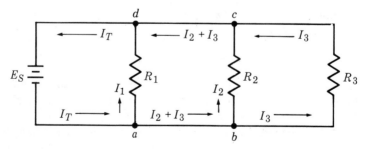

Figure 6–2

Current paths in a parallel circuit

6–3
R_T of Parallel Resistance

In Section 6–2, you learned that the total current in a parallel circuit is the sum of the branch currents. This fact suggests that the total resistance of the circuit must be less than the ohmic value of the smallest resistor, since $R_T = E/I_T$.

Consider the simple parallel combination shown in Figure 6–3a. With 60 V as a source voltage, the current through R_1 is 1 A and through R_2 is 2 A. As indicated, the total current is 3 A. From Ohm's law, we calculate the total resistance as follows:

$$R_T = \frac{E}{I_T} = \frac{60 \text{ V}}{3 \text{ A}} = 20 \ \Omega$$

Note that this total resistance is *lower than* the lowest resistance in the combination. The equivalent circuit for Figure 6–3a is shown in Figure 6–3b.

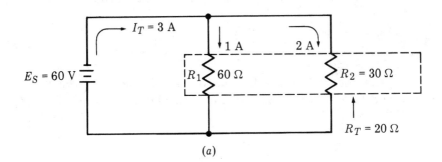

(a)

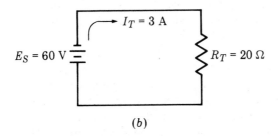

(b)

Figure 6–3

Two parallel resistors: (a) Original arrangement. (b) Equivalent single resistor.

6–4

Calculating R_T by Conductance Method

When three or more resistances are paralleled, we use the *conductance method* to calculate R_T. The line current equals the sum of the branch currents:

$$I_T = I_1 + I_2 + I_3 + \cdots + I_n$$

But each branch current equals E divided by that branch's resistance. For example:

$$I_2 = \frac{E}{R_2}$$

Then

$$I_T = \frac{E}{R_1} + \frac{E}{R_2} + \frac{E}{R_3} + \cdots + \frac{E}{R_n}$$

Therefore

$$\frac{E_T}{R_T} = \frac{E}{R_1} + \frac{E}{R_2} + \frac{E}{R_3} + \cdots + \frac{E}{R_n}$$

Dividing both sides of the equation by E, we obtain the following formula:

$$\frac{1}{R_T} = \frac{1}{R_1} + \frac{1}{R_2} + \frac{1}{R_3} + \cdots + \frac{1}{R_n} \qquad \textbf{6–3}$$

Equation 6–3 states that the total conductance (Section 2–6) of the circuit is equal to the *sum* of the parallel conductances of R_1, R_2, and R_3:

$$G_T = G_1 + G_2 + G_3 \qquad \textbf{6–4}$$

From this equation, we see that each resistance represents a conducting path through which current flows. Increasing the number of parallel resistances increases the total conductance of the circuit by decreasing the total resistance.

The use of the reciprocal key on a hand-held calculator can be very helpful in solving parallel circuit problems. If we enter the resistance value and then press the 1/x key, we have the reciprocal of the resistance, which is its conductance. Hence, if we *add the reciprocals for each resistance, we have the total circuit conductance G_T*. The reciprocal of this value then gives the *equivalent resistance* of the parallel combination.

Example 6–3

Calculate the equivalent resistance of the three resistors shown in Figure 6–4 by using the reciprocal keys on a calculator. (Assume the calculator has been turned on.)

Solution

Enter	Press Keys	Display	Comments
10	$\boxed{1/x}$ $\boxed{+}$	0.1	
20	$\boxed{1/x}$ $\boxed{+}$	0.15	
30	$\boxed{1/x}$ $\boxed{=}$	0.1833	Result is $1/R_T$
	$\boxed{1/x}$	5.4545	The reciprocal of $1/R_T$ equals the total resistance

When only two resistors in parallel are involved, Equation 6–4 can be reduced to solve for the equivalent resistance directly:

$$G_T = G_1 + G_2 = \frac{1}{R_1} + \frac{1}{R_2} = \frac{R_1 + R_2}{R_1 R_2}$$

Therefore

$$R_T = \frac{R_1 R_2}{R_1 + R_2}$$

6–5

This equation indicates that the product of the two resistances divided by their sum gives the equivalent resistance of the pair. The equation is known as the *product-over-sum formula.*

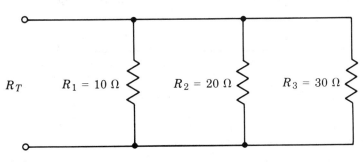

Figure 6–4

Schematic for calculation of R_T

Example 6–4

What is the total resistance of a 10 Ω and a 15 Ω resistor connected in parallel?

Solution

$$R_T = \frac{R_1 R_2}{R_1 + R_2} = \frac{10 \times 15}{10 + 15} = 6\ \Omega$$

We can calculate the unknown value of one of two parallel resistors when the total resistance and the value of one resistor are known. Consider the circuit of Figure 6–5, where R_T and R_1 are given. We can solve for R_2 by using Equation 6–5:

$$R_T = \frac{R_1 R_2}{R_1 + R_2}$$

$$R_2 = \frac{R_1 R_T}{R_1 - R_T} \qquad \textbf{6–6}$$

Example 6–5

Solve for R_2 in the circuit of Figure 6–5.

Solution

$$R_2 = \frac{100 \times 66.67}{100 - 66.67} = 200\ \Omega$$

The ohmic value of R_2 in the circuit of Figure 6–5 can also be determined by the conductance method. The following example illustrates this approach.

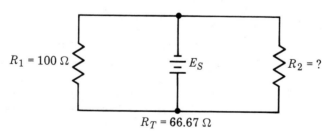

$R_1 = 100\ \Omega$ E_S $R_2 = ?$

$R_T = 66.67\ \Omega$

Figure 6–5

Parallel circuit used in calculating R_2

Example 6-6 ────────────────────────────

Determine the value of R_2 by using the conductance method.

Solution

$$G_T = G_1 + G_2$$

Substitute $\quad\quad 0.015 = 0.01 + G_2$

$$G_2 = 0.015 - 0.01 = 0.005 \text{ S}$$

Therefore $\quad\quad R_2 = \dfrac{1}{G_2} = \dfrac{1}{0.005} = 200 \ \Omega$

6-5
Application of KCL

In some problems, Kirchhoff's current law (KCL) can be an aid in calculating the value of an unknown resistance in a parallel circuit, as demonstrated in the next example.

Example 6-7 ────────────────────────────

From the data given in Figure 6–6, calculate the ohmic value of R_1.

Solution

From KCL, the current entering junction b is the same as the current leaving the junction. Therefore:

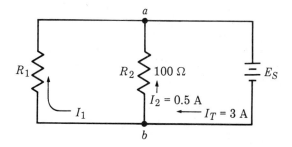

Figure 6-6

Use of KCL in determining the value of R_1

$$I_T = I_1 + I_2$$
$$3\text{ A} = I_1 + 0.5\text{ A}$$
$$I_1 = 3\text{ A} - 0.5\text{ A} = 2.5\text{ A}$$

Now
$$E_{R_2} = I_2R_2 = 0.5\text{ A} \times 100\text{ }\Omega = 50\text{ V}$$
$$E_{R_1} = E_{R_2} = 50\text{ V}$$

Therefore
$$R_1 = \frac{E_{R_1}}{I_1} = \frac{50}{2.5} = 20\text{ }\Omega$$

6–6
Current Divider Formula

In Section 6–1, you learned that the current through each branch can be determined by the ratio E/R and varies *inversely* with branch resistance. Hence, the branch with the smallest resistance will have the greatest current, and the one with the largest resistance will have the least current.

Consider the parallel circuit in Figure 6–7. For this circuit, we have the following relationships:

$$E = \text{voltage across } R_1 = I_1R_1$$
$$E = \text{voltage across } R_2 = I_2R_2$$
$$I_T = I_1 + I_2$$

Therefore
$$I_1R_1 = I_2R_2$$
$$I_1R_1 = (I_T - I_1)R_2 = I_TR_2 - I_1R_2$$

Collect like terms
$$I_1R_1 + I_1R_2 = I_TR_2$$
$$I_1(R_1 + R_2) = I_TR_2$$

Then
$$I_1 = I_T\left(\frac{R_2}{R_1 + R_2}\right) \qquad\qquad \textbf{6–7}$$

Observe that the numerator in Equation 6–7 is R_2 and not R_1. If the same procedure is followed to solve for I_2, we obtain the following equation:

$$I_2 = I_T\left(\frac{R_1}{R_1 + R_2}\right) \qquad\qquad \textbf{6–8}$$

The two expressions in Equations 6–7 and 6–8 are known as the *current divider formulas.*

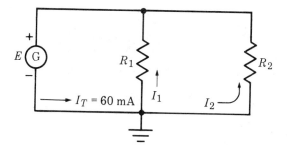

Figure 6–7

Circuit illustrating current divider formula

Example 6–8

From the circuit in Figure 6–7, determine the amount of current through R_1 if $R_1 = 60\ \Omega$ and $R_2 = 90\ \Omega$.

Solution

Apply Equation 6–7:

$$I_1 = \left(\frac{90}{60 + 90}\right) \times 60\ \text{mA} = 0.6 \times 60 = 36\ \text{mA}$$

Example 6–9

Calculate the current through R_2 in Figure 6–7, using the values given in Example 6–8.

Solution

Apply Equation 6–8:

$$I_2 = \left(\frac{60}{60 + 90}\right) \times 60\ \text{mA} = 0.4 \times 60 = 24\ \text{mA}$$

Notice, from Examples 6–8 and 6–9, that I_2 is smaller than I_1 and that the sum of the two load currents equals I_T. Also, it is not necessary to know the supply voltage to find the branch currents when I_T is given. As a reminder, Equations 6–7 and 6–8 apply only to two parallel resistors.

6–7
Total Power in Parallel Circuits

In Section 5–8, you learned that the power expended in a resistor can be determined by dividing the resistance into the square of the impressed voltage:

$$P = \frac{E^2}{R}$$

In a parallel circuit, the total power is the *sum* of the powers expended in each branch. This sum can be expressed as follows:

$$P_T = P_1 + P_2 + P_3 + \cdots + P_n \qquad\qquad \textbf{6-9}$$

Example 6–10 ─────────────────────────────

Determine the total power consumed in the circuit of Figure 6–8.

Solution

$$P_1 = \frac{E^2}{R_1} = \frac{100}{10} = 10 \text{ W}$$

$$P_2 = \frac{E^2}{R_2} = \frac{100}{8} = 12.5 \text{ W}$$

$$P_3 = \frac{E^2}{R_3} = \frac{100}{5} = 20 \text{ W}$$

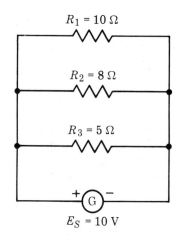

$R_1 = 10\ \Omega$

$R_2 = 8\ \Omega$

$R_3 = 5\ \Omega$

$E_S = 10$ V

Figure 6–8

Circuit for calculation of total power

Substitute this information into Equation 6–9.

$$P_T = 10 \text{ W} + 12.5 \text{ W} + 20 \text{ W} = 42.5 \text{ W}$$

Observe that the branch having the lowest ohmic value dissipates the greatest heat.

6–8
Alternative Methods of Calculating Total Resistance

Assumed-Voltage Method

An easy way to find the total resistance of a parallel-resistive circuit is to *assume* an applied voltage. Any convenient value may be used because it will cancel out in the calculations. Generally, some multiple of 10 is desirable. Calculate the current through each branch as a result of the assumed voltage; this voltage divided by the sum of the branch currents gives the equivalent resistance. The next example demonstrates this technique.

Example 6–11

Calculate the equivalent resistance R_{eq} of the circuit in Figure 6–9.

Solution

Let us assume an applied emf of 100 V. Then, we have the following calculations:

$$I_{R_1} = \frac{100 \text{ V}}{47 \text{ }\Omega} = 2.13 \text{ A}$$

$$I_{R_2} = \frac{100 \text{ V}}{68 \text{ }\Omega} = 1.47 \text{ A}$$

$$I_{R_3} = \frac{100 \text{ V}}{81 \text{ }\Omega} = 1.23 \text{ A}$$

$$I_T = 2.13 + 1.47 + 1.23 = 4.83 \text{ A}$$

$$R_{eq} = \frac{100 \text{ V}}{4.83 \text{ A}} = 20.70 \text{ }\Omega$$

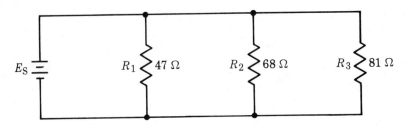

Figure 6–9

Circuit for determining R_T by assumed-voltage method

Equal Parallel Resistors

If several resistors of equal ohmic value are parallel-connected, the equivalent resistance of the combination can be determined by dividing the resistance of one by the total number of resistors in the parallel combination. This rule is expressed by the following formula:

$$R_{eq} = \frac{R}{n}$$ 6–10

where
R = common resistance value
n = number of resistors

Example 6–12

Determine the equivalent resistance of four 100 Ω parallel-connected resistors.

Solution

$$R_{eq} = \frac{100}{4} = 25 \ \Omega$$

Two Parallel Resistors and One Is a Multiple of the Other

There is a simple method of calculating the equivalent resistance of two parallel resistors when one is a multiple of the other. Determine the number of times the smaller goes into the larger (the quotient must be a

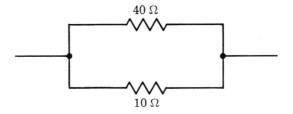

Figure 6–10

Parallel resistors for alternative method of calculating R_{eq}

whole number). Add 1 to this number, and divide it into the larger resistor. The result is the equivalent resistance of the combination.

Example 6–13

Find the equivalent resistance of the parallel combination in Figure 6–10.

Solution

$$40\ \Omega \div 10\ \Omega = 4$$

Add $\qquad 1 + 4 = 5$

Divide $\qquad 40\ \Omega \div 5 = 8\ \Omega$

6–9
Effect of an Open in a Parallel Circuit

The effects of an *open* in a parallel circuit can best be explained with the aid of a schematic diagram such as the one in Figure 6–11. The three lamps are shown connected across a 120 V line, providing normal operation. If an open should develop at point a, there will be no line voltage impressed across any of the lamps. Hence, no current flows, and the entire circuit is considered dead. If the circuit were interrupted at point b, lamp L_3 would not come on, but L_1 and L_2 would function normally. If the lamps were of equal wattage rating, the line current would be reduced to two-thirds of normal value. As you can see, an open in one branch does not adversely affect the operation of the other branches.

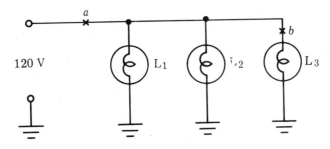

Figure 6–11

Circuit illustrating the effect of an open in a parallel circuit

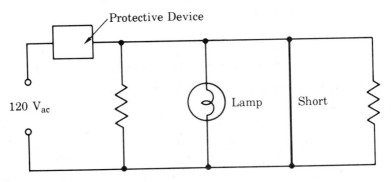

Figure 6–12

Schematic illustrating the effect of a short in a parallel circuit

6–10
Effect of a Short in a Parallel Circuit

A *short* (very low resistance) in a parallel circuit has quite the opposite effect of an open. Should any one of the branches become short-circuited, the total current flowing in the main line will increase tremendously. If the circuit is designed correctly, a *protective device*, such as a fuse or a circuit breaker, will immediately open up and protect the supply.

Figure 6-12 shows a simple parallel circuit having a short. An inspection of the schematic shows that the short is across the supply of 120 V ac, which essentially reduces the branch currents to zero. Current through the short, however, will be limited only by the internal

Figure 6–13

Symbols for protective devices: (a) Fuse. (b) Push-to-reset circuit breaker. (c) Switch type of circuit breaker.

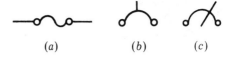

(a)   (b)   (c)

resistance of the voltage supply. If no protective device is in the circuit, the supply will likely be damaged. Fuses or circuit breakers are normally installed in one side of the supply, as indicated in Figure 6–12. Symbols for these devices are shown in Figure 6–13.

⌐—Summary

A parallel circuit contains more than one current path.

The voltage is the same across all branches of a parallel circuit.

The more branches a parallel circuit has, the lower will be the total resistance.

The branch with the least resistance will have the greatest current.

Branch currents are inversely proportional to the branch resistance.

The total current is the sum of the branch currents.

The equivalent resistance R_{eq} will always be lower than the lowest resistance.

For two resistors, the formula for R_T is as follows:

$$R_T = \frac{R_1 R_2}{R_1 + R_2}$$

For three or more resistors, the formula for R_T is as follows:

$$\frac{1}{R_T} = \frac{1}{R_1} + \frac{1}{R_2} + \frac{1}{R_3} + \cdots + \frac{1}{R_n}$$

The total conductance of a circuit is given by the following formula:

$$G_T = G_1 + G_2 + G_3 + \cdots + G_n$$

To find one resistance of a parallel combination when R_T and the other is known, use the following formula:

$$R_2 = \frac{R_1 R_T}{R_1 - R_T}$$

The current leaving a junction must be equal to the current entering the junction (KCL).

The total power in a parallel circuit is the sum of the powers expended in each branch:

$$P_T = P_1 + P_2 + P_3 + \cdots + P_n$$

Total resistance R_T can be determined by calculating I through each branch, using an assumed voltage. The sum of these currents divided into the assumed emf gives the equivalent resistance.

An open in one branch of a parallel circuit causes I_T to be reduced.

A short in a parallel circuit causes I_T to greatly increase.

Fuses and circuit breakers are devices used to protect circuits and supply sources from damage due to overloading.

Progress Test

The bracketed number after each question indicates the section of this chapter where the answer can be found.

    1.   Parallel-resistive circuits are characterized by: (a) equal currents through each branch. (b) equal voltages across each branch. (c) an R_T greater than the lowest resistance. (d) branch currents that are proportional to branch resistances.   [6–1]

    2.   A 68 Ω and an 81 Ω resistor are connected in parallel. The R_T is: (a) 33.65 Ω. (b) 36.97 Ω. (c) 38.41 Ω. (d) 42.18 Ω.   [6–4]

    3.   A 27.3 kΩ and a 51.6 kΩ resistor are paralleled. The R_T is: (a) 19.06 kΩ. (b) 18.23 kΩ. (c) 17.85 kΩ. (d) 16.49 kΩ.   [6–4]

    4.   The total resistance of 500, 1000, and 1500 Ω connected in parallel is: (a) 125 Ω. (b) 179 Ω. (c) 250 Ω. (d) 273 Ω.   [6–4]

    5.   A 180 Ω and a 240 Ω resistor are parallel-connected across a 100 V source. The total circuit current is: (a) 0.83 A. (b) 0.92 A. (c) 0.97 A. (d) 1.02 A.   [6–2]

6. A 55 Ω and a 23 Ω resistor are paralleled. If the current through the source is 2.5 A, what is the source voltage? (a) 36.2 V (b) 40.5 V (c) 46.5 V (d) 52.1 V [6–4]

7. In the circuit of Figure 6-2, assume $I_1 = 5.2$ mA, $I_2 = 3.15$ mA, and $I_T = 10.38$ mA. What is the current through R_3 in milliamperes? (a) 2.03 (b) 2.30 (c) 2.42 (d) 2.63 [6–2]

8. A circuit has three conductances in parallel. If $G_1 = 0.033$ S, $G_2 = 0.027$ S, and $G_3 = 0.008$ S, what is the circuit's resistance? (a) 14.7 Ω (b) 15.5 Ω (c) 16.7 Ω (d) 18.3 Ω [6–4]

9. Two parallel resistors have a total resistance of 400 Ω. If one has 600 Ω, what is the value of the other? (a) 800 Ω (b) 950 Ω (c) 1050 Ω (d) 1200 Ω [6–4]

10. In the circuit of Figure 6-7, if $R_1 = 180$ Ω, $R_2 = 270$ Ω, and $I_T = 24$ mA, what is the current through R_1? (a) 12.9 mA (b) 14.4 mA (c) 14.8 mA (d) 15.4 mA [6–5]

11. In the circuit of Figure 6-2, assume $I_T = 12.3$ A, $R_1 = 100$ Ω, $R_2 = 150$ Ω, and $E_S = 325$ V. Calculate the resistance of R_3. (a) 42.1 Ω (b) 47.2 Ω (c) 56.7 Ω (d) 61.5 Ω [6–5]

12. Four 1250 Ω, 5 W resistors are connected in parallel. What is R_{eq}? (a) 5 kΩ (b) 2.5 kΩ (c) 312.5 Ω (d) 250 Ω [6–8]

13. What total wattage can be dissipated by the resistors described in Question 12? (a) 5 W (b) 10 W (c) 15 W (d) 20 W [6–7]

┌──Problems

Answers to odd-numbered problems are at the back of the book.

1. What is the equivalent resistance of three parallel resistors with values of $R_1 = 430$ Ω, $R_2 = 560$ Ω, and $R_3 = 680$ Ω?

2. Three resistors of 18, 22, and 47 Ω are connected in parallel. Calculate R_{eq}.

3. Three resistors of 20 Ω, 200 Ω, and 2 kΩ are in parallel. What is the equivalent resistance?

4. In the circuit of Figure 6-14, assume $I_2 = 0.6$ A and $I_T = 0.9$ A. Calculate: (a) the value of E_S (b) the ohmic value of R_1.

5. For Problem 4, what is the wattage dissipation of R_1? What is the total dissipation?

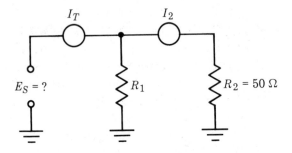

Figure 6–14

Circuit for Problem 4

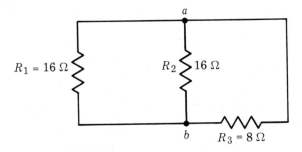

Figure 6–15

Circuit for Problem 8

6. Refer to Figure 6-2. Assume $R_2 = 5.1$ kΩ, $R_3 = 2.7$ kΩ, $I_T = 0.438$ A, and $I_1 = 0.152$ A. Calculate the value of R_1.

7. A 400 Ω, 10 W resistor and a 1.5 kΩ, 50 W resistor are in parallel. What is the maximum voltage that can be connected across them without exceeding the wattage rating of either?

8. Given the circuit of Figure 6-15, what is R_{eq} as measured between a and b?

9. What is R_{eq} of the circuit in Figure 6-16 if $R_1 = 200$ Ω, $R_2 = 400$ Ω, and $R_3 = 600$ Ω?

10. Calculate the values of R_{eq} for the circuits of Figures 6-17a and 6-17b. What effect does the 100 kΩ resistor have?

11. For the circuit of Figure 6-18, determine: (a) I_1 (b) I_2.

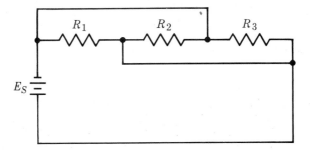

Figure 6–16

Circuit for Problem 9

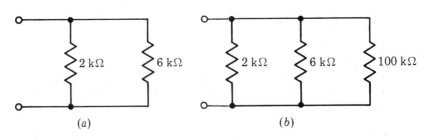

Figure 6–17

Circuit for Problem 10

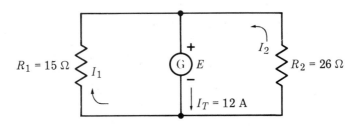

Figure 6–18

Circuit for Problem 11

12. What will the current distribution be in R_2 and R_3 in the circuit of Figure 6–19?

13. When SW is closed (Figure 6–19), what current will flow through R_1?

14. Calculate the power dissipated in the circuit of Figure 6–19 when: (a) SW is open (b) SW is closed.

15. Using the assumed-voltage method, calculate R_{eq} of three parallel resistors with values of 22, 68, and 120 Ω.

16. What is the total power used in the resistors described in Problem 15 if the assumed voltage is 100 V?

17. The following resistors are connected in parallel: 220 Ω, 1.2 kΩ, 1800 Ω, and 47 Ω. Calculate the equivalent total resistance. What is the conductance of the circuit in millisiemens?

18. In the circuit of Figure 6–20, what value of R_1 will make R_T equal to 840 Ω?

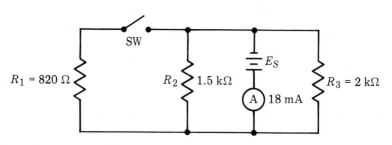

Figure 6–19

Circuit for Problems 12, 13, and 14

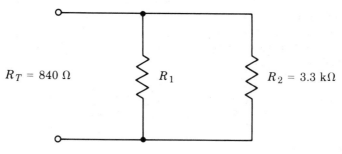

Figure 6–20

Circuit for Problem 18

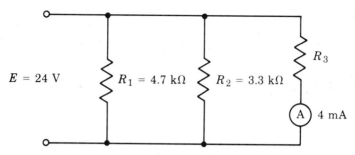

Figure 6–21

Circuit for Problem 19

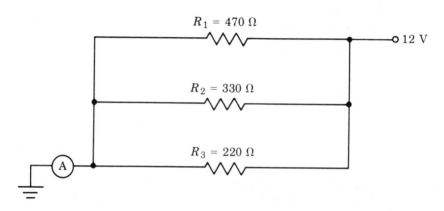

Figure 6–22

Circuit for Problems 20 and 21

19. Refer to Figure 6–21. Calculate the following: (a) R_3 (b) R_T (c) I_T.

20. Refer to the circuit of Figure 6–22. If R_1 opens, how much will the circuit current change?

21. In the circuit of Figure 6–22, if R_2 shorts out, how much of a change, if any, will there be in total current flow to the circuit? (Assume the resistance of the short is 0 Ω.)

22. A current generator provides 500 mA to a parallel circuit. Calculate the amount of current flowing to each resistor in the circuit of Figure 6–23.

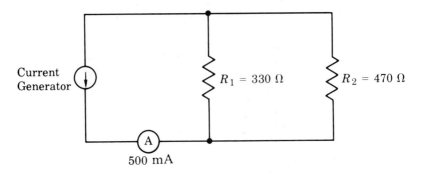

Figure 6–23

Circuit for Problem 22

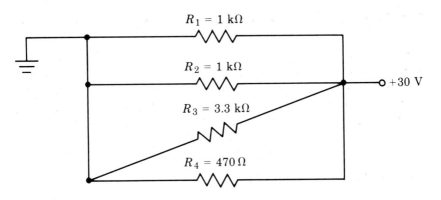

Figure 6–24

Circuit for Problem 23

23. For the circuit of Figure 6-24, calculate: (a) R_T (b) the power consumed by each resistor (c) P_T.

24. Refer to Figure 6-25. Find the value of R_3.

25. Refer to the circuit of Figure 6-26. Calculate: (a) R_T (b) I_T (c) I_2.

26. In the circuit of Figure 6-26, what are G_T and G_2?

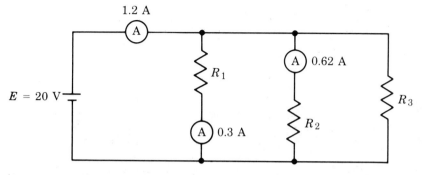

Figure 6–25

Circuit for Problem 24

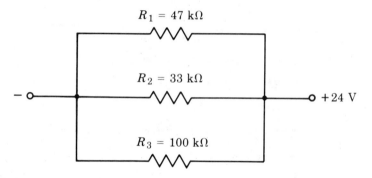

Figure 6–26

Circuit for Problems 25 and 26

7

Series-Parallel Circuit Analysis

Objectives

After studying this chapter, you will be able to:

Recognize series-parallel circuits.

Simplify series-parallel circuits.

Perform simplified series-parallel circuit calculations.

Introduction

The series and parallel circuits previously studied do not account for all of the conditions encountered in electronic circuits. They do, however, provide a good basis for the more complex arrangements to be investigated in this chapter. Many circuits involve combinations of series and parallel connections, which are called *series-parallel circuits*, and they exist in a variety of configurations.

Applications

If two or more loads in a parallel circuit are connected to a power source, it is a form of series-parallel circuit. Although the loads form a parallel connection, the source inevitably has some internal resistance, which is in series with the load. Examples of passive series-parallel circuits with these characteristics are batteries, power supplies, and generators. Active, or amplifying, devices such as transistors, linear integrated circuits, and vacuum tubes also have internal resistance, which may be considered as being in series or parallel (depending on circuit arrangement) with the load to which they are connected.

The circuits associated with these devices themselves contain series-parallel circuits supplying the different current paths used in providing biasing and other operating potentials. Most electric and electronic circuits are made up of series-parallel circuit combinations.

7–1
Simple Series-Parallel Circuits

Series-parallel circuits   Combinations of series and parallel circuits connected to a power source.

Bank   Two or more similar devices parallel-connected across a common power source.

A simple **series-parallel circuit** is shown in Figure 7–1*a*. The parallel connection of R_2 and R_3 is called a **bank**. Now, resistor R_1 is not in series with R_2 and R_3 individually, as may appear at first glance. The circuit current I_1 divides as it enters junction x, and it becomes I_2 and I_3. Neither I_2 nor I_3 is equal to I_1; therefore, neither R_2 nor R_3 can be said to be in series with R_1. From the definition of a *series circuit*, you learned that the *current is the same through every component*. However, R_1 is in series with the parallel combination of R_2 and R_3. By drawing an equiva-

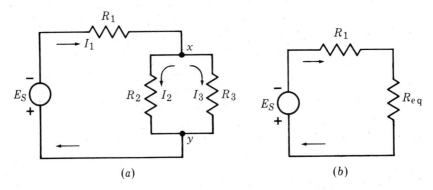

(*a*)   (*b*)

Figure 7–1

Simple series-parallel circuit: (*a*) Original circuit. (*b*) Equivalent circuit.

lent resistance of R_2 and R_3, as in Figure 7-1b, we can illustrate this fact.

> **Equivalent circuit**   For a resistive network, a single resistance with a value equal to the net resistance of the network.

Once a series-parallel circuit has been simplified and an **equivalent circuit** drawn, it is easy to calculate the total resistance. The current division through the several components can then be established. The next example illustrates the technique.

Example 7-1

Given the series-parallel circuit of Figure 7-2, determine the total circuit resistance and current.

Solution

We must first determine R_{eq} of the parallel combination R_1 and R_2.

$$R_{eq} = \frac{24 \times 36}{24 + 36} = 14.4\ \Omega$$

$$R_T = 20\ \Omega + 14.4\ \Omega = 34.4\ \Omega$$

The circuit current I_T is then as follows:

$$I_T = \frac{E}{R_T} = \frac{12\ \text{V}}{34.4\ \Omega} = 0.349\ \text{A}$$

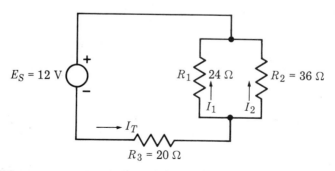

Figure 7-2

Series-parallel circuit

The IR drop across each resistor in the circuit of Figure 7-2 can now be easily determined. Voltage drop $V_{R_1} = V_{R_2}$, which is equal to $I_T R_{\text{eq}}$. Since $I_T = 0.349$ A and $R_{\text{eq}} = 14.4$ Ω, then $V_{R_{\text{eq}}} = 5.025$ V, which is the voltage across R_1 and R_2. Voltage $V_{R_3} = I_T R_3$, which is 6.98 V. The sum of $V_{R_{\text{eq}}}$ and V_{R_3} is equal to the supply voltage (allowing for rounding of numbers).

7-2
Simplifying Series-Parallel Circuits

There are many possible combinations of series-parallel circuits. We will now investigate a few of the simple circuits and formulate a method, or procedure, to facilitate their solution.

Consider the circuit of Figure 7-3a. The lower end of R_1, R_3, and

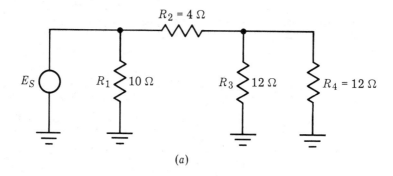

(a)

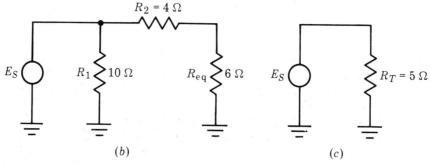

(b) (c)

Figure 7-3

Series-parallel circuit: (a) Original circuit. (b) Partial simplification. (c) Total resistance R_T of 5 Ω.

R_4 and the supply are all connected to ground, which completes the circuits for these elements. The first step is to convert all parallel combinations to equivalent resistances.

In this schematic, R_3 and R_4 are the only such combination, each with a resistance of 12 Ω. In a previous chapter, you learned that the equivalent resistance of two parallel resistors of equal ohmic value is one-half the resistance of one. Therefore, R_{eq} of R_3 and R_4 is 6 Ω.

By redrawing the circuit, as shown in Figure 7–3b, we see that R_{eq} and R_2 are in series. Their series resistance is 6 Ω + 4 Ω = 10 Ω. This equivalent resistance (10 Ω) is in parallel with R_1, which has a value of 10 Ω. Therefore, the total resistance of the circuit is 5 Ω, as shown in Figure 7–3c.

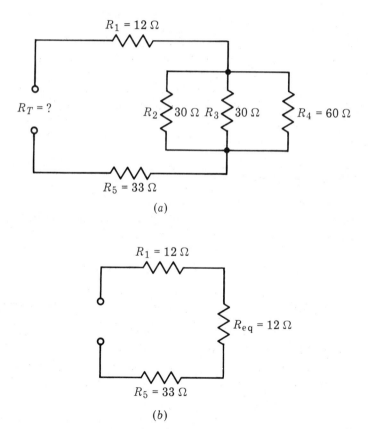

(a)

(b)

Figure 7–4

Series-parallel circuit for Example 7–2: (a) Original circuit. (b) Simplified version.

Example 7-2 ───

Determine the total resistance of the circuit in Figure 7-4a.

Solution

First, we solve for the equivalent resistance of the parallel combination. Let us use the conductance method.

$$G_{eq} = G_2 + G_3 + G_4$$

$$= \frac{1}{R_2} + \frac{1}{R_3} + \frac{1}{R_4} = \frac{1}{30} + \frac{1}{30} + \frac{1}{60} \cong 0.083 \text{ S}$$

$$R_{eq} = \frac{1}{G_{eq}} = \frac{1}{0.083} = 12 \ \Omega$$

By inspection of Figure 7-4b, we see that R_1 and R_5 are in series with R_{eq}. Therefore, R_T is as follows:

$$R_T = R_1 + R_{eq} + R_5 = 12 \ \Omega + 12 \ \Omega + 33 \ \Omega = 57 \ \Omega$$

7-3
Analyzing Series-Parallel Circuits

Let us examine a circuit that contains two parallel combinations, such as the circuit in Figure 7-5. We must begin by determining the equivalent resistance of each. A simplified schematic should then be drawn, showing each parallel network as a single resistor. The sum of these resistors will be R_T. We will perform these operations in the next example.

Example 7-3 ───

For the circuit of Figure 7-5, calculate the equivalent resistance of each branch and R_T.

Solution

Solve for the equivalent resistance between points a and b:

$$R_{ab} = \frac{4 \times 12}{4 + 12} = 3 \ \Omega$$

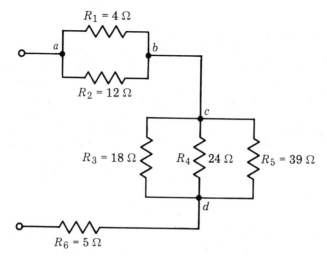

Figure 7–5

Series-parallel circuit for Example 7–3

Similarly, by the conductance method, we can solve for R_{cd}:

$$G_{cd} = \frac{1}{18} + \frac{1}{24} + \frac{1}{39}$$

$$R_{cd} = 8.14 \ \Omega$$

Thus

$$R_T = R_{ab} + R_{cd} + R_6 = 3 \ \Omega + 8.14 \ \Omega + 5 \ \Omega = 16.14 \ \Omega$$

Example 7–4

Calculate R_{eq} of the circuit in Figure 7–6.

Solution

In a more complex circuit like this one, it is generally best to *start combining resistances* at the point *farthest from the source of power.* Therefore, starting with R_8, we observe that it is in parallel with R_6.

$$R_{cd} = \frac{6 \times 3}{6 + 3} = 2 \ \Omega$$

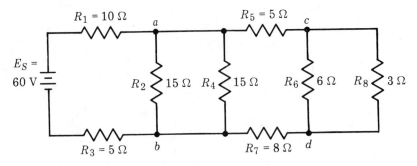

Figure 7-6

Series-parallel circuit for Example 7-4

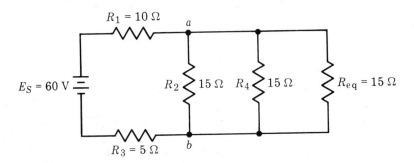

Figure 7-7

Partial simplification of the circuit in Figure 7-6

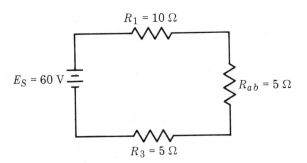

Figure 7-8

Further simplification of the circuit in Figure 7-7

Now, R_{cd} is in series with R_5 and R_7, and the sum of these resistances is

$$5\ \Omega + 2\ \Omega + 8\ \Omega = 15\ \Omega$$

The circuit may be redrawn as in Figure 7–7 to reflect this simplification. Now, R_{eq} is in parallel with R_2 and R_4. Inasmuch as each of these resistances is 15 Ω, the method for calculating equal parallel resistors discussed in Section 6–8 may be used. Thus, their equivalent resistance R_{ab} is as follows:

$$R_{ab} = \frac{R}{n} = \frac{15}{3} = 5\ \Omega$$

If the circuit of Figure 7–7 is redrawn to reflect this simplification, we have the series arrangement shown in Figure 7–8. The total circuit resistance is the sum of these three resistances:

$$R_T = R_1 + R_{ab} + R_3 = 10 + 5 + 5 = 20\ \Omega$$

Thus, we can see that the entire circuit of Figure 7–6 can be represented by a single 20 Ω resistor.

7–4
Voltage and Current Distribution

Now that we have learned how to calculate R_{eq} of a network, we must determine how the supply voltage is distributed throughout the circuit. From this distribution, we can calculate the resulting current through each resistor. The next example illustrates the procedure.

Example 7–5 ——————————————————————————————

Find the voltage and current distribution throughout the network of Figure 7–9.

Solution

We must first calculate R_{eq} of the circuit. Starting with the resistors farthest from the supply, we observe that the series arrangement of R_4 and R_5 is in parallel with R_2. By inspection, we

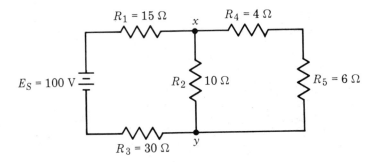

Figure 7-9

Series-parallel circuit for Example 7-5, to show E and I distribution

can see that the equivalent resistance between points x and y is 5 Ω. Therefore:

$$R_{eq} = R_1 + R_{xy} + R_3 = 50 \ \Omega$$

The total circuit current is thus as follows:

$$I_T = \frac{E}{R_{eq}} = \frac{100 \ V}{50 \ \Omega} = 2 \ A$$

Knowing I_T, we now begin to work back toward the resistors farthest from the supply, calculating the current through each resistance as we proceed. Now, I_T must flow through R_1, R_{xy}, and R_3. Therefore, the IR drops across these resistances are as follows:

$$V_{R_1} = I_T R_1 = 2 \ A \times 15 \ \Omega = 30 \ V$$
$$V_{R_{xy}} = I_T R_{xy} = 2 \ A \times 5 \ \Omega = 10 \ V$$
$$V_{R_3} = I_T R_3 = 2 \ A \times 30 \ \Omega = 60 \ V$$

The sum of these voltage drops is, of course, equal to the supply and satisfies KVL. Inasmuch as $V_{R_{xy}} = 10$ V, the current through R_2 will be 10 V/10 Ω = 1 A, and the current through R_4 and R_5 will be 10 V/10 Ω = 1 A. This result satisfies KCL, because the current entering junction x is 2 A, while the current leaving the junction is $I_{R_2} = 1$ A plus $I_{R_{4,5}} = 1$ A, for a total of 2 A. The voltage drop across R_4 is 4 V, and the drop across R_5 is 6 V, which agrees with $V_{R_{xy}} = 10$ V, as calculated.

Example 7–6

Solve for V_o in the circuit of Figure 7–10, using the data supplied.

Solution

Proceed to find R_T as follows (all values expressed in kilohms):

$$R_{cd} = \frac{R_4 \times (R_5 + R_6)}{R_4 + (R_5 + R_6)} = \frac{5 \times 5}{5 + 5} = 2.5 \text{ k}\Omega$$

$$R_{ab} = \frac{R_2 \times (R_3 + R_{cd})}{R_2 + (R_3 + R_{cd})} = \frac{5 \times 5}{5 + 5} = 2.5 \text{ k}\Omega$$

Therefore $R_T = R_1 + R_{ab} = 1 + 2.5 = 3.5 \text{ k}\Omega$

The problem is approximately half-solved. We must now calculate I_T:

$$I_T = \frac{E_S}{R_T} = \frac{24 \text{ V}}{3.5 \text{ k}\Omega} = 6.86 \text{ mA}$$

(*Note:* Dividing kilohms into volts gives an answer in milliamperes.) To find the voltage drop across R_2, we calculate V_{R_1}, multiplying current times resistance, and subtract that result from E_S:

$$V_{R_2} = E_S - V_{R_1} = 24 \text{ V} - (6.86 \text{ mA} \times 1 \text{ k}\Omega)$$
$$= 24 \text{ V} - 6.86 \text{ V} = 17.14 \text{ V}$$

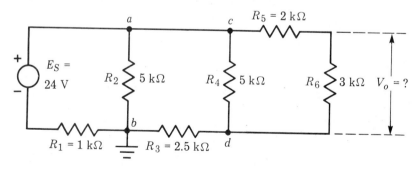

Figure 7–10

Circuit for Example 7–6, with V_o to be calculated

Now, we determine the current through R_2:

$$I_{R_2} = \frac{V_{R_2}}{R_2} = \frac{17.14 \text{ V}}{5 \text{ k}\Omega} = 3.43 \text{ mA}$$

The difference between I_T and I_{R_2} is the current flowing through R_3. Knowing this current, we can determine V_{R_3}, which, when subtracted from V_{R_2}, gives V_{R_4}:

$$I_{R_3} = I_T - I_{R_2} = 6.86 \text{ mA} - 3.43 \text{ mA} = 3.43 \text{ mA}$$

Therefore $\quad V_{R_3} = I_{R_3} \times R_3 = 3.43 \text{ mA} \times 2.5 \text{ k}\Omega = 8.58 \text{ V}$

Then $\quad V_{R_4} = V_{R_2} - V_{R_3} = 17.14 \text{ V} - 8.58 \text{ V} = 8.56 \text{ V}$

Now, V_o can be determined from the voltage divider formula:

$$V_o = V_{R_6} = \frac{R_6}{R_5 + R_6} \times 8.56 \text{ V}$$

$$= \frac{3}{2 + 3} \times 8.56 = 5.14 \text{ V}$$

From Example 7–6, we can see how the voltage dropped across each resistance *diminishes* as we move farther away from the source. If the power supply voltage fluctuated, the output V_o would *fluctuate proportionately*. That is, if the supply increased 10%, the output would increase by the same percentage, and vice versa, which implies that the currents through the several resistors would fluctuate accordingly. This rule points out the imperative need for *stable* source voltage if the output is expected to be stable.

7–5
Voltage and Current Distribution in a Simple Transistor Circuit

A special type of series-parallel circuit is shown in Figure 7–11; it uses a transistor. You do not have to have a complete understanding of how a transistor operates to solve for the voltage and the current distribution in this circuit. Simply stated, a transistor has three connections: emitter (E), base (B), and collector (C). When the circuit is connected to a voltage source, current flows from emitter to base (a diode; see Figure

2–18d), where it divides. Most of the current goes on to the collector; a very small amount leaves the base via R_B. This small amount of base current I_B controls how much collector current I_C flows. (Thus, the transistor is a current-controlled device – a type of electronic valve or amplifier.) If the resistance values are properly selected, there will be approximately 0.6 V between the base and emitter (V_{BE}) connections for a silicon transistor (S_i).

Let us make a voltage and current analysis of the transistor circuit shown in Figure 7–11, assuming an emitter current I_E of 3 mA. This current produces an IR drop across R_E as follows:

$$V_{R_E} = I_E R_E = 3 \text{ mA} \times 1 \text{ k}\Omega = 3 \text{ V}$$

With this result, we can calculate the voltage drop across R_B (where $V_{BE} = 0.6$ V):

$$V_{R_B} = 24 \text{ V} - (V_{BE} + V_{R_E}) = 24 \text{ V} - (0.6 \text{ V} + 3 \text{ V}) = 20.4 \text{ V}$$

The base current through R_B is as follows:

$$I_B = \frac{V_{R_B}}{R_B} = \frac{20.4 \text{ V}}{800 \text{ k}\Omega} = 0.0255 \text{ mA} = 25.5 \text{ }\mu\text{A}$$

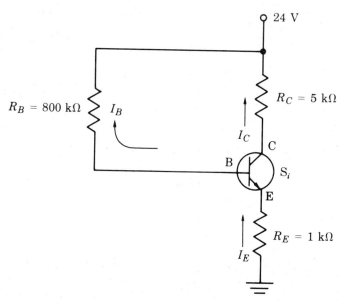

Figure 7–11

Circuit using a transistor as a type of series-parallel arrangement

The collector current is equal to the emitter current minus the base current. Substitute values:

$$I_C = I_E - I_B = 3 \text{ mA} - 0.0255 \text{ mA} = 2.9745 \text{ mA}$$

Then
$$V_{RC} = I_C R_C = 2.9745 \text{ mA} \times 5 \text{ k}\Omega = 14.8725 \text{ V} \cong 15 \text{ V}$$

The voltage drop across the emitter-collector terminals can be determined by using KVL:

$$V_{CE} = 24 \text{ V} - (V_{RC} + V_{RE}) = 24 \text{ V} - (15 \text{ V} + 3 \text{ V}) = 6 \text{ V}$$

Add voltages around the circuit (use KVL):

$$24 \text{ V} - (V_{RC} + V_{CE} + V_{RE}) = 0$$
$$24 \text{ V} - (15 \text{ V} + 6 \text{ V} + 3 \text{ V}) = 0$$

and
$$24 \text{ V} - (V_{RB} + V_{BE} + V_{RE}) = 0$$
$$24 \text{ V} - (20.4 \text{ V} + 0.6 \text{ V} + 3 \text{ V}) = 0$$

Example 7–7

Refer to Figure 7–11. Calculate the voltage and current distribution around the circuit if $R_C = 3.3 \text{ k}\Omega$, $R_E = 470 \ \Omega$, $I_B = 25 \ \mu\text{A}$, and $I_E = 2.5 \text{ mA}$. (For practical purposes, we can assume that $I_C = I_E$.) Assume a supply voltage of 16 V.

Solution

$$V_{RE} = I_E R_E = 2.5 \text{ mA} \times 0.47 \text{ k}\Omega = 1.175 \text{ V}$$
$$V_{RC} = I_C R_C = 2.5 \text{ mA} \times 3.3 \text{ k}\Omega = 8.25 \text{ V}$$
$$V_{CE} = 16 \text{ V} - (V_{RC} + V_{RE})$$
$$= 16 \text{ V} - (8.25 \text{ V} + 1.175 \text{ V}) = 6.575 \text{ V}$$
$$V_{RB} = 16 \text{ V} - (V_{BE} + V_{RE})$$
$$= 16 \text{ V} - (0.6 \text{ V} + 1.175 \text{ V}) = 14.225 \text{ V}$$

Then
$$R_B = \frac{V_{RB}}{I_B} = \frac{14.225 \text{ V}}{25 \ \mu\text{A}} = 0.569 \text{ M}\Omega$$

7-6

Loaded Voltage Dividers

Bleeder current   Current drawn continuously from a power supply by a resistive circuit.

Voltage dividers are used to supply two or more potentials to various parts of a circuit from the same power supply. A simple loaded voltage divider is shown in Figure 7-12, with two separate loads connected to it. The current I_T entering the grounded junction of R_3, load 1, and load 2 divides into I_{Bldr} and $I_{L_1} + I_{L_2}$. The current flowing up through R_3 is called the **bleeder current** and is designated as I_{Bldr}. The bleeder current represents a load imposed on the power supply by the resistance of the unloaded voltage divider arrangement. Current through the bleeder resistor usually ranges from 10% to 20% of the load current.

Regulation   The ability of a power source to maintain a constant output voltage under changing loads.

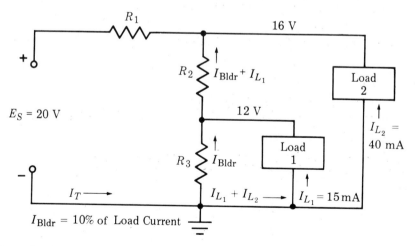

Figure 7-12

Loaded voltage divider

The bleeder current serves two basic functions. First, it allows the filter capacitors in the filter network of the power supply to discharge when the supply is turned off. Second, it places a light load on the supply even if no other current should be drawn. The bleeder current improves the **regulation** of the supply, or the ability of a supply to maintain a constant voltage under changing loads.

> **Current-limiting resistor**   A resistor inserted into an electric circuit to limit the current to some preestablished value.

Inspection of the voltage divider arrangement shown in Figure 7-12 indicates that it is a series-parallel circuit. In some cases, the maximum potential desired is less than the power supply's output. In this case, an additional voltage-dropping resistor, or **current-limiting resistor**, is used to reduce the power supply voltage as represented by R_1.

The procedure used in determining the values of the several voltage divider resistors is given in the next example.

Example 7-8

Using the data supplied in the circuit of Figure 7-12, calculate the ohmic values of R_1, R_2, and R_3 necessary to provide the desired load voltages.

Solution

We must begin by solving for the resistance of R_3. The voltage across R_3 must be 12 V to satisfy the requirements of load 1. The bleeder current is to be 10% of the total load current (that is, $I_{L_1} + I_{L_2}$):

$$I_{\text{Bldr}} = (I_{L_1} + I_{L_2}) \times 0.1 = (15 \text{ mA} + 40 \text{ mA}) \times 0.1 = 5.5 \text{ mA}$$

Calculate R_3:

$$R_3 = \frac{V_{R_3}}{I_{\text{Bldr}}} = \frac{12 \text{ V}}{5.5 \text{ mA}} = 2.18 \text{ k}\Omega$$

To calculate R_2, we must determine its IR drop and the current through it. The voltage drop across R_2 is as follows:

$$V_{R_2} = 16 \text{ V} - 12 \text{ V} = 4 \text{ V}$$

The current through R_2 is the sum of I_{Bldr} and I_{L_1}, since they have joined together at the junction of R_2 and R_3. We may now determine the value of R_2:

$$R_2 = \frac{V_{R_2}}{I_{\text{Bldr}} + I_{L_1}} = \frac{4 \text{ V}}{5.5 \text{ mA} + 15 \text{ mA}} \cong 200 \text{ }\Omega$$

To calculate the required value of R_1 necessary to satisfy the specifications of the divider, we must first determine the voltage drop across R_1:

$$V_{R_1} = E_S - (V_{R_2} + V_{R_3}) = 20 \text{ V} - (4 \text{ V} + 12 \text{ V}) = 4 \text{ V}$$

Now, the current through R_1 will be the sum of the load currents plus the bleeder current:

$$I_{R_1} = I_{\text{Bldr}} + I_{L_1} + I_{L_2} = I_T$$
$$= 5.5 \text{ mA} + 15 \text{ mA} + 40 \text{ mA} = 60.5 \text{ mA}$$

Therefore $\quad R_1 = \dfrac{V_{R_1}}{I_{R_1}} = \dfrac{4 \text{ V}}{60.5 \text{ mA}} = 66 \text{ }\Omega$

To complete the solution of this voltage divider network, let us calculate the wattage dissipated by each resistor:

$$P_{R_1} = \frac{V_{R_1}^2}{R_1} = \frac{(4 \text{ V})^2}{66 \text{ }\Omega} = 0.24 \text{ W}$$

$$P_{R_2} = \frac{V_{R_2}^2}{R_2} = \frac{(4 \text{ V})^2}{200 \text{ }\Omega} = 0.08 \text{ W}$$

$$P_{R_3} = \frac{V_{R_3}^2}{R_3} = \frac{(12 \text{ V})^2}{2.18 \text{ k}\Omega} = 0.066 \text{ W}$$

It is good engineering practice to *double* the calculated *wattage dissipation* and *then go to the next higher commercially available wattage rating*, because the wattage rating is based on the resistor being mounted in free space – that is, with *adequate ventilation*. Taking into account that resistors frequently are mounted under a chassis where ventilation is restricted, we use the following ratings for the divider resistors:

$$R_1 = 1 \text{ W} \quad R_2 = 0.25 \text{ W} \quad R_3 = 0.25 \text{ W}$$

These values are considered the minimum safe values.

7–7
Bridge Circuits

> **Bridge circuit**   A network arranged so that, when an emf is present in one branch, the response of some form of indicating device in another branch may be zeroed by proper adjustments of components in still other branches.

Bridge circuits find wide application in electronics. They are used in many types of measuring instruments and industrial control circuits as well. A bridge circuit is a network so arranged that, when an electromotive force is present in one branch, the response of a suitable detecting device in another branch may be zeroed by adjustment of the electrical constants of still other branches. The elements of the bridge, or *legs*, as they are called, may include resistance, inductance, capacitance, or combinations of all three of these passive elements; sometimes, they include active devices. (*Passive devices* are electronic components that do not amplify, such as resistors, capacitors, and inductors. *Active devices* provide amplification, such as transistors.) Only the relatively simple *Wheatstone bridge* will be explained here, although the general concepts apply to all bridge circuits.

The easiest way to understand the operation of a bridge is to begin with the circuit shown in Figure 7–13. In this circuit, we have two parallel strings of resistors connected across a supply. For the sake of simplicity, let us assume all resistors are of the same value. Then:

$$V_{R_1} = V_{R_2} = V_{R_3} = V_{R_4}$$

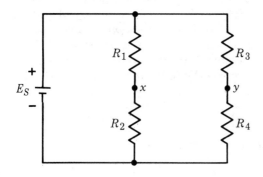

Figure 7–13

Basic elements of a bridge circuit

The voltage drops are equal because the impressed voltage will divide according to the resistance. Since $V_{R_2} = V_{R_4}$, a voltmeter connected between points x and y will read 0 V, because there is *no potential difference between them*. It also follows that if $V_{R_1} = V_{R_3}$, the same condition exists. The circuit in Figure 7–13 is basically a bridge. Under the conditions described here, we have assumed that the bridge is *balanced*.

All legs of a bridge need not have the same resistance to effect balance. Refer to Figure 7–13; suppose $R_1 = 100\ \Omega$, $R_2 = 200\ \Omega$, $R_3 = 300\ \Omega$, $R_4 = 600\ \Omega$, and $E_S = 9$ V. Is the bridge balanced? Let us calculate V_{R_2} and V_{R_4} and find out.

Use the voltage divider formula:

$$V_{R_2} = \frac{R_2}{R_1 + R_2} \times E_S = \frac{200\ \Omega}{100\ \Omega + 200\ \Omega} \times 9\ V = 6\ V$$

$$V_{R_4} = \frac{R_4}{R_3 + R_4} \times E_S = \frac{600\ \Omega}{300\ \Omega + 600\ \Omega} \times 9\ V = 6\ V$$

The bridge is balanced because there would be no potential difference between x and y.

The more conventional way to draw a bridge circuit is shown in Figure 7–14. Note the similarity here with the arrangement in Figure 7–13. In this case, the fourth leg of the bridge is designated R_x. The meter labeled G is a sensitive *galvanometer* or current indicator. When switch SW is closed, the meter will not move when balance is achieved.

The foregoing discussion has indicated the correlation between the resistance in each leg of the bridge and the voltage across each resistance. We can express this correlation by the following ratios:

$$\frac{R_1}{R_2} = \frac{R_3}{R_4} \qquad\qquad 7\text{–}1$$

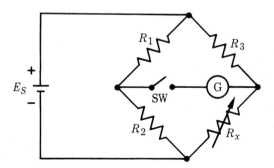

Figure 7–14

Wheatstone bridge

In a Wheatstone bridge, R_1, R_2, and R_3 are *precision variable resistors* attached to calibrated dials, and R_x is the resistor that has an undetermined resistance. By manipulating the three calibrated dials and simultaneously closing SW, we can determine the value of R_x when balance is achieved by reading the calibrated dials. The next example illustrates the procedure.

Example 7–9

What is the ohmic value of R_x in a balanced Wheatstone bridge when $R_1 = 123\ \Omega$, $R_2 = 479\ \Omega$, and $R_3 = 819\ \Omega$?

Solution

Substitute the given resistances into Equation 7–1:

$$\frac{R_1}{R_2} = \frac{R_3}{R_4} \quad \text{so} \quad \frac{123}{479} = \frac{819}{R_x}$$

Cross multiply

$$123\,R_x = 392{,}301$$
$$R_x = 3189.4\ \Omega$$

The unknown resistor could be connected in any leg of the bridge and its resistance determined by using Equation 7–1.

7–8
Troubles in Series-Parallel Circuits

The troubles encountered in series-parallel circuits are generally the result of *shorts* or *opens*. When a short is present, excessive current flows. An open causes the opposite effect. If excessive current flows through a resistor as a result of a shorted component, the resistor is likely to burn out. This situation results in nearly infinite resistance and looks like an open circuit.

Sometimes, the additional current through a resistor, as a result of a short in some other part of the circuit, will be insufficient to cause burnout. However, the added heat generated causes the resistance to change. It may increase or decrease, depending on the resistive material. The result is that the normal distribution of voltage and current throughout the circuit is changed. The technician's responsibility is to analyze the circuit and determine which component is defective.

The technician will analyze the circuit by measuring voltages around the circuit and comparing them with normal readings. It is usually inconvenient to measure currents with current meters, since the circuit would have to be opened wherever the measurement was to be made. Because voltage readings are an analog of current flow, it is advisable to use them in troubleshooting these circuits.

The following examples illustrate how faults in a series-parallel circuit can be localized.

Example 7–10

The voltages indicated in Figure 7–15 are voltages for normal operation. Suppose R_4 becomes open. How will the voltages be affected around the circuit?

Solution

The original current through R_3 and R_4 will be reduced to zero, since an open is an infinite resistance, and the load on the generator consists of R_1 and R_2 only. Their respective IR drops are as follows:

$$V_{R_1} = \frac{40 \ \Omega}{40 \ \Omega + 100 \ \Omega} \times 90 \text{ V} = 25.7 \text{ V}$$

$$V_{R_2} = \frac{100 \ \Omega}{40 \ \Omega + 100 \ \Omega} \times 90 \text{ V} = 64.3 \text{ V}$$

The voltage between b and ground will measure the same as the voltage between a and ground, because there is *no* IR drop across R_3. This voltage will be equal to V_{R_2}, or 64.3 V.

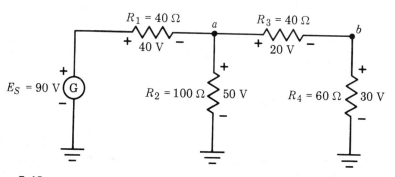

Figure 7–15

Circuit to illustrate troubleshooting techniques

Example 7–11

In the circuit of Figure 7–15, if R_2 is opened, what voltages would exist around the circuit?

Solution

In this situation, the generator would see only R_1, R_3, and R_4 in series.

$$R_T = 40 \ \Omega + 40 \ \Omega + 60 \ \Omega = 140 \ \Omega$$

$$I_T = \frac{90 \text{ V}}{140 \ \Omega} = 0.643 \text{ A}$$

Therefore $\quad V_{R_1} = V_{R_3} = 40 \ \Omega \times 0.643 \text{ A} = 25.7 \text{ V}$
$\qquad\qquad V_{R_4} = 60 \ \Omega \times 0.643 \text{ A} = 38.57 \text{ V}$

Example 7–12

In the circuit of Figure 7–15, if R_3 shorted, what effect would this short have on the rest of the circuit?

Solution

This short would effectively place R_4 in parallel with R_2. Their combined resistance (R_{eq}) would be 37.5 Ω. The R_T of the entire circuit would then be 77.5 Ω, which would result in the following I_T:

$$I_T = \frac{E_S}{R_T} = \frac{90 \text{ V}}{77.5 \ \Omega} = 1.16 \text{ A}$$

Then $\quad V_{R_1} = 1.16 \text{ A} \times 40 \ \Omega = 46.45 \text{ V}$

and $\quad V_{R_2} = V_{R_4} = E_S - V_{R_1} = 90 \text{ V} - 46.45 \text{ V} = 43.55 \text{ V}$

Hence, we see that the voltages around the circuit have changed substantially from normal.

Using the procedures of Example 7–12, we can calculate the various voltages that would exist around the circuit of Figure 7–15 if either R_1, R_2, or R_4 shorted.

┌─Summary

To facilitate the solution of a series-parallel circuit, begin by converting all parallel combinations to equivalent resistances.

To determine R_T of the entire circuit, begin at the farthest point and work toward the supply.

The voltage and current distribution can be calculated, after R_T has been determined, by working from the source back to the farthest resistor.

Any tendency of the supply to fluctuate will cause proportional changes in voltage and current throughout the entire series-parallel circuit.

Voltage dividers are used to supply two or more potentials to various parts of a circuit from the same power source.

The bleeder current in a voltage divider is designed to be 10%–20% of the total load current.

The bleeder current serves to stabilize or improve the regulation of the supply.

The bleeder network serves to discharge the filter capacitors in a power supply.

Bridge circuits are used in many types of electric measuring instruments and industrial control circuits.

When a bridge is balanced, there is no difference of potential between the opposite points of the bridge from where the voltage is applied.

When a Wheatstone bridge is balanced, the four legs of the bridge are related as follows:

$$\frac{R_1}{R_2} = \frac{R_3}{R_4}$$

A short or open in any part of a series-parallel network will change the voltage and current distribution throughout the other parts of the circuit.

⌐——Progress Test

The bracketed number after each question indicates the section of this chapter where the answer can be found.

1. With reference to Figure 7-1a: (a) R_1 is in series with R_2. (b) R_1 is in series with R_3. (c) both a and b are correct. (d) none of the above are correct. [7-1]

2. In Figure 7-3a: (a) R_1 is in parallel with R_3 and R_4. (b) R_1 is in series with R_2. (c) R_4 is in parallel with R_3. (d) R_1 and R_4 are in series. [7-2]

3. In Figure 7-4a, if 100 V were applied at the input of the circuit, the current through: (a) R_1 would be in series with R_5. (b) R_4 would be twice the current through R_3. (c) R_2 and R_3 would equal the current through R_4. (d) R_4 would be one-half the current through $R_2 + R_3$. [7-2]

4. In Figure 7-5, if R_3 shorts, R_T becomes: (a) 10 Ω. (b) 24.63 Ω. (c) 22.86 Ω. (d) 8 Ω. [7-3 and 7-8]

5. Which of the following statements is true regarding Figure 7-6? (a) Resistors R_1, R_2, and R_3 are in series. (b) Resistors R_2 and R_4 are in parallel. (c) Resistors R_5 and R_7 are in parallel. (d) Resistors R_1 and R_5 are in series. [7-3]

6. In the circuit of Figure 7-6, if R_6 shorted: (a) I_{R_8} would increase. (b) I_{R_7} would increase. (c) I_{R_5} would decrease. (d) I_{R_2} would increase. [7-3 and 7-8]

7. In the circuit of Figure 7-6, which of the following statements is correct if R_4 opens? (a) Voltage V_{R_1} will increase. (b) Voltage V_{R_2} will decrease. (c) Current I_{R_1} will decrease. (d) Current I_{R_6} will decrease. [7-3 and 7-8]

8. In Figure 7-10, what is the value of V_o if E_S increases 10%? (a) 5.65 V (b) 5.73 V (c) 5.85 V (d) 6.05 V [7-4]

9. Which of the following is true concerning the voltage divider in Figure 7-12? (a) Resistor R_2 is in series with R_3. (b) Resistor R_1 is in series with R_2. (c) Load 1 is in parallel with load 2. (d) Resistor R_3 is in parallel with load 1. [7-6]

10. In the bridge circuit of Figure 7-14, what is the value of R_x if $R_1 = 10.5$ Ω, $R_2 = 23.1$ Ω, $R_3 = 319$ Ω, and the bridge is balanced? (a) 667.5 Ω (b) 685.8 Ω (c) 693.2 Ω (d) 701.8 Ω [7-7]

⌐──Problems

Answers to odd-numbered problems are at the back of the book.

1. For the circuit of Figure 7–16, calculate: (a) V_{R_2} (b) V_{R_1}.

2. Calculate I_{R_2} and I_{R_3} for the circuit in Figure 7–16.

3. Suppose 36 V is applied to the circuit in Figure 7–4a. Calculate: (a) I_T (b) I_{R_2} (c) I_{R_4}.

4. Refer to Problem 3. If R_4 in the circuit opens, what is the value of I_{R_3}?

5. If 50 V is applied to the circuit of Figure 7–5, what is I_{R_2}?

6. Refer to Problem 5. What power is dissipated in R_4?

7. For the circuit of Figure 7–6, determine the current through the following resistors: (a) R_2 (b) R_8.

8. For Figure 7–17, calculate the following: (a) I_T (b) I_{R_2} (c) V_{R_4}.

9. For Figure 7–17, suppose E_S increases 5%. Calculate the following: (a) V_{R_4} (b) I_{R_2}.

10. In the circuit of Figure 7–18, assume $I_E = I_C = 1.5$ mA. From the data given, calculate: (a) V_{R_E} (b) V_{R_C} (c) V_{CE}.

11. Refer to Figure 7–18. Suppose $I_E = 1.5$ mA and $I_B = 0.01$ I_E. Calculate: (a) I_B (b) V_{R_B} (c) R_B. (Assume $V_{BE} = 0.6$ V.)

12. Refer to Figure 7–19. Suppose load 1 = 24 mA, load 2 = 60 mA, and the bleeder current is 15% of the total load. Calculate: (a) R_3 (b) R_2 (c) R_1.

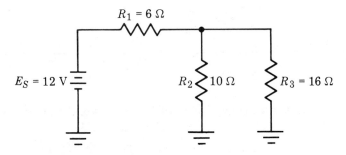

Figure 7–16

Circuit for Problems 1 and 2

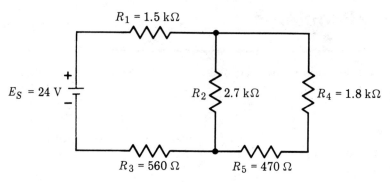

Figure 7–17

Circuit for Problems 8 and 9

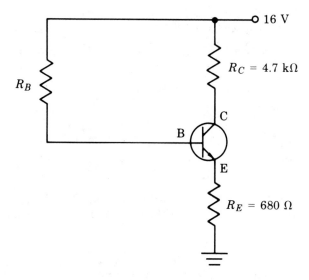

Figure 7–18

Circuit for Problems 10 and 11

13.   In the voltage divider of Figure 7–19, if load R_{L_2} opened, what voltage would exist at the junction of R_2 and R_3?

14.   For the Wheatstone bridge shown in Figure 7–14, determine the value of R_x if $R_1 = 60\ \Omega$, $R_2 = 139\ \Omega$, $R_3 = 1.5\ k\Omega$, and the bridge is balanced.

15.   A Wheatstone bridge balances when $R_1 = 910\ \Omega$, $R_3 = 79\ \Omega$, and $R_4 = 53\ \Omega$. What is the value of R_2?

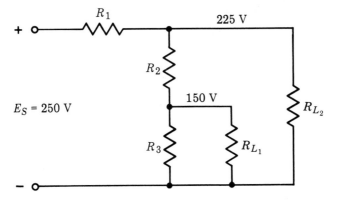

Figure 7–19

Circuit for Problems 12 and 13

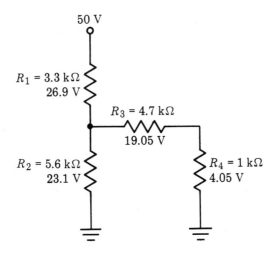

Figure 7–20

Circuit for Problem 16

16.   In the circuit of Figure 7–20, the normal voltages around the circuit are as shown. Later, a technician makes some voltage measurements and records the following: $V_{R_1} = 18.54$ V, $V_{R_4} = 31.46$ V, and $V_{R_3} = 0$ V. Which component has malfunctioned, and what happened to it?

17.   For the circuit shown in Figure 7–21, calculate the following values: (a) V_1 (b) V_3 (c) V_6 (d) I_2 (e) I_4.

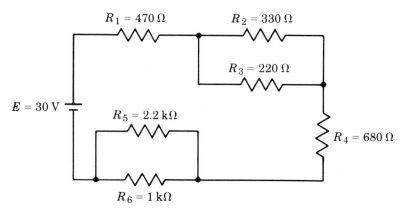

Figure 7–21

Circuit for Problem 17

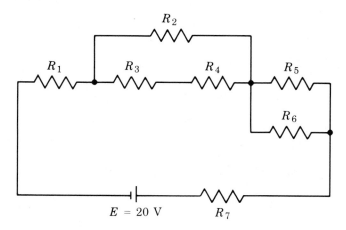

Figure 7–22

Circuit for Problem 18

18. Refer to the schematic shown in Figure 7–22. (*Note:* All resistors are 1 kΩ.) Calculate the following values: (*a*) V_2 (*b*) V_5 (*c*) V_7 (*d*) I_3 (*e*) I_6.

19. In the circuit accompanying Problem 18, if R_6 opens, what will be the new value of V_5?

20. For the schematic of Figure 7–23, all resistors are 500 Ω. Find the following values: (*a*) V_2 (*b*) V_5 (*c*) V_6 (*d*) I_1 (*e*) I_2 (*f*) I_7.

21. In Problem 20, if R_3 of the accompanying circuit shorts out, what will be the new values of V_5 and I_7?

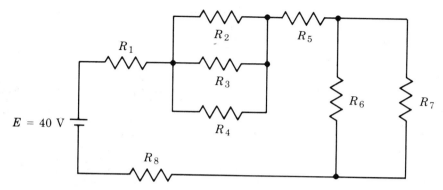

Figure 7–23

Circuit for Problem 20

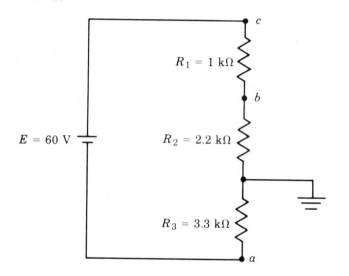

Figure 7–24

Circuit for Problem 22

22. For the unloaded voltage divider circuit shown in Figure 7-24, determine V_a, V_b, and V_c from common ground. Indicate polarities.

23. For the bridge circuit shown in Figure 7-25, determine the voltage from a to b. Indicate polarity.

24. For the schematic shown in Figure 7-26, calculate all the circuit parameters called for in the accompanying data table.

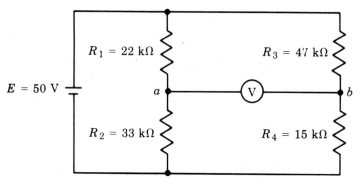

Figure 7–25

Circuit for Problem 23

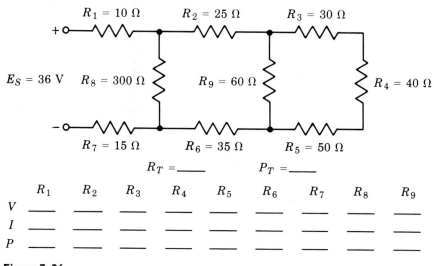

Figure 7–26

Circuit and data table for Problem 24

8

Network Analysis of dc Circuits

Objectives

After studying this chapter, you will be able to:

Determine the current and voltage drop through any resistance.

Determine the current and voltage drop in complex circuits.

Introduction

An introduction to Kirchhoff's laws was given in Chapter 3. They were applied, in a simple way, in the solution of all the series, parallel, and series-parallel circuits you investigated. Every time voltages were added in a series circuit to find the source voltage – or subtracted from known voltages to determine unknown component potentials – you were essentially applying Kirchhoff's voltage law. Kirchhoff's current law was also used every time you added branch currents to find the total current in parallel circuits, or when several branch currents were subtracted from the total current to determine the unknown current in a particular branch.

Many electrical circuits are not simple series, parallel, or series-parallel arrangements. For example, some contain *more than one voltage source* feeding the network from different points; therefore, current paths exist for each supply. Such arrangements are called *networks* or *complex circuits*. Solutions to the current and voltage distribution in such circuits are not easily found by such common techniques as Ohm's law, but other techniques are available. The most frequently encountered methods will be explained in this chapter.

Applications

Complex circuits and networks exist in most electronic equipment. For example, when analyzing the operation of such active devices as transistors, field effect transistors, and most linear integrated circuits, we experience circuits that have more than one voltage present. The incoming signal to be amplified converts some of the dc operating potential into an ac voltage. Thus, both dc and ac voltages act on the circuit elements associated with the device(s). When these conditions exist, the circuit operation can be more easily understood with the aid of any one of several different theorems that will be explained in this chapter.

The use of Kirchhoff's laws provides a complete current distribution analysis throughout the network and designated load resistance. The voltage across each resistor can be calculated from this analysis.

The superposition theorem can be used when the goal is to find the current through any resistor in the network without resorting to simultaneous equations. While not difficult, the process is lengthy.

Thévenin's theorem is best used when the voltage, current, and power in the load only is all we want to know. Presumably, we would not need to know the individual voltage drops or current distribution within the network. It is especially valuable when we wish to analyze a circuit's performance under different load conditions.

Norton's theorem is used in those circumstances where a network is simplified in terms of currents rather than voltages. It is used principally when only the current through the load is desired.

Millman's theorem provides a convenient method for finding the common voltage across a number of parallel branches when different voltage sources are used.

8-1
Kirchhoff's Laws

> ─Network or complex circuit   A combination of series, parallel, and/or series-parallel circuits, frequently with more than one voltage source.

From the introduction to *Kirchhoff's laws*, you learned that the *currents into and out of a node, or junction, are equal to zero.* Similarly, the voltage distribution around a loop, or simple series circuit, is equal to zero. This section will expand on these basic concepts and show how **networks**, or **complex circuits**, can be solved for current and voltage values more easily than is possible with Ohm's law alone.

In Chapter 3, you learned that the algebraic sum of the voltage drops around a complete electric circuit equals zero (KVL). We can verify this rule by applying KVL to the simple circuit of Figure 8-1:

$$50 \text{ V} - IR_3 - 20\text{V} - IR_2 - IR_1 = 0$$

Regardless of the circuit complexity, such an equation can be written for each path in the network.

Any closed path is called a *loop*. The method described here, called *loop analysis*, provides a valuable way of analyzing a circuit's performance in which Kirchhoff's voltage law is the key to the solution. With this method, two or more equations are formed and are then *solved simultaneously.*

The circuit shown in Figure 8-2, a network of four resistors and a voltage source, consists of two loops. In solving this circuit by loop analysis, we assume we have two currents, as shown. Each current is indicated in a direction based on the polarity of the power supply.

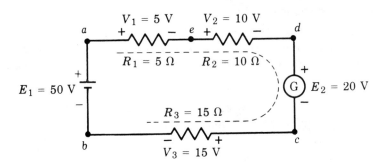

Figure 8–1

Series circuit with series-opposing voltage sources to illustrate Kirchhoff's voltage law

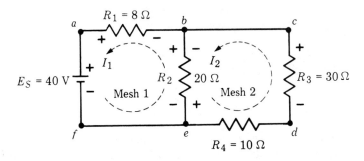

Figure 8–2

Circuit illustrating Kirchhoff's laws

(The true direction is unimportant at this time.) Each of the two closed current paths is called a *mesh*. The individual circuit components (resistors in this case) in each mesh are called *elements*.

From KVL, we know that the sum of the voltage *gains* supplied to a circuit from a voltage source always equals the sum of the voltage *drops* (V_1, V_2, V_3, and so on) in the circuit. Therefore, the sum of the *gains* plus the sum of the drops must equal zero. These voltages must be added algebraically in order to include their polarities. As we go around a circuit adding voltages, we must be careful to indicate voltage *gains* as *plus* values and voltage *drops* as *minus* values. The sign of each voltage can be determined as follows:

1. Observe the direction of current flow around the circuit. In the circuit of Figure 8–1 electrons leave the negative terminal of the supply and return to the positive terminal. The current is ccw since the 50 V source is greater than the opposing 20 V source.

2. We must combine all voltages as we move around the circuit. We start at any convenient point and algebraically add each voltage until we come back to the starting point.

3. Source voltages are considered positive (+) when we move through them from the plus to the minus terminal. The reverse is true if we move through a voltage source from minus to plus.

4. The voltage across any resistance will be negative if we go through it in the direction of the assumed electron flow. Therefore, in Figure 8–1, if we go around the circuit in the direction *abcde*, the voltage drops across the resistors will be $-V_3$, $-V_2$, and $-V_1$. If we go around the circuit in the opposite direction, the voltage drops across the resistors will be positive.

Let us begin at point *a* in Figure 8–1 and proceed counterclockwise around the circuit:

$$E_1 - V_3 - E_2 - V_2 - V_1 = 0$$

Substitute $\qquad 50\,\text{V} - 15\,\text{V} - 20\,\text{V} - 10\,\text{V} - 5\,\text{V} = 0$

Now let us consider the circuit of Figure 8–2, which contains two loops. In order to solve for current and voltage values, we need two equations, one for each loop. The number of equations required is always equal to the number of loops.

It has been assumed that the current in each mesh is in a counterclockwise direction, as indicated. Polarities are assigned to each resistor according to the direction of current. Observe that resistor R_2 carries two currents in opposite directions. A separate set of polarity signs is used for the voltage drops caused by each of these currents. A loop equation can now be written for each mesh. Beginning at point f, the equation for loop *feba* is as follows:

$$-20I_1 + 20I_2 - 8I_1 + 40 = 0 \qquad \textbf{8-1}$$

Combine terms $\qquad -28I_1 + 20I_2 + 40 = 0$

Transpose $\qquad -28I_1 + 20I_2 = -40 \qquad \textbf{8-2}$

Since two opposing currents pass through R_2, two voltage drops of opposite polarity appear across R_2. Both of these drops ($-20I_1$ and $+20I_2$) must be included in the equation with their proper polarities.

The second loop *edcb* does not contain a voltage source. When a loop does not have a voltage source, the algebraic sum of the IR drops alone must total zero. For instance, for the loop just mentioned, going counterclockwise from point e, the loop equation is as follows:

$$-10I_2 - 30I_2 - 20I_2 + 20I_1 = 0 \qquad \textbf{8-3}$$

Simplify $\qquad 20I_1 - 60I_2 = 0 \qquad \textbf{8-4}$

Again, note that two opposing voltage drops ($+20I_2$ and $-20I_1$) were included for R_2.

Equations 8-2 and 8-4 are now solved simultaneously to determine the values of I_1 and I_2:

$$-28I_1 + 20I_2 = -40 \qquad \textbf{8-2}$$
$$20I_1 - 60I_2 = 0 \qquad \textbf{8-4}$$

To eliminate I_2, we multiply Equation 8-2 by 3 and add the resulting equation to Equation 8-4:

$$-84I_1 + 60I_2 = -120 \qquad \textbf{8-5}$$
$$\underline{20I_1 - 60I_2 = 0}$$
$$-64I_1 \qquad\quad = -120$$

Dividing both sides by -64 gives the value of I_1:

$$I_1 = 1.875 \text{ A}$$

To obtain the value of I_2, we substitute 1.875 A into Equation 8-4 in place of I_1:

$$20 \times 1.875 - 60I_2 = 0$$
$$37.5 - 60I_2 = 0$$
$$-60I_2 = -37.5$$
$$I_2 = 0.625 \text{ A}$$

From these calculations, we see that $I_1 = 1.875$ A and $I_2 = 0.625$ A. The voltage drops across each resistor can now be calculated by Ohm's law as follows:

$$V_1 = I_1R_1 = 1.875 \times 8 = 15 \text{ V}$$
$$V_2 = (I_1 - I_2)R_2 = (1.875 - 0.625)20 = 25 \text{ V}$$
$$V_4 = I_2R_4 = 0.625 \times 10 = 6.25 \text{ V}$$
$$V_3 = I_2R_3 = 0.625 \times 30 = 18.75 \text{ V}$$

We can verify the accuracy of these calculations by adding all the voltages in *each loop* and seeing that they equal zero:

$$V_2 + V_1 - E_S = 0$$
$$25 \text{ V} + 15 \text{ V} - 40 \text{ V} = 0$$

and

$$V_4 + V_3 - V_2 = 0$$
$$6.25 \text{ V} + 18.75\text{V} - 25 \text{ V} = 0$$

The following example illustrates the use of KVL in solving for the unknown voltages and currents in a circuit.

Example 8-1

Refer to the circuit of Figure 8–3. Calculate I_1 and I_2 and the voltage across each resistor.

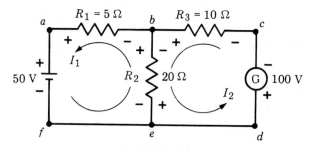

Figure 8–3

Circuit for Example 8–1

Solution

A counterclockwise current has been assumed in each mesh and polarities assigned as shown. Don't forget that R_2 has two opposing voltage drops! KVL for loop *feba* is as follows:

$$-20I_1 + 20I_2 - 5I_1 + 50 = 0 \qquad \text{8-6}$$

Simplify $\qquad -25I_1 + 20I_2 = -50 \qquad \text{8-7}$

And KVL for loop *edcb* is as follows:

$$+100 - 10I_2 - 20I_2 + 20I_1 = 0 \qquad \text{8-8}$$

Simplify

$$+20I_1 - 30I_2 = -100 \qquad \text{8-9}$$

Multiply Equation 8-7 by -3, and multiply Equation 8-9 by -2:

Add $\qquad 75I_1 - 60I_2 = 150 \qquad \text{8-10}$

$$\underline{-40I_1 + 60I_2 = 200} \qquad \text{8-11}$$

$$35I_1 \qquad\quad = 350$$

Therefore $\qquad I_1 = 10 \text{ A} \qquad \text{8-12}$

Substitute $I_1 = 10$ A into Equation 8-9:

$$+200 - 30I_2 = -100 \qquad \text{8-13}$$

$$-30I_2 = -300$$

$$I_2 = 10 \text{ A} \qquad \text{8-14}$$

Had any of the currents been negative, we would know that the assumed direction was incorrect. However, the *magnitude* of current would be correct.

Having calculated each current, we can now proceed to determine the IR drops across each resistor:

$$V_{R_1} = I_1 R_1 = 10 \times 5 = 50 \text{ V}$$
$$V_{R_2} = (I_1 - I_2) R_2 = 0 \times 20 = 0$$
$$V_{R_3} = I_2 R_3 = 10 \times 10 = 100 \text{ V}$$

Let us consider one more example so that we become more familiar with Kirchhoff's law.

Example 8-2 ─────────────────────────────

Determine the current and voltage distribution around the circuit shown in Figure 8-4.

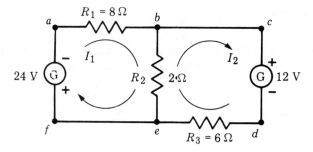

Figure 8–4

Circuit for Example 8–2

Solution

A clockwise current has been assumed in each loop (based on the indicated supply polarities). So KVL for loop *fabe* is as follows:

$$+24 - 8I_1 - 2I_1 + 2I_2 = 0 \qquad \text{8–15}$$

Simplify $\qquad -10I_1 + 2I_2 = -24 \qquad \text{8–16}$

And KVL for loop *cdeb* is as follows:

$$+12 - 6I_2 - 2I_2 + 2I_1 = 0 \qquad \text{8–17}$$

Simplify

$$+2I_1 - 8I_2 = -12 \qquad \text{8–18}$$

Multiply Equation 8–16 by 4 and add the resulting equation to Equation 8–18:

$$-40I_1 + 8I_2 = -96 \qquad \text{8–19}$$
$$+2I_1 - 8I_2 = -12$$

Add $\qquad \overline{-38I_1 \qquad\quad = -108} \qquad \text{8–20}$

$$I_1 = 2.84 \text{ A} \qquad \text{8–21}$$

Substitute $I_1 = 2.84$ A into Equation 8–16:

$$-28.4 + 2I_2 = -24 \qquad \text{8–22}$$
$$+2I_2 = +4.4$$
$$I_2 = 2.2 \text{ A} \qquad \text{8–23}$$

Since currents I_1 and I_2 each have positive values, we know that the direction of the assumed currents was correct. Knowing the values of these currents, we can easily calculate the voltage drops across each resistor:

$$V_{R_1} = I_1 R_1 = 2.84 \times 8 = 22.72 \text{ V}$$
$$V_{R_2} = (I_1 - I_2) R_2 = 0.64 \times 2 = 1.28 \text{ V}$$
$$V_{R_3} = I_2 R_3 = 2.2 \times 6 = 13.2 \text{ V}$$

8–2
Superposition Theorem

The *superposition theorem* provides an alternative method for solving linear, multisource circuits without the need for simultaneous equations. The technique is fairly simple. All of the supply sources but one are replaced by their own internal resistance or by a short circuit if the internal resistance is negligible. The current distribution around the circuit is then solved by using Ohm's law. This process is repeated, using each voltage source as the only supply. The final current through each component is the *algebraic sum of the currents produced* when each source acted alone. To show that we obtain the same results with the superposition theorem that we obtained with Kirchhoff's laws, we will use the circuit values of Figure 8–4, which is redrawn in Figure 8–5.

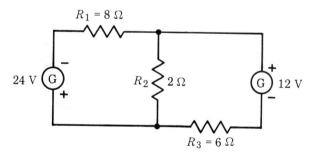

Figure 8–5

Circuit for Example 8–3, illustrating the superposition theorem

Example 8–3 ───────────────────────────────────────

Using the circuit values indicated in Figure 8–5, calculate the current through each resistor and the associated *IR* drops.

Solution

Let us replace the 12 V supply by a *short circuit*, as shown in Figure 8–6a. This arrangement places R_2 and R_3 in parallel, and the combination is in series with R_1. The total resistance is calculated as follows:

$$R_T = R_1 + \frac{R_2 R_3}{R_2 + R_3} = 8 + \frac{2 \times 6}{2 + 6} = 9.5 \ \Omega$$

Calculate the resulting current flow:

$$I_T = \frac{24}{9.5} = 2.53 \ \text{A}$$

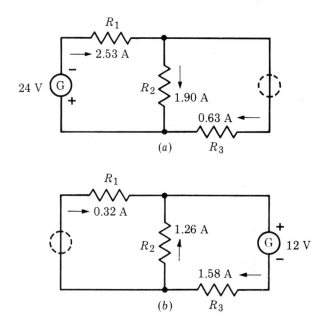

Figure 8–6

Circuits equivalent to the circuit of Figure 8–5: (a) With 12 V supply shorted. (b) With 24 V supply shorted.

This current flows through R_1 and divides into I_{R_2} and I_{R_3}. With the *current divider formula*, these currents can be calculated as follows:

$$I_{R_2} = \frac{R_3}{R_2 + R_3} \times I_T = \frac{6}{2 + 6} \times 2.53 = 1.90 \text{ A}$$

$$I_{R_3} = \frac{R_2}{R_2 + R_3} \times I_T = \frac{2}{2 + 6} \times 2.53 = 0.63 \text{ A}$$

The direction and magnitudes of these currents are shown in Figure 8–6a.

We must now repeat the process with the other voltage source shorted. The circuit becomes as shown in Figure 8–6b. Solve for R_T:

$$R_T = R_3 + \frac{R_1 R_2}{R_1 + R_2} = 6 + \frac{8 \times 2}{8 + 2} = 7.6 \text{ } \Omega$$

Calculate the resulting current distribution:

$$I_T = \frac{12}{7.6} = 1.58 \text{ A}$$

This current flows through R_3 and divides into I_{R_1} and I_{R_2}. Again, with the current divider formula, these currents are calculated as follows:

$$I_{R_1} = \frac{R_2}{R_1 + R_2} \times I_T = \frac{2}{8 + 2} \times 1.58 = 0.32 \text{ A}$$

$$I_{R_2} = \frac{R_1}{R_1 + R_2} \times I_T = \frac{8}{8 + 2} \times 1.58 = 1.26 \text{ A}$$

The direction and magnitudes of these currents are shown in Figure 8–6b.

Now, to find the resulting current flow through each resistor, we must *superimpose* the currents shown in Figures 8–6a and 8–6b and add them algebraically. The final current flow is as indicated in Figure 8–7.

For the resulting currents in Figure 8–7, the voltage drops across each resistor are as follows:

$$V_{R1} = I_1 R_1 = 2.85 \times 8 = 22.8 \text{ V}$$
$$V_{R2} = I_2 R_2 = 0.64 \times 2 = 1.28 \text{ V}$$
$$V_{R3} = I_3 R_3 = 2.21 \times 6 = 13.26 \text{ V}$$

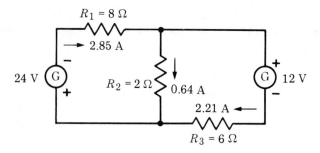

Figure 8–7

Resulting current flow of the circuit in Figure 8–5

Kirchhoff's voltage law says that the sum of the voltage drops around a circuit (loop) should equal zero. Therefore, $V_{R_1} + V_{R_2}$ should equal the 24 V supply:

$$22.8 + 1.28 = 24.08 \text{ V}$$

This result is very nearly equal to 24 V; the difference is due to rounding off some of the previous calculations. Similarly, $V_{R_3} - V_{R_2}$ should equal the 12 V supply:

$$13.26 - 1.28 = 11.98 \text{ V}$$

Again, this result is very nearly equal to 12 V.

8–3
Thévenin's Theorem

It is not always necessary to calculate the current through each resistor and all the voltage drops. Often, we are concerned only with the voltage across a particular load, the current through it, and its power. Thévenin, a French physicist, developed a theorem in 1883 whereby the laborious work involved in solving complex circuits could be substantially reduced. One feature of Thévenin's approach is that it presents an easy way for us to calculate a circuit's behavior *under different load conditions.* The theorem says that *any network, no matter how complex, can be replaced by an equivalent circuit consisting of a constant-voltage source in series with an equivalent resistance.* This constant-voltage source is designated as V_{TH} and the equivalent series resistance as R_{TH}.

The value of V_{TH} is found by calculating (or measuring) the voltage that would appear across the load if it were removed. In other words, V_{TH} is the *open-circuit voltage* across whatever resistor is assumed to be the load.

The value of R_{TH} is calculated by looking back into the network from the open-circuit terminals (the load is not yet connected) with all voltage sources short-circuited. It is possible to measure R_{TH} with an ohmmeter if all voltage sources are replaced with shorts. The next example illustrates this technique.

Example 8–4

Using Thévenin's theorem and the circuit of Figure 8–8, calculate the current through R_L.

Solution

First, remove R_L and calculate the Thévenin voltage V_{TH} across terminals a to b. Voltage V_{TH} will be the same as V_{R_2} because there will be no IR drop across R_3. Calculate V_{R_2}:

$$V_{TH} = V_{R_2} = \frac{R_2}{R_1 + R_2} \times 30 \text{ V} = \frac{12}{8 + 12} \times 30 = 18 \text{ V} \qquad \textbf{8–24}$$

Next, replace all internal voltage sources with a short, and determine R_{TH} by looking into the network from terminals a to b. The circuit now looks like the circuit shown in Figure 8–9. Calculate R_{TH}:

$$R_{TH} = \frac{R_1 R_2}{R_1 + R_2} + R_3 = \frac{8 \times 12}{8 + 12} + 4 = 8.8 \text{ }\Omega \qquad \textbf{8–25}$$

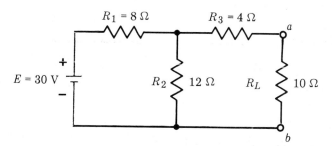

Figure 8–8

Circuit for Example 8–4, illustrating Thévenin's theorem

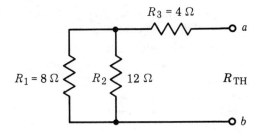

$R_3 = 4\,\Omega$

a

$R_1 = 8\,\Omega$ R_2 $12\,\Omega$ R_{TH}

b

Figure 8–9

Calculating R_{TH} for the circuit of Figure 8–8.

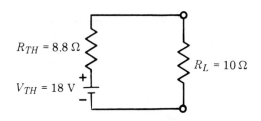

$R_{TH} = 8.8\,\Omega$

$R_L = 10\,\Omega$

$V_{TH} = 18\text{ V}$

Figure 8–10

Thévenized circuit for the circuit of Figure 8–8

The circuit in Figure 8–8 is shown in its Thévenized form in Figure 8–10. The current through R_L is as follows:

$$I_L = \frac{V_{\mathrm{TH}}}{R_{\mathrm{TH}} + R_L} = \frac{18}{8.8 + 10} = 0.96\text{ A} \qquad \textbf{8–26}$$

By knowing the current through the load, we can determine V_{R_L}:

$$V_{RL} = I_L R_L = 0.96 \times 10 = 9.6\text{ V} \qquad \textbf{8–27}$$

Example 8–4 was relatively easy, since only one voltage source was employed. The following example uses two voltage sources.

Example 8–5

Using Thévenin's theorem and the circuit in Figure 8–11, calculate the current through R_2 (the assumed load, in this case).

Solution

First, remove R_2 and calculate R_{TH} (looking into terminals a to b with the two voltage sources shorted):

$$R_{\mathrm{TH}} = \frac{R_1 R_3}{R_1 + R_3} = \frac{6 \times 10}{6 + 10} = 3.75\,\Omega$$

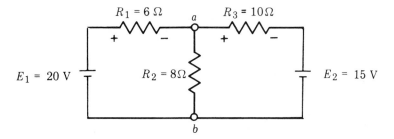

Figure 8–11

Circuit for Example 8–5, illustrating Thévenin's theorem for two voltage sources

Before we can calculate V_{TH} we must determine the net voltage acting in the circuit and then the circuit current:

$$E_{\mathrm{net}} = 20 \text{ V} - 15 \text{ V} = 5 \text{ V}$$
$$I_{\mathrm{circuit}} = \frac{5}{R_1 + R_3} = \frac{5}{16} = 0.3125 \text{ A}$$

Therefore, the circuit current of 0.3125 A will produce the following voltage drops across R_1 and R_3:

$$V_1 = 0.3125 \text{ A} \times 6 \text{ } \Omega = 1.875 \text{ V}$$
$$V_3 = 0.3125 \text{ A} \times 10 \text{ } \Omega = 3.125 \text{ V}$$

The polarity of these IR drops is indicated in Figure 8–11. The voltage V_{ab} is equal to the series-aiding potentials:

$$V_{ab} = E_2 + V_3 = 15 \text{ V} + 3.125 \text{ V} = 18.125 \text{ V}$$

We can solve for V_{ab} by using an alternative approach, such as the following:

$$V_{ab} = E_1 - V_1 = 20 \text{ V} - 1.875 \text{ V} = 18.125 \text{ V}$$

Notice that the polarity of V_1 is opposing E_1 (series-opposing). The Thévenin circuit of Figure 8–11 can now be drawn as in Figure 8–12 and the load current calculated:

$$I_2 = \frac{V_{\mathrm{TH}}}{R_{\mathrm{TH}} + R_2} = \frac{18.125}{11.75} = 1.54 \text{ A}$$

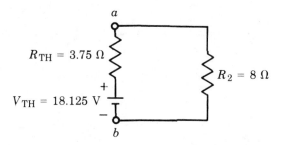

Figure 8–12

Thévenized circuit for the circuit of Figure 8–11

The voltage across R_2 is as follows:

$$V_{R_2} = I_2 R_2 = 1.54 \times 8 = 12.34 \text{ V}$$

Once a circuit has been Thévenized, the following formula will facilitate the calculation of voltage across the desired load:

$$V_{RL} = \frac{V_{TH} R_L}{R_{TH} + R_L} \qquad \textbf{8-28}$$

8–4
Norton's Theorem

An analysis of some complex circuits can be made easier when considered in terms of *current distribution*. This analysis is frequently helpful, for example, in networks having a high internal resistance, such as in certain transistor applications. *Norton's theorem* can be used in these situations. The theorem states that *a network can be reduced to a simple parallel circuit in shunt with a current source.* Current from this source *divides between the equivalent internal resistance of the network and the load.*

Current versus Voltage Sources

Current source   A two-terminal circuit element with an output current essentially independent of the voltage between its terminals and characterized by a high internal impedance.

A **current source** is a two-terminal circuit element with an output current that is *essentially independent of the voltage between its terminals.* The symbol for such a device is shown in Figure 8–13a; the arrow indicates the direction of current. Resistor R represents the equivalent internal resistance of the network. It is obvious that the current supplied by the generator will divide between the load R_L and the internal resistance R, since they are parallel. Typical examples of a current generator are transistors, tetrodes, and pentode vacuum tubes, all of which have high internal resistances. Current through these devices is somewhat independent of the applied voltage. In their equivalent circuits, these active devices can be represented by this two-terminal symbol.

Voltage source   A two-terminal circuit element having an output voltage essentially independent of the current through it and characterized by a low internal impedance.

A **voltage source** is a two-terminal circuit element with an output voltage that is *essentially independent of the current through it.* The symbol for a device of this kind is indicated in Figure 8–13b, and it may be either dc or ac. It is characterized by a relatively *low internal series resistance*, represented by R in the drawing. Since the load R_L is in series with the generator's internal resistance, it is apparent that the output voltage will be distributed between R and R_L, since they are in series. Typical examples of voltage generators are a triode vacuum tube and some kinds of transistors. The equivalent circuit of these devices is the two-terminal symbol shown.

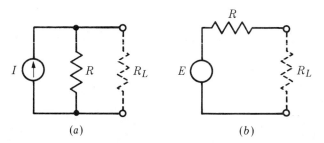

(a)   (b)

Figure 8–13

Electrical symbols: (a) For a current generator. (b) For a voltage generator.

Norton Equivalent Circuit

Norton's theorem states that any linear network that is energized by one or more sources of energy, if viewed from any two points in the network, can be replaced by a single constant-current source in parallel with an equivalent resistance. An illustration of this theorem is given in Figure 8–14; here, an entire network connected to terminals x and y can be replaced by the single current source I_N in parallel with a resistance R_N. The value of I_N is equal to the short-circuit current through the terminals x and y. In other words, I_N is the current that the network can send through the shorted terminals x and y.

To find the value of R_N, the internal resistance of the network, we must look back from terminals x and y. For this calculation, terminals x and y are assumed to be open (load removed), as in calculating R_{TH} for Thévenin's theorem. Consequently, the Norton equivalent resistance is the *same* as the Thévenin equivalent resistance:

$$R_N = R_{\text{TH}} \qquad\qquad \textbf{8-29}$$

To calculate the short-circuit current, or Norton current, I_N, we first determine the open-circuit voltage appearing across the desired load terminals, that is, the same voltage that is calculated for a Thévenized circuit. In fact, there is a close relationship between Norton's and Thévenin's theorems, as evidenced by the manner in which the Norton current is calculated:

$$I_N = \frac{V_{\text{TH}}}{R_{\text{TH}}} \qquad\qquad \textbf{8-30}$$

This current, represented by I_N in Figure 8–14, will divide according to the current divider formula. Some of the current will go through R_N, and the balance will go through whatever load is connected across terminals

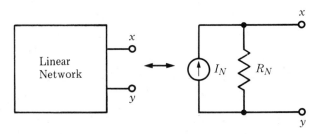

Figure 8–14

Any linear network (in the block at the left) and the equivalent Norton circuit (at the right)

x and y. The division of current I_N will be inversely proportional to the resistances R_N and R_L (load connected between terminals x and y) or directly proportional to their respective conductances.

As an aid in the application of Norton's theorem, the following current divider equation can be used in determining the current through the desired load:

$$I_{RL} = \frac{I_N R_{TH}}{R_{TH} + R_L} \quad \text{or} \quad I_{RL} = \frac{I_N R_N}{R_N + R_L} \qquad \text{8–31}$$

The following examples demonstrate the use of Norton's theorem.

Example 8–6

For the circuit of Figure 8–15, calculate the current through R_L and the voltage across it.

Solution

The basic steps involved in the solution are illustrated in Figure 8–16.

1. Assume R_L is removed. Calculate the voltage across R_2 by using the voltage divider formula:

$$V_{R_2} = V_{TH} = \frac{R_2}{R_1 + R_2} \times E_S = \frac{15}{30 + 15} \times 90 \text{ V} = 30 \text{ V}$$

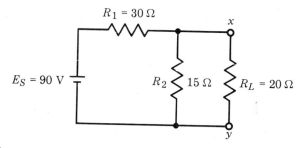

Figure 8–15

Circuit for Example 8–6, demonstrating Norton's theorem

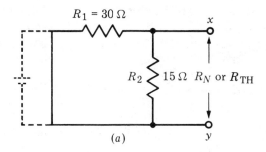

(a)

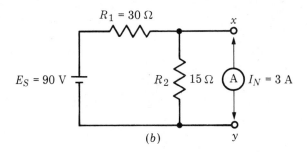

(b)

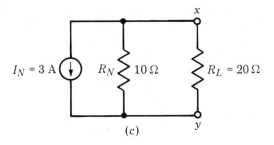

(c)

Figure 8–16

Evolution of Norton's equivalent circuit, using the circuit of Figure 8–15: (a) Calculating R_N. (b) Determining I_N. (c) Equivalent circuit.

2. Assuming the voltage source E_S is shorted, calculate R_N:

$$R_{TH} = R_N = \frac{R_1 R_2}{R_1 + R_2} = \frac{30 \times 15}{30 + 15} = 10\ \Omega$$

3. Calculate I_N:

$$I_N = \frac{V_{TH}}{R_N} = \frac{30\ \text{V}}{10\ \Omega} = 3\ \text{A}$$

4. Calculate the load current:

$$I_{R_L} = \frac{I_N R_N}{R_N + R_L} = \frac{3 \times 10}{10 + 20} = 1 \text{ A}$$

5. Calculate the voltage across R_L:

$$V_{R_L} = I_{R_L} R_L = 1 \text{ A} \times 20 \text{ } \Omega = 20 \text{ V}$$

Should it be desired to develop Thévenin's equivalent circuit, V_{TH} can be determined as follows:

$$V_{TH} = I_N R_N \qquad\qquad \text{8–32}$$

Example 8–7

Using Norton's theorem, determine the current through R_L in the circuit of Figure 8–17.

Solution

1. Remove R_L and calculate V_N (V_{xy}):

$$V_N = E_2 + I_c R_2$$

$$\text{where } I_c = \frac{E_1 - E_2}{R_1 + R_2} = \frac{8 - 6}{0.2 + 0.1} = \frac{2}{0.3} = 6.67 \text{ A}$$

$$\text{So} \quad V_N = E_2 + I_c R_2 = 6 + 6.67 \times 0.1 = 6.67 \text{ V}$$

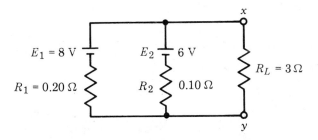

Figure 8–17

Circuit for example 8–7

2. Calculate R_N:

$$R_N = \frac{0.2 \times 0.1}{0.2 + 0.1} = \frac{0.02}{0.3} = 0.067 \ \Omega$$

3. Calculate I_N:

$$I_N = \frac{V_N}{R_N} = \frac{6.67}{0.067} = 100 \ \text{A}$$

4. Calculate the load current:

$$I_{R_L} = \frac{I_N R_N}{R_N + R_L} = \frac{100 \times 0.067}{0.067 + 3} = \frac{6.7}{3.067} = 2.19 \ \text{A}$$

Conversion Formulas

The equivalent resistance for Norton's theorem is the *same* as the equivalent resistance for Thévenin's theorem. There is also a close relationship between Norton's and Thévenin's theorems in the calculation of current. The comparison of the two theorems is summarized as follows:

Thévenin to Norton	Norton to Thévenin
$R_{\text{TH}} = R_N$	$R_N = R_{\text{TH}}$
$V_{\text{TH}} = I_N R_N$	$I_N = \dfrac{V_{\text{TH}}}{R_{\text{TH}}}$

8–5
Millman's Theorem

Finding the *common voltage across any network* containing a number of parallel voltage sources can be simplified by using *Millman's theorem*, which is a combination of Thévenin's and Norton's theorems. A typical example of its use is shown in Figure 8–18. In any network like this one, there will be a common voltage appearing across output terminals x and y. This voltage V_{xy} is the net effect of all the sources acting in the circuit and can be calculated from the following formula:

$$V_{xy} = \frac{E_1/R_1 + E_2/R_2 + E_3/R_3}{1/R_1 + 1/R_2 + 1/R_3} \qquad \textbf{8–33}$$

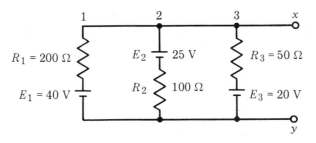

Figure 8–18

Calculating V_{xy} with Millman's theorem

Observe that the terms in the numerator are *expressions of current*. Thus, the current in branch 1 is expressed as E_1/R_1; in branch 2, as E_2/R_2; and so on. The terms in the denominator are *conductances*. The sum of the branch currents, $\Sigma\, I$, divided by the sum of the conductances, $\Sigma\, G$, yields the terminal voltage V_{xy}. In other words, $V_{xy} = I_T/G_T = I_T R_{eq}$.

Example 8–8

Calculate the terminal voltage V_{xy} for the circuit in Figure 8–18, using Millman's theorem.

Solution

$$V_{xy} = \frac{E_1/R_1 + E_2/R_2 + E_3/R_3}{1/R_1 + 1/R_2 + 1/R_3} = \frac{40/200 + 25/100 + 20/50}{1/200 + 1/100 + 1/50}$$

$$= \frac{0.2 + 0.25 + 0.4}{0.005 + 0.01 + 0.02} = \frac{0.85}{0.035} = 24.286 \text{ V}$$

If one branch of a circuit contains no voltage source, as in Figure 8–19, the same procedure is used, except that the current in that branch will be zero. The procedure is illustrated in the following example.

Example 8–9

Determine the common voltage V_{xy} and the load current in the circuit of Figure 8–19, using Millman's theorem.

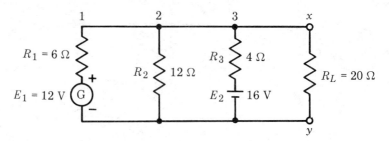

Figure 8–19

Circuit for Example 8–9, illustrating Millman's theorem

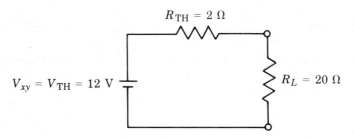

Figure 8–20

Equivalent circuit for the circuit of Figure 8–19

Solution

$$V_{TH} = V_{xy} = \frac{12/6 + 0/12 + 16/4}{1/6 + 1/12 + 1/4}$$

$$= \frac{2 + 0 + 4}{0.167 + 0.083 + 0.25} = \frac{6}{0.5} = 12 \text{ V}$$

Calculate R_{TH}:

$$R_{TH} = \frac{1}{1/6 + 1/12 + 1/4} = 2\ \Omega$$

The circuit is now as shown in Figure 8–20. Calculate the load current:

$$I_{R_L} = \frac{V_{TH}}{R_{TH} + R_L} = \frac{12 \text{ V}}{2\ \Omega + 20\ \Omega} = 0.545 \text{ A}$$

Summary

Complex circuits are not simple series, parallel, or series-parallel arrangements.

Complex circuits frequently contain more than one voltage source feeding the network from different points.

Solutions to the current and voltage distributions in complex networks are not easily found by such common techniques as Ohm's law.

Kirchhoff's laws provide a complete analysis of the current and voltage distributions throughout a network.

Kirchhoff's current law (KCL) states that the algebraic sum of all currents entering a junction must equal the sum of the currents leaving the junction.

Kirchhoff's voltage law (KVL) states that the algebraic sum of all voltages around any closed path must equal zero.

The superposition theorem states that, in any linear network having more than one voltage source, the current and voltage in any one component of the network can be determined by algebraically adding the effect of each voltage source separately while all other sources are temporarily shorted.

Thévenin's theorem states that any network having two open terminals (x and y) can be replaced by a single voltage source V_{TH} in series with a single resistance R_{TH}. Voltage V_{TH} drives any load connected across these terminals with R_{TH} in series.

Norton's theorem states that any linear network can be reduced to a simple parallel circuit whose resistance R_N is in shunt with a current source I_N.

Millman's theorem provides a method of calculating the common voltage across parallel branches having different voltage sources. The formula used in this calculation is given in Equation 8–33.

Progress Test

The bracketed number after each question indicates the section of this chapter where the answer can be found.

1. In the circuit of Figure 8–21, if $E_1 = 30$ V and $E_2 = 20$ V, the voltage across R_2 is: (a) 6.6 V. (b) 3.3 V. (c) 2.8 V. (d) 2.3 V.   [8–1]

2. If voltage source E_2 were reversed in the circuit of Figure 8–21, the voltage across R_1 would be: (a) 33.3 V. (b) 24.6 V. (c) 16.7 V. (d) 11.7 V.   [8–1]

3. The loop equation for the circuit in Figure 8–1, beginning at point b is: (a) $-IR_2 + 20$ V $- IR_3 - 50$ V $- IR_1 = 0.$ (b) $IR_2 - 20$ V $+ IR_3 + 50$ V $+ IR_1 = 0.$ (c) $-IR_2 - 20$ V $- IR_3 + 50$ V $- IR_1 = 0.$ (d) $-IR_2 + 20$ V $- IR_3 - 50$ V $- IR_1 = 0.$ [8–1]

4. Which of the following statements apply to the superposition theorem? (a) It can be used with nonlinear networks. (b) It does not require the use of simultaneous equations. (c) The total current through each resistor is the arithmetic sum of the individual currents produced by each voltage source. (d) Only one voltage source is shorted at a time when calculating currents.   [8–2]

5. In the circuit of Figure 8–22, the total current through R_2 is (use the superposition theorem): (a) 0.11 A. (b) 0.54 A. (c) 0.69 A. (d) 0.8 A.   [8–2]

6. Voltage V_{TH} is the: (a) average value of all voltage sources in the network. (b) short-circuit voltage across R_{TH}. (c) equivalent voltage of all series voltage sources. (d) open-circuit potential across the load.   [8–3]

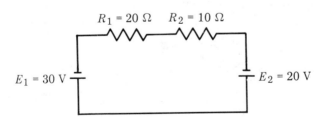

$R_1 = 20\ \Omega$ $R_2 = 10\ \Omega$

$E_1 = 30$ V $E_2 = 20$ V

Figure 8–21

Circuit for Questions 1 and 2

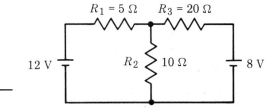

Figure 8–22

Circuit for Question 5

7. Resistance R_{TH} is: (a) the net circuit resistance in parallel with V_{TH}. (b) the equivalent network resistance looking back from the open load terminals with all voltage sources shorted. (c) in parallel with R_L with all voltage sources shorted. (d) the network's internal resistance in parallel with R_L. [8–3]

8. A current generator: (a) has high internal series resistance. (b) has low internal series resistance. (c) delivers a somewhat constant current regardless of normal loads. (d) delivers a current proportional to the applied voltage. [8–4]

9. A constant-voltage generator is characterized by a: (a) low internal resistance in series with a voltage supply. (b) high internal resistance in parallel with the internal supply. (c) low internal resistance in parallel with the internal supply. (d) low internal resistance in parallel with the supply. [8–4]

10. Which of the following statements applies to Norton's theorem? (a) Current I_N is the current through the load. (b) Current I_N divides between the load and the internal resistance. (c) The Norton resistance is the reciprocal of the Thévenin resistance. (d) Current I_N is calculated with all voltage sources short-circuited. [8–4]

┌──── **Problems**

Answers to odd-numbered problems are at the back of the book.

1. Using Kirchhoff's law, calculate the current through R_3 in the circuit of Figure 8–23.

2. Calculate the current through R_1 in the circuit of Figure 8–23.

3. Determine the current through R_2 in the circuit of Figure 8–23.

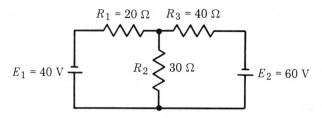

Figure 8–23

Circuit for Problems 1–3

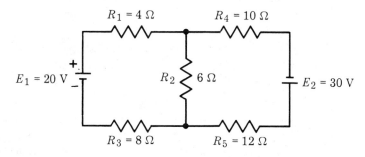

Figure 8–24

Circuit for Problems 4–6

Problems 4 through 6 refer to Figure 8–24.

    4.   Calculate the mesh current I_2 produced by E_2.

    5.   What is the value of mesh current I_1?

    6.   Determine the voltage drop across R_2.

Problems 7 through 10 refer to Figure 8–25.

    7.   What is the magnitude of current through R_3?

    8.   Calculate I_{R_1}.

    9.   What net current passes through R_2?

    10.   Verify that the sum of the voltage drops around loop b equals 14 V.

    11.   Using the superposition theorem, calculate the net current through each resistor in the circuit of Figure 8–26.

    12.   Determine the IR drops across each resistor in the circuit of Figure 8–26.

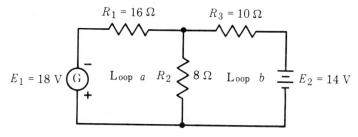

Figure 8–25

Circuit for Problems 7–10

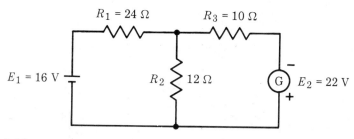

Figure 8–26

Circuit for Problems 11–13

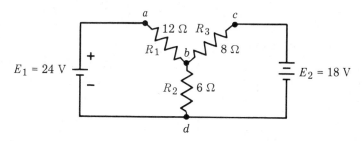

Figure 8–27

Circuit for Problem 14

13.　Verify that V_{R_1} plus V_{R_2} equals E_1 in the circuit of Figure 8–26.

14.　Determine the magnitude and direction of the current through each resistor in the circuit of Figure 8–27, using the superposition theorem.

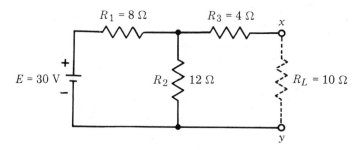

Figure 8–28

Circuit for Problems 16 and 17

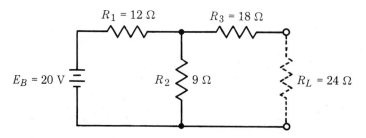

Figure 8–29

Circuit for Problems 18 and 19

15. From the currents determined in Problem 14, calculate the voltage across each resistor.

16. From the data given in Figure 8–28, calculate the Thévenin voltage V_{TH} that would appear across terminals x to y with R_L removed.

17. In the circuit of Figure 8–28, when load R_L is connected, what voltage appears across it? What current passes through it?

18. Calculate V_{TH} and R_{TH} from the data given in Figure 8–29.

19. What is the current through and the voltage across R_L in the circuit of Figure 8–29?

Problems 20 and 21 refer to the bridge circuit in Figure 8–30.

20. Using Thévenin's theorem, you are to calculate the current through R_L. Determine V_{TH} and R_{TH}.

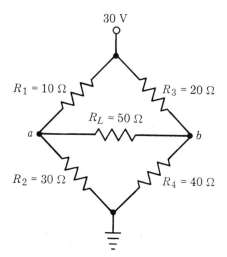

Figure 8–30

Circuit for Problems 20 and 21

21. From the information obtained in Problem 20, calculate the voltage across and the current through R_L.

22. Find the Norton current in the circuit of Figure 8–31.

23. Determine the current through R_L in the circuit of Figure 8–31, using Norton's theorem.

24. Refer to Figure 8–29. Determine: (a) the Norton current (b) I_{R_L}.

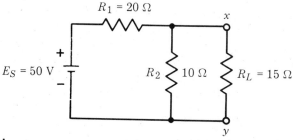

Figure 8–31

Circuit for Problems 22 and 23

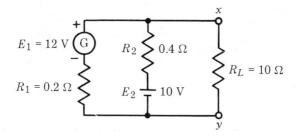

Figure 8–32

Circuit for Problem 25

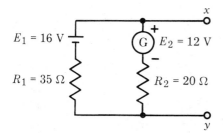

Figure 8–33

Circuit for Problem 26

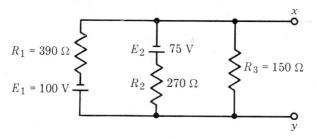

Figure 8–34

Circuit for Problem 27

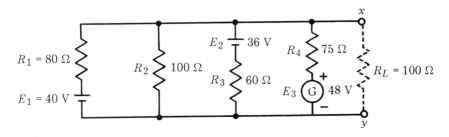

Figure 8–35

Circuit for Problem 28

25. Refer to Figure 8–32. Determine: (a) the Norton current (b) the current passing through R_L.

26. Using Millman's theorem, determine the terminal voltage V_{xy} of the circuit in Figure 8–33.

27. What is the Millman voltage of the circuit in Figure 8–34?

28. From the data presented in Figure 8–35, determine the following: (a) the Millman voltage across terminals x and y (b) the current through a 100 Ω load.

29. Thévenize the circuits in Figures 8–36a, 8–36b, 8–36c, and 8–36d, and determine R_{TH} and V_{TH} for each.

30. Using the V_{TH} and R_{TH} values calculated for each part of the figure accompanying Problem 29, draw the equivalent circuits, and determine the voltage across each load resistor.

31. Nortonize the circuits in Figures 8–37a, 8–37b, 8–37c, and 8–37d, and determine R_N and I_N for each.

32. Using the I_N and R_N values calculated for each part of the figure accompanying Problem 31, draw the Norton equivalent circuits and determine the current through each load.

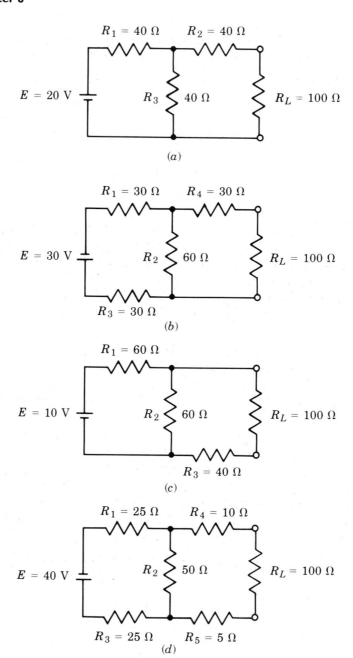

Figure 8–36

Circuits for Problem 29

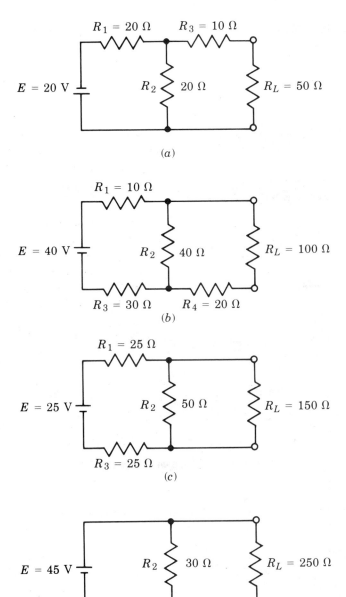

(a)

(b)

(c)

(d)

Figure 8–37

Circuits for Problem 31

9

Conductors and Insulators

Objectives

After studying this chapter, you will be able to:

Identify the factors that determine the effectiveness of a conductor in carrying an electric current.

Identify the factors that constitute a good insulator.

Introduction

The various components of an electronic circuit are connected by means of *conductors*. Generally speaking, conductors are small insulated wires. In the case of printed circuit boards, the conductive pattern forms the conductors. They should have very low resistance so that no appreciable voltage drop exists between source and load. Any voltage drop along the conductor constitutes a loss of energy.

Insulators should have very high resistance, ranging into hundreds or thousands of megohms. Examples of common insulator types are mica, glass, ceramic, Teflon, and air. Some insulating materials perform well at low but not at high frequencies. Temperature also has a significant effect on the characteristics of insulators and conductors.

Applications

Without conductors, we could neither generate nor conduct electric currents. Copper wire is an efficient conductor; within this metal, electrons move freely when under the influence of an electromotive force. However, should some material such as rubber, mica, glass, quartz, or waxed paper be substituted for the copper wire, electron movement between the two charges will be virtually nonexistent, because these materials have a very high resistance, which makes them valuable as insulators.

A knowledge of conductor sizes and of a conductor's ability to carry currents is also essential to the design and construction of electric equipment. If the conductor is too small, it may overheat; if it is too large, it may weigh and/or cost too much.

9–1
Circular Mils

For comparison of the resistance and the size of one conductor with another, some convenient unit must be established. This unit is the *mil-foot (mil-ft)*. A conductor will have this unit size if it has a *diameter of one mil (0.001 in.) and is one foot long.*

The *circular mil (c.mil)* is the standard unit of wire cross-sectional area. Because the diameters of round conductors are usually only a small fraction of an inch, it is convenient to express them in mils, equal to 1/1000 of an inch, to avoid the use of decimals. For example, the diameter of a conductor is expressed as 25 mils rather than 0.025 in. A circular mil is the *area of a circle having a diameter of one mil*, as illustrated in Figure 9–1. The area in circular mils of a round conductor is obtained by *squaring the diameter* measured in mils. This rule is an engineering agreement dating back many decades and is *not related* to the area of a circle being equal to πr^2.

Example 9–1 ————————————————————————

What is the circular mil area of a round conductor with a diameter of 25 mils?

Solution

$$A = d^2 = 25^2 = 625 \text{ c.mils}$$

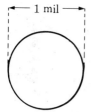

├── 1 mil ──┤

Figure 9–1

Circular mil

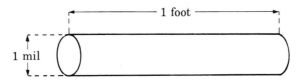

Figure 9–2

Circular mil-foot

Example 9–2

A wire has a diameter of 80.81 mils. What is its area in circular mils?

Solution

$$A = d^2 = 80.81^2 = 6530 \text{ c.mils}$$

A *circular mil-foot*, as shown in Figure 9–2, is actually a unit of volume. It *is one foot long and has a cross-sectional area of one circular mil*. Because it is considered a unit conductor, the circular mil-foot is useful in making comparisons between conductors made of different metals. For example, a comparison of the resistivity of various types of conductors may be made by determining the resistance of a circular mil-foot of each of the conductors.

9–2
Specific Resistance

Specific resistance or resistivity   The resistance of a conductor expressed in ohms per unit length per unit area, that is, per circular mil-foot.

Specific resistance, or **resistivity,** is the resistance in ohms offered by a unit volume (the circular mil-foot) of a material to the flow of electric current. A substance that has *high resistivity* will have *low conductivity*, and vice versa. For example, the specific resistance of copper is 10.4 Ω/mil-ft. In other words, a copper wire 1 c.mil in cross-sectional area and 1 ft long has 10.4 Ω resistance.

A list of specific resistivities of several different types of materials is given in Table 9–1. The values indicated are based on 20°C.

The resistance of a conductor of uniform cross-sectional area is *directly proportional to length* and *inversely proportional to cross-sectional area*. Thus, if the length of a particular conductor were unchanged but its cross-sectional area doubled, the resistance would be reduced by one-half.

Table 9–1 Specific Resistivities
(Ω/c.mil-ft at 20°C)

Material	Resistivity (ρ)
Silver	9.56
Copper (annealed)	10.4
Gold	14.0
Aluminum	17.0
Tungsten	34.0
Brass	42.0
Iron	61.0
Nichrome	675.0

9–3
Calculation of Resistance

The relationship of specific resistance, length, and cross-sectional area is given by the following equation:

$$R = \rho \frac{l}{A} \qquad \text{9–1}$$

where ρ = specific resistance (Greek letter *rho*)

l = length in feet

A = cross-sectional area in circular mils

The next two examples illustrate use of this formula.

Example 9–3 ─────────────────────────────

Calculate the resistance of a piece of copper wire at 20°C if it is 25 ft long and 40 mils in diameter.

Solution

First, calculate the cross-sectional area:

$$A = d^2 = 40^2 = 1600 \text{ c.mils}$$

Substitute $$R = \frac{10.4 \times 25}{1600} = 0.163 \ \Omega$$

Example 9–4 ─────────────────────────────

A piece of nichrome resistance wire is 17 ft long and has a diameter of 20.1 mils. Calculate its resistance at 20°C.

Solution

From Table 9–1, we see that nichrome has a specific resistance of 675 Ω/c.mil-ft. Thus,

$$R = \frac{675 \times 17}{404} = 28.4 \ \Omega$$

9–4
Wire Sizes

┌─ **American Wire Gage**   The system of notation for measuring the
│   size of conductors or wires.
└───

Wires are manufactured in sizes numbered according to the **American Wire Gage** (AWG). Some of these numbers appear in Table 9–2. Notice that the wire diameters become smaller as the gage num-

Table 9–2 American Wire Gage (AWG) Wire Sizes

Gage Number	Diameter (mils)	Circular Mil Area	Ohms per 1000 ft at 25°C
00	365.0	133,000.0	0.0795
0	325.0	106,000.0	0.100
1	289.0	83,700.0	0.126
2	258.0	66,400.0	0.159
3	229.0	52,600.0	0.201
4	204.0	41,700.0	0.253
5	182.0	33,100.0	0.319
6	162.0	26,300.0	0.403
7	144.0	20,800.0	0.508
8	128.0	16,500.0	0.641
9	114.0	13,100.0	0.808
10	102.0	10,400.0	1.02
11	91.0	8,230.0	1.28
12	81.0	6,530.0	1.62
13	72.0	5,180.0	2.04
14	64.0	4,110.0	2.58
15	57.0	3,260.0	3.25
16	51.0	2,580.0	4.09
17	45.0	2,050.0	5.16
18	40.0	1,620.0	6.51
19	36.0	1,290.0	8.21
20	32.0	1,020.0	10.4
21	28.5	810.0	13.1
22	25.3	642.0	16.5
23	22.6	509.0	20.8
24	20.1	404.0	26.2
25	17.9	320.0	33.0
26	15.9	254.0	41.6
27	14.2	202.0	52.5
28	12.6	160.0	66.2
29	11.3	127.0	83.4
30	10.0	101.0	105.0
31	8.9	79.7	133.0
32	8.0	63.2	167.0
33	7.1	50.1	211.0
34	6.3	39.8	266.0
35	5.6	31.5	335.0
36	5.0	25.0	423.0

bers increase. Larger and smaller wires, or conductors, are manufactured but have not been included in the table. In typical applications where the current is in milliamperes, a #22 wire would be used. By comparison, #14 wire is customarily used in residential-lighting circuits and #12 for wall plugs. When any conductor is selected, consideration must be given to the maximum current it can safely carry and the voltage its insulation can stand without breakdown.

Copper is most frequently used for wire conductors because it has low resistance per unit length, is less expensive than silver or gold, and is easily solderable. The copper is usually *tinned* (covered with a thin coating of solder) and may be solid or stranded.

Twin-lead transmission line A type of transmission line comprised of two parallel conductors covered by a solid insulation.

Coaxial cable A transmission line in which one conductor is concentric to another and separated by a continuous solid dielectric spacer.

Many electric cables are used in industry to interconnect components. *Cables* consist of two or more conductors within a common covering. Figure 9–3a shows a typical 300 Ω **twin-lead transmission line**, or cable, such as is used in TV to connect the antenna to the receiver. The cable shown in Figure 9–3b is a **coaxial cable**, which is used extensively for conducting high-frequency currents and consists of an inner conductor surrounded by polyethylene or other highly resistive insulation. Over the insulation is a flexible, tinned copper braid, which is in turn enclosed in a vinyl jacket. The inner conductor and braid constitute the two leads.

(a)   (b)

Figure 9–3

Cables: (a) Twin-lead transmission line. (b) Coaxial cable.

Printed Circuits

Modern techniques utilize printed circuit boards (PCBs) to provide necessary wiring between components. Printed circuit boards are fabricated from copper-clad, epoxy-based material that is painted with a light-sensitive coating. A negative is made of the desired circuit pattern and projected onto the sensitized copper. The exposed board is then immersed in a tank of ferric chloride, which etches away the unexposed copper to leave the desired circuit intact. Tiny holes are drilled at appropriate places in the remaining pattern, or circuit, to accommodate the required components, which are then soldered to the circuit. Customarily, the copper-wiring pattern is on one side of the board and the components on the other. More complex circuits use double-sided boards (circuits on each side).

Aluminum Conductors

Although aluminum has only about 60% of the conductivity of copper, it is much lighter in weight than copper and is now frequently used by the electrical power companies. Because aluminum conductors are not easily soldered, lugs, or terminals, are generally fastened to them by special tools.

9–5
Insulators

Dielectric The insulating medium between the plates of a capacitor.

Any substance that has a very high resistance, ranging in the order of megohms, may be classified as an insulator. Insulators are used for two basic purposes: (1) to insulate conductors and (2) to store an electric charge when a voltage is applied (for example, the dielectric in capacitors; see Chapter 11). An insulator that maintains its charge because electrons cannot flow to neutralize it is called a **dielectric**.

For any insulator, there is a certain voltage that will cause *breakdown* of the internal structure. The resulting current flow burns a hole through the material, rendering it useless.

> ─ **Dielectric strength**   The maximum voltage a dielectric can with-
> stand without breaking down, usually expressed in volts per
> mil.

Table 9–3 compares the **dielectric strength**, or breakdown voltage, of several types of insulators. The thickness of the material is assumed to be 1 mil. Obviously, the thicker the dielectric, the greater the
breakdown voltage will be.

Example 9–5

What is the breakdown voltage (V_{BD}) of a sheet of polystyrene
with a thickness of 2.5 mils and a dielectric strength of 650 V/
mil?

Solution

$$V_{BD} = 2.5 \text{ mils} \times 650 \text{ V/mil} = 1625 \text{ V}$$

> ─ **Corona**   A luminous discharge of electricity caused by ionization
> of the air around a conductor carrying a high potential.

Corona is a luminous discharge of electricity due to the ionization of the air. It appears on the surface of a conductor when the potential applied to it exceeds a certain (very high) value. Generally, if the
voltage (usually in kilovolts) is raised to a higher level, a spark discharge occurs. So that the corona effect is reduced, conductors should
be smooth, round, and thick. This configuration tends to equalize the
potential difference from all parts of the conductor to the surrounding

Table 9–3 Breakdown Voltage of Insulators

Material	Dielectric Strength (V/mil)	Material	Dielectric Strength (V/mil)
Air or vacuum	20	Paraffin wax	200–300
Bakelite	300–550	Polystyrene	500–760
Glass	335–2000	Porcelain	40–150
Mica	600–1500	Rubber, hard	450
Paper	1250	Shellac	900

air. Any sharp point will have an intensified field, which increases susceptibility to corona and spark discharge.

Corona discharge may become a serious problem in many TV sets. For example, if the high-voltage leads connected to the picture tube are not properly installed, corona discharge may occur. Soldered connections in these high-voltage circuits should be heavily soldered, to the point of forming a round ball at the point of connection, to prevent corona.

9–6
Conduction in Gases and Liquids

If sufficient potential is placed across two terminals in a gas-filled tube, some electrons will be removed from the gas atoms. These free electrons collide with other atoms, which either release one or more electrons or retain the new electron. An atom (or molecule) that has fewer or more electrons than normal is called an *ion*. A positive ion is one that has lost electrons, and a negative ion is one that has acquired more electrons than normal. This breaking up of the gas atom is called *ionization*. Positive ions tend to travel, or move, to negative potentials, while negative ions move toward positive potentials; these two movements constitute a current. The current that flows between the two electrodes is called *ionization current*.

The releasing of an electron from its atomic orbit is generally accompanied by the emission of light. It is this glow that we see in neon lights, fluorescent lamps, and other gas-filled tubes. Before ionization, the gas acts like an insulator; after ionization, the resistance drops to a significantly lower value.

> **Electrolysis**   The process of changing the chemical composition of a material (called the electrolyte) by sending an electric current through it.

Liquids containing certain acids or salts may become good conductors of electricity because of ionization. Such solutions are known as *electrolytes*. Examples of electrolytes are the battery acid in a storage cell and the conductive paste used in dry cell batteries. The process of producing a chemical change in a material by allowing ionization currents to flow in an electrolyte is called **electrolysis**. Oxygen and hydrogen, for example, are produced by electrolysis in water.

9–7
Effect of Temperature on Resistance

The resistance of most materials is affected by temperature changes. Pure metals such as silver and copper increase in resistance with an increase in temperature; semiconductor materials such as silicon and germanium decrease in resistance with an increase in temperature. Those few alloys, such as constantan and manganin, that change very little are used in measuring instruments requiring constant internal resistance for accurate results.

> **Temperature coefficient**   A factor used in calculating the change in the characteristics of a device or circuit element with changes in temperature.

Every material has what is known as a **temperature coefficient** of resistance, indicated by α (Greek letter *alpha*). This expression (α) indicates how much the material's resistance will change for each degree of temperature (°C). Table 9–4 indicates α for several commonly used materials for a temperature of 20°C. By using the following formula, we can determine the resistance of a material if α and the temperature are known:

$$R_x = R_{20}(1 + \alpha \, \Delta T) \qquad\qquad 9\text{--}2$$

where
R_x = resistance to be determined
R_{20} = resistance at 20°C
α = temperature coefficient of resistance
ΔT = temperature change

Example 9–6 ————————————————————————

A length of copper wire has a resistance of 21.7 Ω at 20°C. Calculate its resistance at 50°C.

Table 9–4 Temperature Coefficients

Material	Temperature Coefficient α (20°C)
Silver	0.0038
Copper	0.00393
Aluminum	0.004
Tungsten	0.0045
Carbon	−0.0003

Solution

$$R_x = R_{20}(1 + \alpha \, \Delta T) = 21.7\,[1 + 0.00393\,(50 - 20)]$$
$$= 21.7 \times 1.1179 = 24.258 \; \Omega$$

9–8
Switches

Switches are used to open and close circuits. The particular type used depends on the circuit complexity. When only one side of the line is to be opened, a *single-pole, single-throw (SPST) switch* is used. The basic design is shown in Figure 9–4a. A more conventional SPST switch appears in Figure 9–4b.

An old-style *triple-pole, double-throw (TPDT) knife switch* is shown in Figure 9–5 for comparison. Industrial power circuits frequently use triple-pole switches (modern, fully enclosed types) to control each leg of a three-phase power line. The differences in style, packaging, and voltage and current ratings of these switches are considerable. For additional information, reference should be made to a manufacturer's catalog.

Another type of switch is the *rotary switch*, usually used in more complex circuits where it is necessary to control a number of different circuits. Electronic instruments and TV receivers (channel selector switch) are typical examples of this kind of switch.

The elementary parts of a rotary switch are indicated in Figure 9–6a. The *wiping contact* makes connections at all times with the metal part (shaded in the figure), which has an extended arm rotated by a shaft; the end of the shaft can be seen in the drawing. As the switch

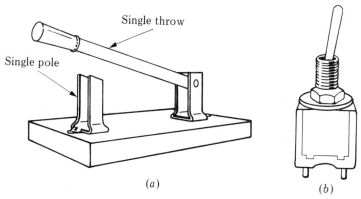

Single throw

Single pole

(a)

(b)

Figure 9–4

Switches: (a) Basic SPST switch. (b) Modern SPST switch.

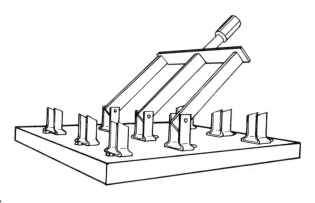

Figure 9–5

Old-style TPDT knife switch illustrating basic parts

rotates, it alternately makes connections with contacts 1 through 10. In the drawing, the wiper contact is connected, via the shaded part, with contact 6. The assembly shown is called a *wafer*. In complex circuits, rotary switches may have many wafers, all driven by a common shaft.

Figure 9-6*b* shows a miniature, fully enclosed rotary switch used in modern equipment. The schematic diagram of a six-position rotary switch appears in Figure 9-6*c*.

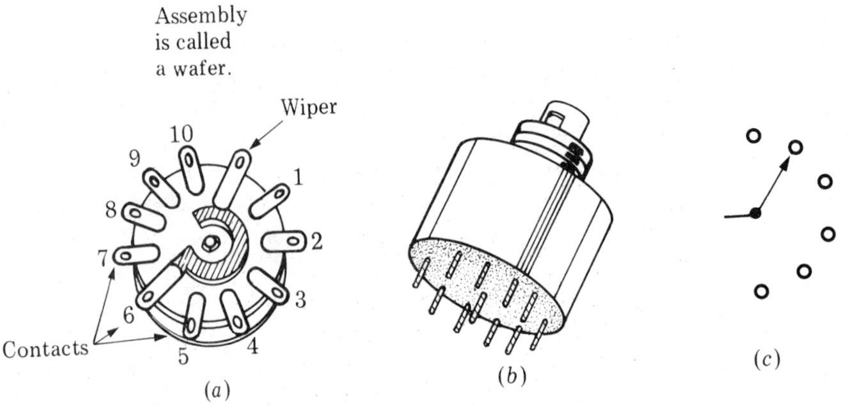

Figure 9-6

Rotary switch: (a) Pictorial drawing. (b) Miniature enclosed switch. (c) Schematic diagram of six-position switch.

9-9
Fuses and Circuit Breakers

Fuse A protective device containing a special wire that melts when excessive current passes through it, thereby opening the circuit.

Circuit breaker An automatic device that opens a circuit (without destroying itself) when an electric current beyond the designed value passes through it.

Fuses and **circuit breakers** protect circuits against *overloads* and *shorts*, thus preventing damage to wiring and components. Excessive current (I^2R) melts the fusible element and opens the circuit. Fuses come in many sizes and current ratings (1/500 A to hundreds of amperes). A typical fuse is shown in Figure 9–7a, with its symbol indicated in Figure 9–7b. Two types of fuse holders are shown in Figures 9–8a and 9–8b; the fuse in Figure 9–8b is located inside the tubular assembly.

Figure 9–7

Fuse: (a) Glass cartridge fuse. (b) Schematic symbol.

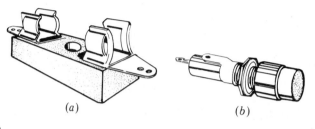

Figure 9–8

Typical fuse holders: (a) Fuse block. (b) Panel-mounted.

Some types of fuses are called *slo-blow* and are designed to handle momentary overloads several times their rating. They are customarily used in applications involving a substantial surge of current when the circuit is first turned on.

The normal resistance of a fuse is very low. When excessive load causes it to open, however, its resistance becomes infinite, and the full-circuit voltage appears across the open. For this reason, fuses have voltage ratings to indicate the maximum potential that can appear across the blown-out element without the electricity jumping across or arcing over.

There are two basic types of circuit breaker: *thermal* and *magnetic*. The thermal breaker contains a thermal element, generally in the form of a spring, that expands with heat and trips open the circuit. After cooling, it may be reset to restore normal operation (assuming the overload has been eliminated).

The magnetic breaker contains an *electromagnet* through which the circuit current must pass. If the current rises beyond the rating of the breaker, the pull of the magnet releases a trip mechanism that allows it to open. The schematic symbols for three different types of breakers are shown in Figure 9–9.

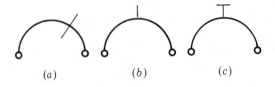

Figure 9–9

Schematic symbols of circuit breakers: (*a*) Switch. (*b*) Push-to-reset. (*c*) Push-to-reset, pull-to-trip.

⌐─**Summary**

A good conductor has very low resistance.

Good insulators have very high resistance, ranging into the hundreds or thousands of megohms.

The resistivity of a conductor is expressed in ohms per mil-foot. Its cross-sectional area is measured in circular mils.

The resistance of a conductor of uniform cross-sectional area varies directly as the product of the length and specific resistance, and inversely as the cross-sectional area.

A cable consists of two or more conductors within a common covering.

Coaxial cables are used for conducting high-frequency currents.

Printed circuit boards are made of copper-clad, epoxy-based material that has the wiring pattern etched on it to interconnect the various components.

The dielectric strength of a material is a measure of how much voltage it can withstand, per 1 mil thickness, before breaking down.

Corona discharge is due to the ionization of gases at very high potentials on an object.

When a gas becomes ionized, it will act as a conductor.

Electrolytic solutions are conductors of electricity.

Pure metals have positive temperature coefficients; semiconductors have negative coefficients.

The temperature coefficient α of a material indicates how much its resistance will change per degree Celsius.

Switches are rated by the amount of current and voltage they can handle without breaking down. They are also categorized as SPST, DPDT, and so on.

Fuses and circuit breakers are devices that protect circuits. They are rated according to the amount of current they can handle before opening.

Progress Test

The bracketed number after each question indicates the section of this chapter where the answer can be found.

    1.   A good electric conductor: (a) has low conductance. (b) will always be made of stranded wire. (c) provides a minimum IR drop. (d) has few free electrons. [Introduction]

    2.   The resistance of a conductor of uniform cross section: (a) varies inversely with length. (b) varies inversely as the square of its diameter. (c) varies directly as its cross-sectional area. (d) decreases as the specific resistance increases. [9-2 and 9-3]

    3.   An electric heater has a heating element made of 4 ft of #30 nichrome wire. What is its resistance at 20°C? (a) 0.27 Ω (b) 3 Ω (c) 27 Ω (d) 168 Ω [9-3]

    4.   What is the diameter of an aluminum wire 2000 ft long with a resistance of 3 Ω at 20°C? (a) 106 mils (b) 160 mils (c) 187 mils (d) 1060 mils [9-3]

    5.   How long is a copper wire 32 mils in diameter if its resistance is 2.54 Ω at 20°C? (a) 7.81 ft (b) 86.1 ft (c) 189.5 ft (d) 250 ft [9-3]

    6.   An output transformer in a transistorized receiver is wound with 80 ft of #32 gage copper wire. What is the resistance of the winding at 20°C? (a) 13.16 Ω (b) 20.85 Ω (c) 23.6 Ω (d) 31.2 Ω [9-4]

    7.   An electrolyte is: (a) a liquid or paste that conducts current because of ionization. (b) an ionized gas that conducts current. (c) an ionized current. (d) always made up of positive ions. [9-6]

    8.   What is the resistance of a copper wire at 60°C if its resistance is 6.15 Ω at 20°C? (a) 6.321 Ω (b) 6.988 Ω (c) 7.117 Ω (d) 7.361 Ω [9-7]

┌──Problems

Answers to odd-numbered problems are at the back of the book.

1. The diameter of a copper conductor is 0.032 in. What is its circular mil area?

2. An aluminum conductor has a diameter of 0.258 in. Calculate its cross-sectional area in circular mils.

3. A certain piece of copper wire has a circular mil area of 642. What is its diameter in mils?

4. A certain wire has a resistance of 7.8 Ω. Another copper wire is twice as long but has three times the cross-sectional area. What is its resistance?

5. What is the resistance of 150 ft of #22 copper wire at 20°C?

6. Calculate the resistance of a piece of #27 nichrome wire 11.5 ft long at 20°C.

7. Determine the length of a piece of tungsten wire with a resistance of 57 Ω at room temperature and with a diameter of 8 mils.

8. A certain wire has a diameter of 5 mils, a length of 4 in., and a resistance of 0.185 Ω at 20°C. What is its specific resistance?

9. Determine the resistance of 87 ft of #18 copper wire at 25°C.

10. A power transformer wound with #18 copper wire has 2.37 Ω resistance at 20°C. What will the resistance be at 65°C?

11. A #0 aluminum cable 2500 ft long has 0.4250 Ω resistance at 20°C. What is the cable's resistance at 50°C?

12. What is the resistance of 600 ft of #20 copper wire?

13. A length of #28 gage copper wire has a resistance of 43.5 Ω. How long is the wire?

14. What is the resistance of a 47 W soldering iron heating element made of #28 gage nichrome wire?

15. An electric circuit in a home is wired with 350 ft of #12 gage copper wire. Determine the resistance of the wire.

16. If the electric circuit described in Problem 15 were to be wired with the same gage aluminum wire, what would be the circuit resistance?

17.   Suppose the electric circuit described in Problems 15 and 16 were powered by 120 V ac and connected to a coffeepot with a heating element of #20 gage nichrome wire 6 ft long. Determine the actual voltage at the coffeepot for the following: (a) aluminum wire (b) copper wire.

18.   Suppose the circuit described in Problem 17 has a 20 A fuse. Would the fuse blow when the coffeepot is connected to the 120 V line?

19.   What modifications would have to be made to the element of the coffeepot heating element described in Problem 18 in order for it to operate without blowing a 20 A fuse when connected to 120 V?

10

Basic Electric Measuring Instruments

244

Objectives

After studying this chapter, you will be able to:

Accurately read analog voltmeter, ammeter, and ohmmeter scales.

Properly install a current meter in a circuit.

Identify the precautions to be followed with the use of ohmmeters.

Explore the versatility of an oscilloscope and some of its uses.

Introduction

Voltage, resistance, and current measurements are often required for checking a circuit for normal operation or for localizing a defective component. Many different kinds of instruments, extending over a wide price range, are available to perform these tests. However, several basic instruments are most commonly used by technicians and engineers. These instruments are voltmeters, ohmmeters, current-reading meters, and oscilloscopes. This chapter will provide a basic understanding of the use of these instruments.

245

Applications

Two basic kinds of instruments are used in designing, testing, and troubleshooting electronic equipment: analog and digital. *Analog meters* include a needle (pointer) that moves across a meter scale, like the speedometer in an automobile. *Digital meters* provide a numeric readout (numbers) on liquid crystal displays (LCDs). Earlier instruments used light-emitting diodes (LEDs), but they required more power to operate, and their display tended to wash out under bright ambient light.

10–1
Digital Multimeters

Many instruments are known as *multimeters*, so called because they are designed to measure voltage (dc and ac), current, and resistance. Selection of the desired function is provided by a switch. Another selector switch permits us to choose the correct *range* for the measurements to be made. Multimeters may provide either analog or digital readout. Digital multimeters are referred to as DMMs.

The DMM has certain advantages over the analog meter. Among these advantages are speed, increased accuracy, and reduction of operator errors, such as pointer-to-scale interpretation or parallax errors.

> **Parallax error**   Erroneous indications resulting from not reading an analog meter at right angles to the meter face.

The meter should always be read at right angles to the meter face. Since the meter divisions are small and the pointer is raised above the scale, reading the needle position from another angle will result in an inaccurate reading, or **parallax error.**

Digital multimeters display measurements as discrete numerals rather than as pointer deflections on a continuous scale. One disadvantage of DMMs is that the reading will not stabilize if the measured values are constantly changing. An analog instrument will reflect these variations.

Some confusion exists regarding the most fundamental of all DMM characteristics: its number of digits. The number of digits available on a DMM equals the number of 9s it can display, which indicates the number of full digits. But DMMs are usually rated as 3½ or 4½ digits. What does this rating mean? Since most instruments can *overrange*, this rating means a digit is added. However, the overrange digit is not a full digit.

Let us consider an example. A display of 1999 really has only three full digits; the 1 indicates overrange. This overrange digit permits the user to read beyond the normal full scale without loss of accuracy. Overranging is important since it extends the instrument's usefulness. For example, suppose a voltage changes from 9.99 to 10.1 V. A 3-digit DMM without overranging can only measure up to 9.99 V. Therefore, the 0.01 and 0.1 V (from 9.99 to 10.1 V) would be lost in the process. With overranging, a 3½ digit instrument could measure 10.1 V without loss of sensitivity or the need to switch to the next higher range.

Digital multimeters consist of two sections. One section is a voltage-sensitive circuit that accepts low-level dc inputs (usually 2 V max) and displays the measured value on a numeric readout. This section may be a series of LCDs or LEDs. The other section is a signal conditioner, or front end, that converts and scales the input signal into a dc voltage proportional to its amplitude and to the range of the instrument.

A popular, high-quality, hand-held DMM is shown in Figure 10–1. Figure 10–2 shows a typical analog multimeter.

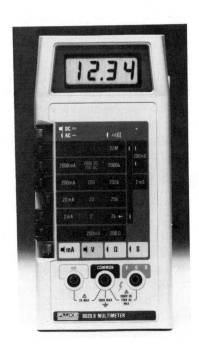

Figure 10–1

Typical, professional, hand-held digital multimeter (Courtesy of John Fluke Manufacturing Company)

Figure 10–2

Typical analog multimeter (Courtesy of Heath Co.)

10–2
Voltmeters

Meters are delicate instruments and should be handled with care. They are connected to the circuit or device being measured by means of test leads. One test lead has a red probe, while the other has a black probe. These probes represent the positive and negative connections, respectively, to the meter. They are well insulated so that the operator will not receive an electric shock when measuring higher voltages.

Voltmeters (VMs) are connected *across the circuit element* whose voltage is being measured. They are designed to have *very high input resistance*, typically 10 MΩ or more. An ideal voltmeter would have *infinite resistance* and thus present no load to the circuit to which it is connected. Real voltmeters always draw some current, however small, and therefore appear to the circuit as an additional element in parallel with the element across which voltage is being measured.

Figure 10–3 illustrates the loading effect that an older type of VM can have. Figure 10–3a is a simple voltage divider with 50 V across each resistor. When the VM is connected across R_2 as in Figure 10–3b, the division of voltages changes. It changes because the resistance R_m (100 kΩ) of the meter is in parallel with R_2, giving an R_{eq} of 50 kΩ. This new R_{eq} causes the voltage drops to change, as shown in Figure 10–3c.

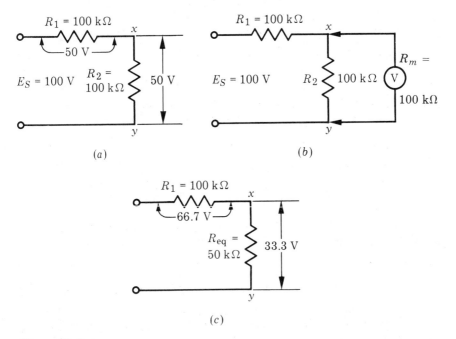

Figure 10–3

Loading effect of voltmeter on a circuit: (a) No meter connected. (b) Voltmeter with 100 kΩ internal resistance connected across R_2. (c) Resulting voltage drops with voltmeter connected.

10–3
Reading the Scales

The scales on different instruments will vary somewhat, but they all follow the basic arrangement shown in Figure 10–4. There are four scales shown in Figure 10–4. They are as follows, from top to bottom:

1. The top scale measures ohms from 0 to 1K ohm.
2. The second scale measures from 0 to 1.
3. The third scale measures from 0 to 3.
4. The bottom scale measures decibels from −20 to +2 dB.

Note that the middle two scales are for both dc and ac measurements.

A *function switch* selects the measurement function: dc, ac, or ohms. The *range switch* selects the full-scale volts, current, and ohms

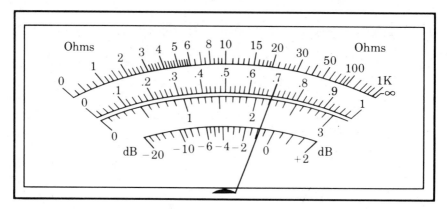

Figure 10–4

Basic scales of a typical analog multimeter

multiplier. For example, if this switch is positioned to the 1 V range, then a maximum of 1 V (dc or ac) can be measured, as read on the 0–1 scale. This same scale is used when the range switch is positioned to any multiple of 1 (such as 0.1, 10, 100, 1000) by using the *appropriate scale multiplier.*

Example 10–1

What voltage is indicated on the meter in Figure 10–4 if the range switch is on 10?

Solution

We must read on the 0–1 scale and multiply the needle reading by 10. Since the needle reads 0.7, then 0.7 × 10 = 7 V.

Example 10–2

What voltage is indicated by the needle in Figure 10–5 if the range switch is on 30 V?

Solution

Use the 0–3 scale, since this scale is a submultiple of 30 (that is, 0.1 × 30). Since the needle reads 1.4, the voltage is 1.4 × 10 = 14 V.

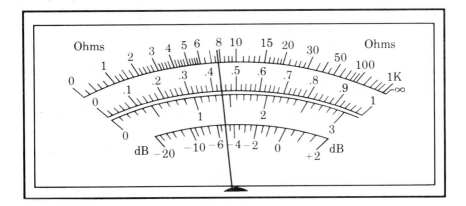

Figure 10–5

Scale reading for Example 10–2

(*Note*: This reading could be a dc or an ac voltage, depending on the position of the selector switch.)

You may often have to use your multimeter to check, maintain, and repair electronic equipment that contains *dangerously high voltages*. Because of this danger, you should always observe the following safety procedures:

1. Always handle the test probe by the insulated housing only. Be careful not to touch the exposed tip.
2. When you measure high voltages, turn off the power to the equipment to be tested *before* you connect the test leads. If you cannot turn off the power, be very careful to avoid accidental contact with any object that could provide a ground return (circuit completion) path.
3. If it is at all possible, use only one hand when you test energized equipment. Keep one hand in your pocket or behind your back to minimize the possibility of accidental shock.
4. If possible, insulate yourself from ground while making measurements. Stand on a properly insulated floor or floor covering.
5. Before you connect the test leads for a resistance measurement, turn off the power to the equipment to be tested, and discharge any capacitors that may have stored a charge.

10–4
Measuring ac Voltages

Not all ac voltages can be measured with uniform accuracy. For this reason, the following information is presented.

When a dc voltage is applied to a resistor, it produces a measurable temperature increase. If an ac voltage is applied to the same resistor and produces the same temperature increase, then the ac voltage must be producing the same amount of power. The ac voltage that produces this power is proportional to the square root of the mean power, and it is called the *root mean square (rms)* voltage. Usually, ac meters are calibrated in rms voltage. For a *sine wave* (see Figure 10–6a), the

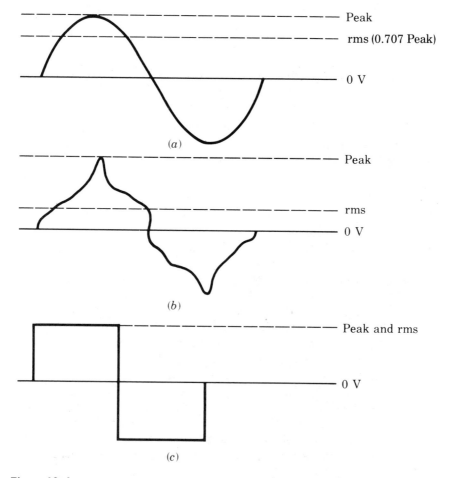

(a)

(b)

(c)

Figure 10–6

Three basic types of ac voltages: (a) Sine wave. (b) Complex wave. (c) Square wave.

most common ac voltage waveform, the rms value of each half cycle is 0.707 times the peak of the waveform. Only sine waves give correct readings on conventional ac voltmeters. For more information on the characteristics of sine waves, refer to Chapter 15.

If a *nonsinusoidal* (not a sine wave) waveform, such as a square wave, sawtooth wave, or pulse, is being measured, the indicated reading on the scale will not be accurate. For example, the complex waveform shown in Figure 10-6*b* contains a spike (peak) that may be several times as large as the rms value of the waveform. Since the spike is of such short duration, the rms value of the overall waveform is barely affected. On the other hand, the symmetrical square wave (a square wave having positive and negative portions of equal amplitude and time duration) shown in Figure 10-6*c* would indicate an rms value higher than its peak value.

The preceding discussion indicates an important caution. Always examine any nonsinusoidal waveform with an oscilloscope or a true rms meter if you want a highly accurate measurement.

10-5
Ammeters

Ammeters are very low resistance instruments designed to measure the current in a circuit. (Theoretically, they should have 0 Ω internal resistance, but we don't know how to make such instruments.) They must be connected in *series* with the circuit element. That is, the circuit must be opened to allow the meter to be connected. The positive side of the current meter must be connected so that, when you trace through the circuit, you come to the positive terminal of the voltage supply without going through the meter. Likewise, the negative terminal of the current meter must lead to the power source's negative terminal without going through the meter.

Remember, you must open the circuit to install the current meter. *Never connect the meter across any circuit element.* Since ammeters have *very low internal resistance*, any voltage applied across them (like the *IR* drop across a resistor) will cause excessive current, and the meter will likely be damaged.

A typical ammeter connection and an ammeter's schematic symbol are illustrated in Figure 10-7. *Ammeters should have very low internal resistance* so that the current in the circuit is the same with or without the meter. It doesn't matter where the meter is inserted in the circuit, since current is the same in each component. Observe the meter polarity. The terminal where the current enters the meter is *negative*; the other terminal is *positive*.

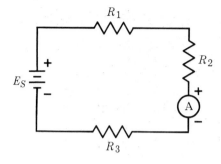

Figure 10–7

Ammeter symbol and connection in
a series circuit

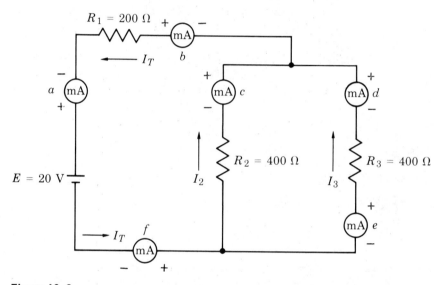

Figure 10–8

Simple series-parallel circuit showing possible milliammeter connections

For most electronic circuits, currents are in the milliampere (mA) or microampere (μA) range. You will have to select an appropriate meter to measure the current. When multimeters are used, it is good practice to set the instrument to a *high current range* and then switch down to lower ranges until you reach the proper range. This procedure will prevent accidental damage.

The circuit shown in Figure 10–8 illustrates a number of possible current meter connections. Every meter is in series with one of the circuit elements. Carefully observe all meter polarities. Meters at *a*, *b*, and f all measure I_T and therefore read the same. The meter at c measures I_2, and the meters at d and e measure I_3. The sum of the meter readings at c and d, for example, equals the total current I_T, measured at either a, b, or f.

Actually, technicians and engineers find it more convenient and much faster to measure the current in a circuit by reading the voltage drop across a resistor and dividing that value by the ohmic value of the resistor (Ohm's law). Since most resistors now being used have 5% or even 2% tolerance, and since voltmeters are also quite accurate, the resulting "measured" current is obtained quickly and accurately.

Example 10-3

Determine the current in the circuit shown in Figure 10-9.

Solution

If a voltmeter connected across R_2 reads exactly 3 V, then circuit current is as follows:

$$I = \frac{V_{R_2}}{R_2} = \frac{3.0 \text{ V}}{3 \text{ k}\Omega} = 1.0 \text{ mA}$$

The accuracy of this measurement is limited only by the resistor and meter tolerances.

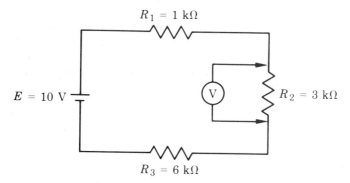

Figure 10-9

Circuit for Example 10-3, showing how current through R_2 can be measured by using Ohm's law

10-6
Ammeter Shunts

—**Shunt**   A precision resistor connected in parallel with a current-
indicating meter for the purpose of bypassing a specific fraction
of current around the meter.

When an ammeter is to be used in a circuit that has a current
greater than the meter is designed to handle, a **shunt** must be used. A
shunt is a precision resistor *connected in parallel* with the meter for the
purpose of *shunting*, or *bypassing*, a specific fraction of the current
around the meter. Resistor R_{sh} in Figure 10-10 is the ammeter shunt;
shunts are usually mounted across the meter's terminals. Current I_T approaching the meter divides into I_m and I_{sh}, with the currents being inversely proportional to the resistance (current divider law).

To calculate the shunt resistance for a particular meter, use the
following formula:

$$R_{sh} = \frac{R_m I_m}{I_{sh}}$$

10-1

where
R_{sh} = shunt resistance
R_m = meter resistance
I_m = meter current for full-scale deflection
I_{sh} = shunt current

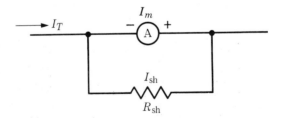

Figure 10-10

Ammeter with shunt connected

Example 10–4 ────────────────────────────────

What shunt resistance is required to extend the range of a 0–1 mA meter movement, having 27 Ω internal resistance, to read a total of 10 mA?

Solution

Because the meter movement can handle only 1 mA, the other 9 mA must flow through the shunt.

$$R_{sh} = \frac{R_m I_m}{I_{sh}} = \frac{27 \times 1 \times 10^{-3}}{9 \times 10^{-3}} = 3 \ \Omega$$

Because the shunt is 1/9 the value of the meter resistance, it will have 9 times more current than the meter. If the meter is connected in a circuit carrying 5 mA, the current will divide in a ratio of 1 to 9 between meter and shunt. In this case, the shunt current is 4.5 mA, and the meter current is 0.5 mA. The needle would deflect only about half scale.

10–7
Ohmmeters

Ohmmeters are instruments that measure resistance. They are also used to make *circuit continuity* checks and to locate *shorted* or *open* circuits or components. Since resistance values vary from fractions of an ohm to megohms, ohmmeters must have a selector switch for choosing the appropriate range. With analog meters, a range should be selected that will give a reading in the center two-thirds of the scale. It is difficult to obtain accurate readings on the higher end of the scale, where the divisions are crowded. *Note:* An ohmmeter scale is *nonlinear.*

Ohmmeters have internal voltage sources (batteries) to provide for their operation since they must force a small current through the circuit or element being tested. *A word of caution: Avoid connecting an ohmmeter across any circuit in which a voltage is present.* Damage to the meter may result. Standard operating practice is to de-energize a circuit (turn off the power) before making resistance measurements.

On a typical ohmmeter, the *range switch* has resistance ranges as follows: ×1, ×10, ×100, ×1K, ×10K, ×100K, and ×1M. If, for ex-

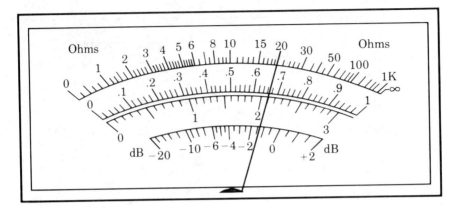

Figure 10–11

Typical ohmmeter scale

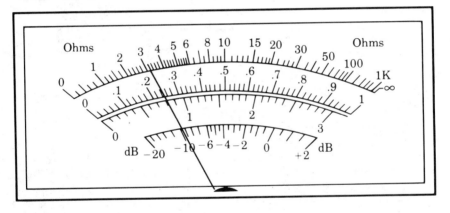

Figure 10–12

Ohmmeter scale reading for Example 10–5

ample, the needle points to 20 and the range switch is positioned to ohms ×100, then the resistance being measured is 20 × 100 = 2000 Ω. A typical ohmmeter scale is shown in Figure 10–11. The needle points exactly to 20.

In Figure 10–11, observe the *nonlinearity* of the ohms scale (top). The left part of the scale is expanded, and the right part is compressed. Notice that the space between the main numbers diminishes as you move from left to right. Also, observe that 10 is approximately midscale. For most accurate resistance readings, select a resistance range, with the range switch, that will cause the needle to read in the center two-thirds of the scale.

Example 10-5

Refer to Figure 10–12. What resistance is indicated if the range switch is on ×1K?

Solution

The needle indicates 3.2 on the ohms scale. Therefore, the resistance being measured is $3.2 \times 1K = 3.2 \text{ k}\Omega$.

10-8
Continuity Testing

Ohmmeters are frequently used to check for *continuity* in electric conductors. A continuous conductor has very low resistance. An ohmmeter connected to the ends of such a conductor that has no breaks will read practically 0 Ω. A common failure in multiconductor cables is that one of the conductors is open somewhere inside the cable. This open indicates an infinite resistance and shows up as an *open circuit*.

Continuity testing is frequently used to check for opens (breaks) in multiconductor electric cables, as illustrated in Figure 10–13. This check is made by testing for continuity between terminal x and the leads coming from the other end of the cable. The conductor that has 0 Ω to terminal x is the one connected to it (wire 4 in Figure 10–13). This same technique is used to test continuity of coils, trans-

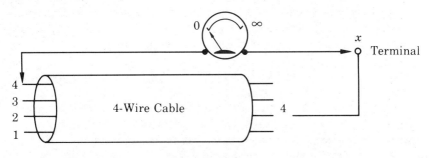

Figure 10-13

Testing a cable for continuity, using an ohmmeter on the ×1 scale

former windings, and relays. The range switch should be on ×1 for continuity testing.

10–9
Introduction to Oscilloscopes

The single instrument most widely used in electronic engineering and technology is the *cathode-ray oscilloscope* (often abbreviated as scope). This instrument displays not only the *amplitude* and *period* of the signal (frequency) but also the *shape* of the wave. (For more information on ac waves and voltages, see Chapter 15.) The signal being viewed is projected on the face of a *cathode-ray tube (CRT)*. It is characterized by a high input resistance (that is, impedance) and, consequently, does not appreciably load the circuit being tested.

Almost anything can be measured on the two-dimensional graph drawn by an oscilloscope. Normally, voltage is displayed on the vertical, or Y, axis and time on the horizontal, or X, axis. This type of display presents far more information than do other test and measurement instruments such as multimeters and frequency counters. For example, with a scope, you can determine the amplitude and shape of an ac signal, whether there is any dc component, how much noise is present (if any), and the frequency of the signal. With a scope, you can see *everything at once*; you do not have to make several different tests with various pieces of test equipment.

In addition to measuring electric signals, scopes measure nonelectrical phenomena when appropriate *transducers* are used. Transducers change one kind of energy into another. Microphones and loudspeakers are two familiar examples. In microphones, sound waves are converted to electric signals; loudspeakers convert electric signals to sound waves. There are many kinds of transducers. Others transform pressure, light, temperature, or mechanical stress into electric signals.

The record and reproduce heads used with cassette players and other magnetic recorders are further examples of electric transducers. The record head records the audio or video electric signals as magnetic patterns on ferromagnetic tape; the reproduce head converts these patterns to electric signals. With proper transducers, scopes provide very extensive test and measurement capabilities.

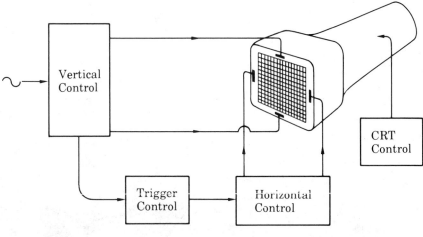

Figure 10–14

Display and control systems of an oscilloscope

10–10
Basic Circuits of an Oscilloscope

The basic control systems of an oscilloscope are shown in the functional block diagram of Figure 10–14. Each of the blocks is named for its function. A very brief explanation of the function of each block follows. In subsequent sections, more detailed explanations will be given.

The *vertical system* controls the vertical axis of the CRT. Any time the electron beam, which draws the graph, moves up or down, it does so under the control of the vertical control circuit. This system is the Y axis control. The *horizontal system* controls the left-to-right movement of the electron beam. This system is the X axis control.

The *trigger circuitry* determines the beginning of the horizontal sweep. It is responsible for creating stable CRT displays. A signal that is selected to do the triggering should be time-related to the displayed signal. The *display system* contains the CRT, which shows the signal that is being observed. The *CRT control system* includes those circuits that provide control for intensity (brightness), focusing of the electron beam for sharpness, and vertical and horizontal positions. More expensive oscilloscopes also have controls for trace rotation, beam finding, and automatic intensity and focus.

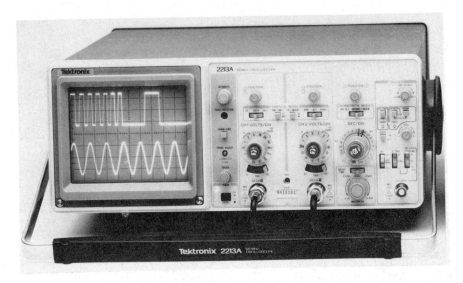

Figure 10-15

Tektronix 2213 dual-trace scope (Courtesy of Tektronix, Inc.)

One very widely used instrument is the Tektronix Model 2213 dual-beam oscilloscope, which is shown in Figure 10-15. It has the capability of displaying two separate signals simultaneously on the CRT. This display is very informative when you are comparing *phase relationships*, *waveshapes*, or *amplitudes* of two different signals.

10-11
Display System of an Oscilloscope

The functional elements of a CRT (a vacuum tube device) are shown in Figure 10-16. Some of these elements, the *cathode*, the *control grid*, and the *anode*, serve the same functions that they serve in a conventional vacuum tube, even though they are constructed as cylinders. The surface of the cathode is coated with special materials that permit it to boil off large quantities of electrons when it is raised to the correct temperature. The *heater* provides the energy to heat the cathode, and it gets its power from a low-voltage, 60 Hz source, usually a step-down power transformer winding.

The control grid voltage (relative to the cathode) determines the intensity of the beam current and therefore the brightness of the displayed pattern. This system is referred to as *Z* axis control. The flow of electrons is from the cathode to the positive potentials of the anodes (*accelerating grid* and *anode wall coating*). High voltages, up to several thousand volts, are used for the last anode; the voltage depends on the size of the CRT.

The electron beam must pass between two sets of *deflection plates* before reaching the CRT. Voltages applied to the *vertical deflection plates* cause the beam to move up and down according to the intensity of the potentials applied to them. The output of the *vertical control amplifier* provides these voltages. Similarly, the beam is deflected from left to right across the CRT from voltages applied to the *horizontal deflection plates*. These voltages are supplied by the *horizontal control circuits*. Because of the very small mass of the electrons in the beam, the voltages connected to the deflection plates can display events on the face of the CRT that represent time elements in nanoseconds. The entire cylinder assembly, consisting of the elements just described, is referred to as an *electron gun*.

On the inside surface of the face of the CRT is a phosphor coating, which fluoresces when struck by the electron beam from the gun. The result is a glow seen through the glass for a very short time afterward, tracing the path of the beam. Various colors can be produced, depending on the chemical composition of the phosphor. Green phosphors are generally used for the CRTs in oscilloscopes because they are more efficient. Figure 10–17 shows the face of a typical CRT; the display is a sine wave.

A TV picture tube is a CRT. Though much larger screens are usually used with TV, these tubes possess the same basic elements as

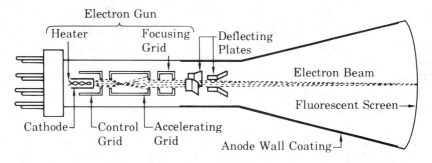

Figure 10–16

Basic components of a CRT with electrostatic focusing and deflection

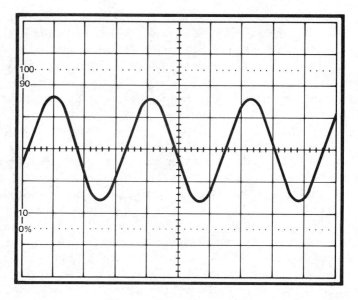

Figure 10–17

Face of a typical CRT displaying a sine wave

scopes, with one major difference. Picture tubes use *magnetic deflection* because of their larger size. Two sets of coils are used to provide horizontal and vertical deflection. They constitute a *yoke*, which is slipped over the neck of the CRT up against the bell section.

Etched on the inside of the faceplate is a grid of lines that serves as a reference for measurements. This grid is called the *graticule*. On older scopes, the graticule was a piece of Plexiglas with a ruled scale inscribed on one of its surfaces; this Plexiglas was placed over the face of the CRT. Some CRTs had the graticule etched directly on the outside surface of the CRT faceplate. On newer CRTs, the graticule is etched or silk-screened on the inside surface of the CRT faceplate. This technique eliminates measurement inaccuracies called *parallax errors*. These errors occur when the trace and the graticule are on different planes and the observer is shifted slightly from the direct line of sight. A typical CRT faceplate with a graticule is shown in Figure 10–18. Notice the major and minor divisions.

Though different-sized CRTs may be used, graticules are usually laid out in an 8 × 10 pattern. Each of the 8 vertical and 10 horizontal lines block off major divisions of the screen. These divisions are often in centimeters. The *tick marks* on the center graticule lines represent minor divisions, or subdivisions. Since rise time measurements are very common in pulse or square wave circuits, graticules often include

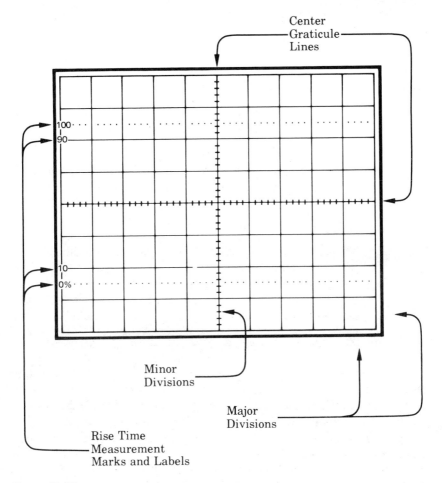

Figure 10–18

Typical CRT faceplate with a graticule

rise time measurement markings: dashed lines for 0% and 100% points and labeled lines for 10% and 90% (see Figure 10–18).

10–12
Vertical System of an Oscilloscope

The vertical system of a scope takes the input signals to be examined and develops deflection voltages to control the *Y* axis on the CRT. This system provides the following basic controls: vertical positioning, input coupling, and vertical sensitivity. Scopes with two beams have these three controls for each input channel.

Input Coupling

The *input-coupling switch* lets you control how the signal being examined is coupled to the vertical channel. Three selections are usually provided:

1. The dc coupling allows all of an input signal to be displayed.
2. The ac coupling inserts a capacitor between the input and the vertical amplifier to block any dc component of the signal. Thus, it allows only the ac components to reach the vertical amplifier.
3. The GND (ground) coupling disconnects the input signal from the vertical system and causes the triggered display to show the scope's chassis ground. The trace displayed is the ground reference level. Switching from ac or dc to ground and back is a convenient way to measure signal voltage levels with respect to ground. *Note*: Using the GND position *does not* ground the signal in the circuit you are probing.

Figure 10–19 illustrates the display of a signal that has both dc and ac components. The ground reference is shown in Figure 10–19*a*. The dc component (Figure 10–19*b*) does not change the shape of the ac wave (Figure 10–19*c*) but causes it to be displaced from the ground reference line by an amount equal to the dc present. The vertical-sensitivity switch (explained in the following paragraph) is set at 0.2 V per division for each signal shown.

Vertical Sensitivity

A *volts-per-division* (VOLTS/DIV) rotary switch controls the *sensitivity* of the *vertical* channel or amplifier. Having different sensitivities extends the range of an oscilloscope's applications. Typical ranges provided by this switch are from 2 mV/DIV to 100 V/DIV.

Using the VOLTS/DIV switch to change sensitivity also changes the scale factor, or the value of each major division on the graticule. For example, with a setting of 2 V/DIV, each of the eight major vertical divisions represents 2 V, and the entire screen can show 16 V from top to bottom. The center knob of this control must be in the CAL (calibration) position (full clockwise) for the settings on the

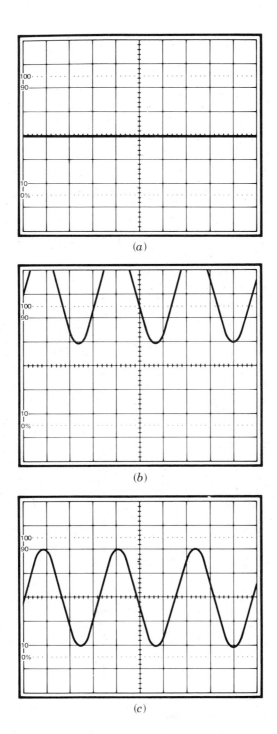

(a)

(b)

(c)

Figure 10–19

Effect on the CRT display of an ac signal with a dc voltage component: (a) Ground coupling. (b) dc coupling. (c) ac coupling.

VOLTS/DIV switch to be correct. When this control (potentiometer) is out of calibration, it provides a continuously variable change in the scale factor. The VOLTS/DIV control is useful when you are making quick amplitude comparisons on a series of signals.

Example 10–6

What is the peak-to-peak voltage (E_{pp}) of the wave shown in Figure 10–20 if the VOLTS/DIV switch is on 10 mV/DIV?

Solution

Since the wave fills four complete vertical divisions, the signal's amplitude is as follows:

$$E_{pp} = 10 \text{ mV/DIV} \times 4 \text{ DIV} = 40 \text{ mV}$$

Dual-trace scopes have provisions for displaying more than one signal at a time. For example, channel 1 can be shown on one trace and channel 2 on the other. Dual-trace scopes allow easy comparison of different signals. For information on scopes having more elaborate vertical-switching functions, refer to the manufacturer's instruction manual.

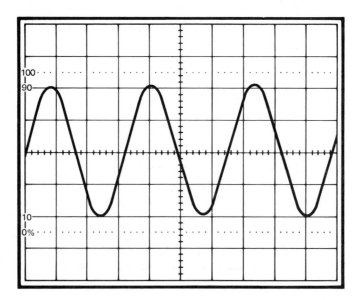

Figure 10–20

Waveform for Example 10–6

10–13
Horizontal System of an Oscilloscope

The horizontal circuitry in an oscilloscope provides the deflection voltages required to deflect the beam across the X axis of the CRT. It also contains a *sweep generator*, often called a *time-base generator*, which provides a *sawtooth waveform*, or *ramp*, that is used to control the scope's sweep rates. This generator lets you select the time units, observing the signal for either very short times, measured in micro- or nanoseconds, or relatively long times, measured in seconds.

Figure 10–21 details the waveform that produces the horizontal sweep. The *ramp* is the most critical part of the wave. It must be *extremely linear*; otherwise, the electron beam would not be deflected across the screen at a uniform rate. For example, if the seconds-per-division (SEC/DIV) switch is set to 1 μs/DIV, then the beam must sweep through *each* of the 10 horizontal divisions in exactly 1 μs. If it did not, then the viewed waveform would be subject to distortions produced by the oscilloscope. The circuit that made the *rate of rise* in the ramp *linear* was one of the most important advances in oscillography. A linear ramp meant that the horizontal beam movement could be calibrated directly in *units of time*.

The falling edge of the waveform shown in Figure 10–21 is called the *retrace*. It is during this time interval that the beam returns to the left side of the screen.

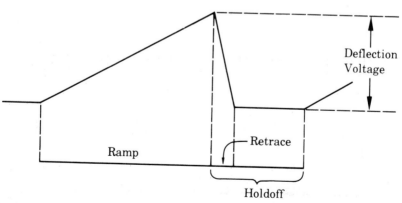

Figure 10–21

Details of the sawtooth waveform

Example 10-7

Calculate the period of the wave shown in Figure 10-22 if the SEC/DIV switch is on 50 μs/DIV. From this information, determine its frequency.

Solution

This measurement can be facilitated by lining up one rising edge of the square wave with the graticule that is second from the left-hand side of the screen. This alignment can be done with the horizontal position control. Now, count the major and minor divisions across the center horizontal graticule line from left to right for one complete cycle. In Figure 10-22, there are 7.2 divisions.

$$\text{period} = 7.2 \text{ DIV} \times 50 \ \mu\text{s/DIV} = 360 \ \mu\text{s}$$

The frequency (f) of the wave can be calculated from the formula $f = 1/T$, where T is the period of the wave:

$$f = \frac{1}{360 \times 10^{-6}\text{s}} = 2.778 \text{ kHz}$$

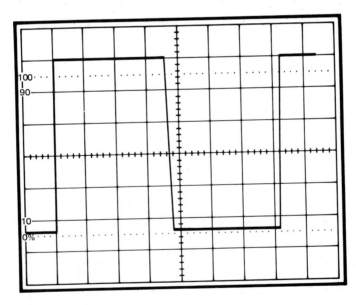

Figure 10-22

Square wave for Example 10-7

10–14
Trigger System of an Oscilloscope

Triggering is responsible for creating stable CRT displays. Proper setting of the trigger controls guarantees that each sweep starts at the correct time. Selecting a particular trigger point on a repetitive waveform means that each sweep of the beam will overlay the previous waveform. When triggering is derived directly from the signal being displayed, the display remains stable as long as there is an adequate signal present. Figure 10–23 illustrates the effect of triggering.

Oscilloscopes have a *trigger selector switch* that permits the operator to choose the particular trigger method desired. Three positions are available:

1. In the INT (internal) position, triggering is derived from the signal being displayed.
2. In the LINE position, the ac power line is used to derive the triggering. This trigger source is very useful when you are troubleshooting any circuit containing power line frequencies.

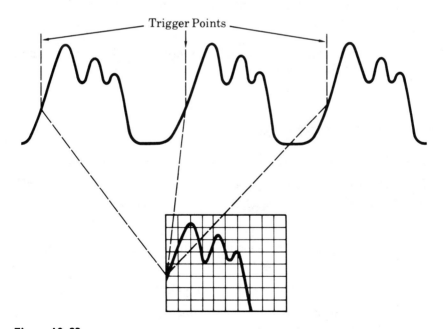

Figure 10–23

Triggering on an oscilloscope

3. In the EXT (external) position, triggering is derived from an external signal. It should be time-related to the display signal for stability.

Depending on the nature of the waveform being analyzed, it is possible to have the scope display begin on either the rising or the falling edge of the signal. The *slope switch* allows for this selection, with + meaning a positive-going signal and − meaning a negative-going signal. The *level control* selects the specific point (amplitude) on the waveform where triggering is to begin.

Most oscilloscopes have a *mode selector switch* that determines how triggering is to be accomplished. The switch usually has three positions:

1. In the NORM (normal) mode, when the triggering signal is inadequate or absent, or when the level control is misadjusted, there is no display. However, the NORM position is one of the most useful because it can handle a wider range of trigger signals than any other triggering mode.
2. In the AUTO (automatic) mode, the trigger level control will be matched to the trigger signal. This mode lets you trigger on signals with changing amplitudes or waveshapes without making adjustments to the level control.
3. In the TV mode, triggering is matched to a TV frame or line. This mode is useful when you are examining signals found in a television system.

Summary

Multimeters are designed to measure voltage, current, and resistance.

Digital multimeters (DMMs) provide speed, accuracy, and reduction of operator errors.

Overranging in a digital meter allows you to read beyond the normal full scale without loss of accuracy.

Voltmeters are designed to have very high input resistance.

Care should be exercised when measuring high voltages with a voltmeter.

Most ac voltmeters only provide accurate readings when you are measuring sine waves.

Current meters have very low internal resistance.

Current meters must always be connected in series with the circuit element whose current is being measured.

A fast and convenient way to measure the current in a circuit is to read the voltage drop across a resistor and divide that value by the ohmic value of the resistor.

Shunts are used to extend the range of a current meter.

Ohmmeter scales are nonlinear.

Ohmmeters can be used for continuity checks.

Oscilloscopes are one of the most widely used measuring instruments.

With an oscilloscope, you can determine the amplitude and shape of a wave.

The trigger circuitry of an oscilloscope determines the beginning of the horizontal sweep.

The electron gun of a scope consists of the cathode, control grid, focusing and accelerating anodes, and deflection plates.

The graticule of a scope serves as a reference for measurements.

⌐Progress Test

The bracketed number after each question indicates the section of this chapter where the answer can be found.

1. Which of the following statements is true regarding DMMs? (a) They provide analog readout. (b) They create parallax errors. (c) They offer increased accuracy. (d) Their readings quickly stabilize when measured values are constantly changing. [10–1]

2. Overranging in a DMM: (a) provides very high input resistance. (b) allows readings beyond the normal full scale. (c) reduces the number of accurate digits by one. (d) applies only to the ohms scale. [10–1]

3. Voltmeters should be connected: (a) in series with the circuit element. (b) across the element whose voltage is being measured. (c) in series or in parallel because of their high resistance. (d) so as not to load the circuit. [10–2]

4.   The same voltmeter scale can be used to measure full-scale voltages of : (a) 0.1, 1, 0.3, and 3 V. (b) 0.3, 3, 30, and 300 V. (c) 1, 10, 30, and 100 V. (d) 0.1, 1, 10, and 0.3 V.   [10–3]

5.   When measuring high voltages, you should: (a) only handle the insulated test leads. (b) begin on low-voltage ranges and work up. (c) never measure high voltages. (d) leave the power on when connecting test leads.   [10–3]

6.   Which of the following statements is true concerning ac voltmeters? (a) They can read all ac voltages accurately. (b) They can measure the shape of the wave. (c) They usually require different scales than dc voltmeters. (d) They are only accurate when you are measuring sine waves.   [10–4]

7.   The internal resistance of a typical ammeter is: (a) zero. (b) low. (c) high. (d) infinity.   [10–5]

8.   Current-measuring instruments must always be connected in: (a) series with the circuit. (b) parallel with the circuit. (c) series-parallel with a circuit. (d) shunt with the device whose current is being measured.   [10–5]

9.   Ammeter shunts are used to: (a) increase its current-handling capabilities. (b) increase the meter's accuracy. (c) make the instrument more rugged. (d) limit the meter's current-measuring ability.   [10–6]

10.   Ohmmeters can be used to: (a) measure voltage. (b) read current. (c) make continuity checks. (d) measure batteries.   [10–7 and 10–8]

11.   Ohmmeter scales are: (a) linear. (b) square law. (c) nonlinear. (d) uniformly divided.   [10–7]

12.   For the most accurate readings, the ohmmeter range switch should be positioned so that the resistance being measured provides a pointer reading that is in the: (a) center of the scale. (b) first third of the scale. (c) center two-thirds of the scale. (d) last third of the scale.   [10–7]

13.   Which of the following signal characteristics can be displayed by an oscilloscope? (a) amplitude and period only. (b) period and shape only. (c) amplitude, period, and shape. (d) shape and amplitude only.   [10–9]

14.   Voltages that can be displayed on a scope are limited to: (a) dc only. (b) ac only. (c) ac or dc. (d) ac with no dc component.   [10–9]

15.   Which of the following statements is correct regarding

the trigger circuit of a scope? (a) It determines the beginning of the horizontal sweep. (b) It controls the stopping point of the trace. (c) It determines when the Y axis displacement begins. (d) It works on non-repetitive waveforms. [10–10]

16. The function of the cathode in a CRT is to: (a) focus the electron beam. (b) control the flow of electrons to the phosphor coating on the CRT. (c) supply large quantities of electrons. (d) deflect the electron beam. [10–11]

17. The function of the graticule is to: (a) protect the face of the CRT. (b) eliminate parallax error. (c) improve the reliability of the CRT. (d) provide a reference for measurements. [10–11]

18. Which of the following statements is correct regarding the vertical system of a scope? (a) It develops the horizontal deflection voltages. (b) It modifies the input signal to match the input characteristics of the scope. (c) It controls the length of the ramp voltage in the horizontal deflection circuit. (d) It develops deflection voltages to control the Y axis. [10–12]

19. The time-base generator: (a) is a part of the vertical deflection system. (b) controls the duration of the ramp voltage. (c) affects the Z axis control. (d) is controlled by the period of the input wave. [10–13]

20. The ramp generator: (a) is also called the time-base generator. (b) is part of the vertical display system. (c) causes nonlinear time measurements. (d) is controlled by the trailing edge of the input signal. [10–13]

21. Which of the following statements is correct with regard to triggering? (a) It controls the start of the horizontal sweep. (b) It is controlled by the period of the wave. (c) It determines the sweep speed of the electron beam. (d) It controls the starting point of vertical deflection. [10–14]

Problems

Answers to odd-numbered problems are at the back of the book.

1. Draw the schematic symbol for a voltmeter.

2. Draw the schematic symbol for an ammeter.

3.   What instrument is used to measure electromotive force? (a) ohmmeter (b) galvanometer (c) ammeter (d) voltmeter.

4.   What is the maximum voltage a 3½ digit DMM can read when set to the 200 V range?

5.   Which of the following statements concerning voltmeters is correct? (a) It has very high internal resistance. (b) It must be connected in series with the device being measured. (c) It has low internal resistance. (d) Its low and high ends must be calibrated before it is used.

6.   Which of the following statements regarding the use of electronic voltmeters is correct? (a) The polarity of the measured voltage must be observed. (b) Polarity is unimportant. (c) When you are measuring unknown voltages, you start on low scales and work up. (d) The position of the range control is not significant.

7.   Which schematic (a, b, c, or d) in Figure 10–24 shows the correct connection for a voltmeter?

8.   What voltage is indicated on the meter in Figure 10–25 if the needle is at position B and the range switch is on 30 V? (a) 2.2 V (b) 72 V (c) 23 V (d) 7.2 V.

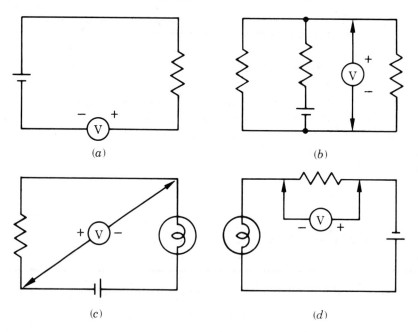

(a)

(b)

(c)

(d)

Figure 10–24

Circuits for Problem 7

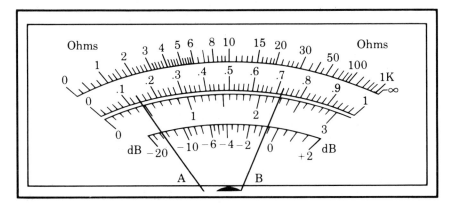

Figure 10–25

Meter readings for Problems 8–10

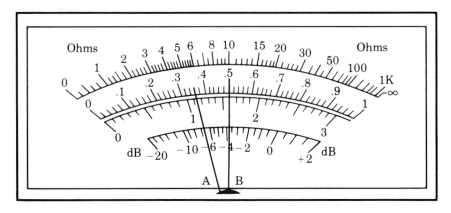

Figure 10–26

Meter readings for Problems 11 and 12

9. What voltage is indicated on the meter in Figure 10–25 if the needle is at position A and the range switch is on 100 V? (a) 1.4 V (b) 12 V (c) 72 V (d) 14 V.

10. Refer to Figure 10–25. What is the measured voltage if the range switch is on 0.1 V and the needle is at position B? (a) 0.0072 V (b) 0.072 V (c) 0.22 V (d) 0.71 V.

11. What voltage is indicated by the needle in Figure 10–26 if it is at position A and the range switch is on 0.1 V? (a) 0.36 V (b) 0.0115 V (c) 0.036 V (d) 0.36 V.

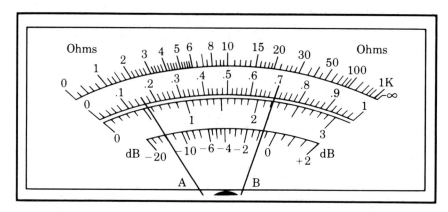

Figure 10–27

Meter readings for Problems 13 and 14

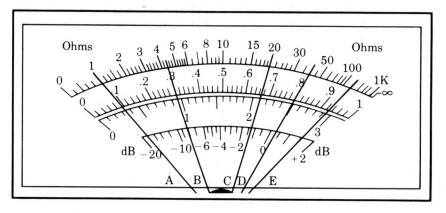

Figure 10–28

Meter readings for Problem 16

12. What voltage is indicated by needle position B in Figure 10–26 if the range switch is set to 30 V? (a) 16 V (b) 1.6 V (c) 50 V (d) 5 V.

13. Refer to Figure 10–27. If the instrument is set to the 10 mA range and the needle is at position A, what is the current reading?

14. Refer to Figure 10–27. If the range switch is on 0.01 mA, what current is being read when the needle is at position B?

15. Which of the following ohmmeter scales should be used for continuity testing? (a) $R \times 1$ (b) $R \times 10$ (c) $R \times 1000$ (d) $R \times 1M$.

16. Refer to Figure 10–28. For each of the following functions, list the value of the indicated meter reading.

	Range	Function	Value
A	$R \times 1K$	Ω	_____
B	10 V	dc+	_____
C	$R \times 1$	Ω	_____
D	3 V	ac	_____
E	100 V	dc	_____
A	1000 V	dc	_____
B	$R \times 100$	Ω	_____
C	30 V	dc	_____
D	$R \times 1K$	Ω	_____

17. Refer to Figure 10–29. For each of the following functions, list the value of the indicated meter reading.

	Range	Function	Value
A	10 V	dc	_____
B	$R \times 100K$	Ω	_____
C	0.1 V	dc	_____
D	0.3 V	dc	_____
E	$R \times 1K$	Ω	_____
A	$R \times 1$	Ω	_____
B	30 V	dc	_____
C	$R \times 100$	Ω	_____
D	1000 V	dc	_____

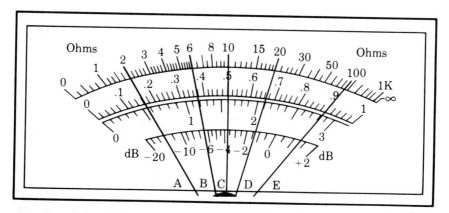

Figure 10–29

Meter readings for Problem 17

18. Calculate the shunt resistance required to extend the range of a 0–1 mA meter with a resistance of 27 Ω to the following: (a) 5 mA (b) 50 mA (c) 500 mA.

19. What value of shunt resistance is required to convert a 2 kΩ, 0–50 μA meter movement to read the following? (a) 100 μA (b) 1 mA (c) 10 mA.

20. A certain 0–1 mA meter is shunted with a 0.707 Ω resistance. Calculate the meter resistance if 99 mA passes through the shunt when the meter reads full scale.

21. A 0–50 μA meter, with R_m = 2 kΩ, is shunted with a 222.2 Ω resistance. What current passes through the shunt when the meter reads full scale?

22. In Problem 21, what current passes through the shunt if the meter reads a 0.25 scale deflection?

23. Match the following oscilloscope control names to the correct oscilloscope control functions in the accompanying list: (a) horizontal position (b) intensity (c) dc/ac (d) VOLTS/DIV (e) vertical position (f) slope (g) SEC/DIV (h) trigger selector (i) focus.

_____ Selects capacitive or direct coupling of vertical input signal.
_____ Controls sharpness of beam.
_____ Selects whether triggering will be on LINE, INT, or EXT.
_____ Determines the voltage value of each vertical division.
_____ Allows manual left-right adjustment of beam.
_____ Allows manual up-down adjustment of beam.
_____ Controls brilliance of trace.
_____ Controls repetition rate of sawtooth deflection voltage.
_____ Tells sweep whether to start on positive- or negative-going waveform.

24. Describe the major functions of an oscilloscope.

25. What is the grid of lines on the front of a CRT called?

26. Oscilloscopes have controls for the X, Y, and Z axes. Briefly describe the function of each.

27. What are the three major elements of a CRT?

28. Which of the following statements are true regarding the trigger circuitry of a scope? (a) It controls the beginning of the vertical deflection. (b) It provides a choice for connecting the input signals. (c) It provides the sawtooth waveform for the horizontal sweep. (d) It se-

lects a point on the signal being displayed to control the start of the horizontal sweep.

29.   Which of the following terms indicates the type of scope deflection system used? (a) electromagnetic (b) electrostatic (c) trigger (d) Z axis.

30.   A certain square wave has a peak-to-peak amplitude of 4.3 divisions as displayed on the graticule. What is the signal's amplitude if the VOLTS/DIV switch is set to 5 mV/DIV?

31.   What is the frequency of a wave shown on a scope whose period is 7.2 divisions when the SEC/DIV switch is on 0.5 ms?

32.   The horizontal time-per-division factor is valid only if the: (a) VOLTS/DIV switch is in CAL (b) trigger level control is in AUTO (c) SEC/DIV switch is in CAL (d) slope switch is in positive.

33.   An oscilloscope displays the _____, _____, and _____ of a wave.

34.   The X axis displays _____, and the Y axis displays _____.

35.   What is the amplitude of the signal displayed in Figure 10–30 if the VOLTS/DIV switch is on 2 mV/DIV?

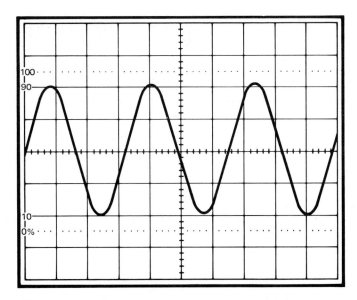

Figure 10–30

Graticule display for Problems 35–37

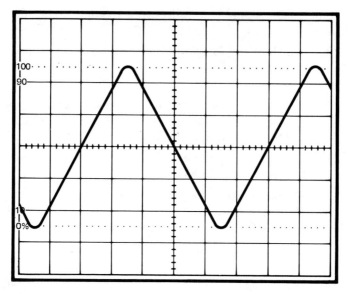

Figure 10–31

Waveform for Problems 38 and 39

36.   What is the amplitude of the displayed waveform in Figure 10–30 if the VOLTS/DIV switch is on 5 V/DIV?

37.   What is the period of the wave shown in Figure 10–30 if the SEC/DIV switch is on 1 ms/DIV? What is its frequency?

38.   Determine the peak-to-peak amplitude of the signal displayed in Figure 10–31 if the VOLTS/DIV switch is on 10 V/DIV.

39.   Calculate the period of the wave displayed in Figure 10–31 if the SEC/DIV switch is on 2 μs/DIV. What is the frequency of the signal?

11

Capacitors

Objectives

After studying this chapter, you will be able to:

Determine which basic type of capacitor to use for a particular application.

Calculate the total capacitance of capacitors that are connected in either series or parallel combinations.

Introduction

The function of a *capacitor*, sometimes called a *condenser*, is to store an electric charge. The storage may be for short or relatively long periods of time. The shorter periods may be in the nanosecond range, while the longer periods are measured in seconds.

Capacitors vary in shape, size, and design, depending on their intended use. For example, filter capacitors used with low frequencies, such as power supplies, have very large capacities so that they can store large charges when the rectifiers are conducting. Most of the charge is then released when the rectifiers are not conducting, thereby maintaining a steady flow of current to the load. On the other hand, capacitors used with high frequencies have much smaller capacities. They do not have a chance to receive as large a charge because of the very short charge and discharge times.

Applications

The capacitor, or condenser, which consists of two conducting surfaces separated by an insulator called a *dielectric*, has the ability to store and release an electric charge over a period of time. Depending on the capacitor and the circuit, the time may vary from many seconds to billionths of a second. This delay in charge and discharge is useful in selecting or rejecting ac currents of particular frequencies.

Condensers may be used in filters where undesired high-frequency currents are effectively shorted out to ground through the condenser while lower-frequency currents bypass the condenser, as they would any high resistance. For example, the hum produced by stray 60 Hz ac currents in a radio, a hi-fi system, or a cassette player can be reduced by filters.

Another electronics industrial heating process, dielectric heating, differs in principle from induction heating in that the major heating element is a condenser rather than a transformer, and the material to be heated is an insulator. Dielectric heating is used in forming plywood, roasting coffee, hardening plastics, and heating microwave ovens.

11-1
Electrostatic Fields

The phenomenon of *electrostatic fields* was investigated in Section 1-2. We discovered the action resulting from placing a charged object in close proximity with another charged body. If the charged bodies possessed the same sign of charge, a repelling force was established between them. If they had unlike signs, an attractive force prevailed. These forces are the result of the electrostatic field surrounding every charged body.

The direction of the electrostatic field can be represented by lines of force drawn perpendicular to the charged surfaces. They are assumed to originate on the positively charged surface. Each line of

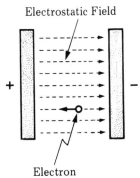

Electrostatic Field

+   −

Figure 11–1

Electrostatic field between two charged surfaces exerting force on electron

Electron

force is represented by an arrow pointing from positive to negative, as represented in Figure 11-1. The electron will be forced to move in a direction opposite to that of the electrostatic lines, or, in other words, toward the positively charged surface.

From Coulomb's law, we learned that *the force existing between two charged bodies is directly proportional to the product of their charges and inversely proportional to the square of the distance separating them.* Hence, the greater the distance between the electrons and the positive charge, the less will be the force of attraction. The repelling force of the negative plate on the electron likewise varies inversely with the square of the distance between them.

11–2
Charging Action in a Capacitor

Capacitor or condenser   A device for storing electrical charges.

In its simplest form, a **capacitor**, or **condenser**, consists of two metal plates separated by an insulating material called a dielectric. Such a capacitor is represented in Figure 11-2.

Let us now analyze what happens when a dc potential is applied to the capacitor plates, as shown in Figure 11-3. The positive potential of the supply attracts many of the free electrons in the left-hand plate,

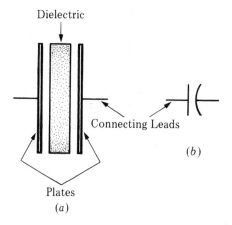

Dielectric

Connecting Leads

(b)

Plates

(a)

Figure 11–2

Capacitor: (a) Simple form. (b) Schematic symbol.

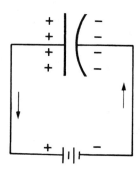

Figure 11–3

Electron motion during charging

while the negative terminal forces electrons into the right-hand plate. Thus, in every part of the circuit, a displacement of electrons occurs. This current is *large* when the capacitor is first charging and *tapers off* near full charge.

As electrons are forced into the right-hand plate and others are removed from the left-hand plate, a *difference of potential* develops across the capacitor. Each electron forced into the right plate makes it more negative, while the left plate becomes more positive as electrons are removed. The charge building up across the capacitor tends to *oppose* the source voltage. That is, as the charge increases, it becomes more difficult for the source to force more electrons around the circuit. As the capacitor continues to charge, the voltage building up across it increases until it equals the supply potential, at which time electron movement in the plate and connecting leads ceases. *Thus, capacitors oppose any change of voltage across them.*

Figure 11–4 shows the charging action of a capacitor. As shown in Figure 11–4a, current is maximum at the instant the charging be-

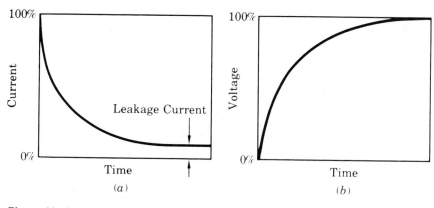

Figure 11–4

Capacitor's charging characteristics: (*a*) Current versus time. (*b*) Voltage versus time.

gins. On the other hand, voltage across the capacitor is minimum when charging begins, as indicated in Figure 11–4*b*. The voltage tapers off (approaches 100%) when the capacitor is nearing its full charge.

> **Displacement current**   The current that appears to flow in a capacitive circuit.

The charging action described above has all the appearances of a true current flow, although it must be emphasized that *no current* flows through the capacitor. Since the insulating material (dielectric) between the plates has no free electrons (the characteristic of any good insulator), no current can flow through the capacitor. The current that appears to flow (see Figure 11–3) is the result of the displacement of electrons in the plates and connecting wires to the voltage source and is referred to as a **displacement current**.

11–3
Storing Charges in a Capacitor

Consider the drawing in Figure 11–5*a*, which represents an atom within a particular dielectric material of a capacitor. Since there is no charge indicated on the plates, the electrons will travel in normal orbits around the nucleus.

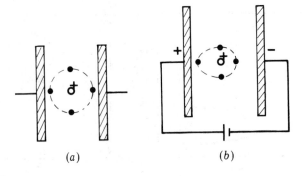

Figure 11–5

Electron orbits of an atom within a dielectric: (a) With no electric field. (b) In the presence of an electrostatic field.

We previously established that a free electron placed in an electrostatic field will be repelled by any negative charge and attracted by a positive one. The same is true, with qualifications, if the electron is in a *bound state*, that is, not free to leave its orbit around the nucleus. In Figure 11–5b, the same atom is represented after a potential has been applied to the plates. Notice that the electron orbits have been distorted. The bound electrons are attracted toward the positive plate and repelled from the negative plate. The resulting orbits in the dielectric are elongated in the direction of the positive charge. The voltage source must supply the energy required for this distortion.

If the voltage applied to the plates of a capacitor is increased, there will be further elongation of the electron orbits in the dielectric around their nuclei, which results from the increased charge. For any given dielectric material and thickness, there is a *maximum potential* that can be applied without forcing electrons out of their orbits. If this maximum is exceeded, the dielectric breaks down (becomes shorted), and the capacitor is destroyed. The characteristics of dielectrics will be investigated in Section 11–6.

When a capacitor is fully charged, its emf will be equal to the applied voltage, and charging current will cease. At full charge, the electrostatic field between the plates is at maximum intensity, and the energy stored in the dielectric is therefore at maximum. If a charged capacitor is disconnected from its charging source, the charge will be retained for considerable time. Just how long depends on the amount of *leakage current* through the dielectric. Since no dielectric material has infinite resistance, there is always some leakage. However, it is extremely small for a high-quality capacitor, and the charge may be retained for days. Because electrical energy is stored in a capacitor, a charged capacitor can act as a *voltage source* until the charge is dissipated.

Reason tells us that the larger the capacitance of the capacitor or the greater the impressed voltage or both, the larger will be the stored energy. This statement is indeed true, as indicated by the following formula:

$$\text{energy} = 1/2\, CE^2 \text{ (joules)} \qquad\qquad \text{11-1}$$

where
C = capacitance in farads (F)
E = capacitor voltage in volts (V)

Note: The letter symbol for capacitance is C.

11-4
Discharging a Capacitor

If a capacitor is to be discharged, the energy stored in the dielectric must be released. Placing a short, such as a heavy-gage wire or the blade of a screwdriver, across the capacitor terminals, although effective for this purpose, is not recommended. Many capacitors are charged to very high voltages and can be lethal if a person accidentally touches both terminals.

A proper method of discharging a capacitor is shown in Figure 11-6. A suitable resistor and switch can be connected as indicated — again using caution if high voltages are involved. When the switch is closed, the excess electrons on the negative plate can flow through the resistor to the positive plate. In doing so, the electrostatic energy in the capacitor will be converted to heat in the resistor. When the capacitor is fully discharged, the distorted orbits of the electrons in the dielectric return to their normal positions.

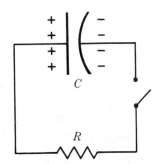

Figure 11-6

Discharging a capacitor

11–5
The Farad

The SI unit of capacitance is the farad, named in honor of the English physicist and chemist Michael Faraday (1791–1867). The unit symbol for the farad is F. Capacitance is the ability of a component to hold a charge. The amount of charge Q that is stored is *proportional to the applied voltage*. The charge is also determined by the capacitance of the capacitor. The relationship of these terms can be expressed as follows:

$$Q = CE \qquad\qquad \text{11–2}$$

where Q = electric charge to be measured in coulombs
E = electric pressure across the capacitor in volts
C = capacitance in farads

In terms of direct current, this equation may be stated as follows: *A capacitor has a capacity of one farad if one volt applied across the plates produces a charge of one coulomb in the capacitor.*

Example 11–1 ——————————————————————————

A 0.05 F capacitor has 15 V dc applied across its plates. What is its charge in coulombs?

Solution

$$Q = CE = 0.05 \text{ F} \times 15 \text{ V} = 0.75 \text{ C}$$

Example 11–2 ——————————————————————————

What is the capacitance of a capacitor that receives a charge of 0.001 C when 200 V is applied?

Solution

$$C = \frac{Q}{E} = \frac{10 \times 10^{-4}}{2 \times 10^{2}} = 5 \times 10^{-6} \text{ F}$$

The unit of capacitance may also be described in terms of an alternating current (one that is constantly changing). *If E changes at the rate of one volt per second and causes one coulomb of charge per second to flow, the capacitance is one farad.* Since 1 C/s is 1 A, we may say that *the capacitance of the device is one farad if one ampere flows through the circuit when the applied emf is changing at the rate of one volt per second.*

The charging (and discharging) of a capacitor takes time. Large capacitors take more time to charge than small ones. Just because a voltage is suddenly applied across a capacitor does not imply that it instantly receives its full charge. The phenomenon of charging and discharging will be explained fully in Chapter 14.

The farad is an extremely large unit of capacitance. In practice, it is more convenient to use the microfarad (μF), which is 1×10^{-6} F. In fact, for very high frequency applications, the nanofarad (nF) and the picofarad (pF) are more frequently used. These units are 1×10^{-9} and 1×10^{-12} F, respectively. In older publications, the term *micromicrofarad* is used in place of the *picofarad*.

Example 11-3

A certain capacitor is rated at 0.0068 F. What is its capacity in microfarads and in nanofarads?

Solution

One farad equals 1×10^6 μF. Therefore:

$$0.0068 \text{ F} \times 10^6 = 6800 \ \mu\text{F}$$

There are 10^3 nF in 1 μF. Therefore:

$$6800 \ \mu\text{F} \times 10^3 = 6,800,000 \text{ nF}$$

Example 11-4

A capacitor is marked 390 pF. What is its capacity in microfarads?

Solution

To convert picofarads to microfarads, we must multiply by 1×10^{-6} (move the decimal point six places to the left). Therefore:

$$390 \times 10^{-6} = 0.000390 \ \mu\text{F}$$

11–6
Dielectrics

Dielectric constant (k)   The ratio of the capacitance of a capacitor with a given dielectric to the capacitance of an otherwise identical capacitor having air or vacuum for its dielectric.

One of the principal factors affecting the capacitance of a capacitor is the type of dielectric material used between the plates. These materials, or insulators, are rated by their ability to produce *dielectric flux* in terms of a parameter called **dielectric constant** (k). Materials having a high dielectric constant can create more capacitance than ones with a low k for the same plate area and separation.

A standard is necessary for comparing the dielectric constant of one material to another. A *vacuum* serves as this standard and is assigned a value of *unity*. Air has essentially the same value. The dielectric constant of all materials is compared with that of a vacuum. Thus, if a particular dielectric material has a k value of 6.5, it will produce 6.5 times more capacity than if a vacuum were used in the same-size capacitor.

Table 11-1 indicates the dielectric constants of several commonly used materials. The dielectric strength of a number of materials is given in Table 9-3.

Table 11-1   Dielectric Constants

Material	k
Air or vacuum	1.0
Paper	2.0–6.0
Plastic	2.1–6.0
Mineral oil	2.2–2.3
Silicone oil	2.7–2.8
Quartz	3.8–4.4
Glass	4.8–8.0
Porcelain	5.1–5.9
Mica	5.4–8.7
Askarel (nonflammable synthetic) oil	5.6–5.9
Aluminum oxide	8.4
Tantalum pentoxide	26
Ceramic	12–400,000

11-7
Factors Affecting Capacitance

Several factors affect the ability of a capacitor to store an electric charge. They are as follows:

1. The area of the plates.
2. The distance between the plates.
3. The dielectric constant of the material between the plates.

The relationship of these parameters can be understood by reference to the elemental parallel-plate capacitor illustrated in Figure 11-7. The following formula also expresses the relationship:

$$C = 0.224 \frac{kA}{d} \text{ (picofarads)} \qquad \text{11-3}$$

where
k = dielectric constant of material between the plates
A = area of one plate in square inches
d = distance between plates in inches
0.224 = conversion factor when using inches

When the dimensions are in centimeters (cm), use the following formula:

$$C = 0.08842 \frac{kA}{d} \qquad \text{11-4}$$

where
0.08842 = conversion factor when using centimeters

An examination of these formulas indicates that *the capacitance varies directly with plate area and inversely with the spacing between plates.* The dielectric constant also *directly affects* the

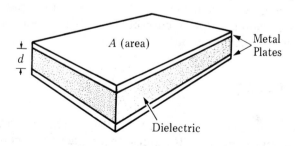

Figure 11-7

Parallel-plate capacitor

capacitance. For example, if mica is substituted for air as the dielectric, the capacitance will increase from 5.4 to 7.4 times.

If the capacitor is composed of more than two parallel plates, its capacitance in picofarads is calculated by the following formula:

$$C = 0.08842 \frac{kA}{d} (n - 1) \qquad \text{11-5}$$

where

n = number of plates

Example 11-5

A simple capacitor has two parallel plates separated by 2 cm. If the area of one plate is 100 cm² and the dielectric is polystyrene, what is the capacitance in picofarads?

Solution

$$C = 0.08842 \times \frac{2.6 \times 100}{2} = 11.49 \text{ pF}$$

11-8
Voltage Ratings

Dielectric Breakdown

The dielectric material must be a good insulator. However, insulators will begin to allow some current to leak through if subjected to a sufficiently high potential. This leakage reduces the capacitor's ability to hold a charge. If the applied voltage is too great, the dielectric will break down, and arcing will occur between the plates. The capacitor is then short-circuited, and the flow of current through it can cause damage to other components in the circuit.

Voltage Rating

Working voltage direct current (WVDC)   The maximum dc voltage that can be steadily applied to a capacitor without danger of arcing.

Capacitors have a *voltage rating* that should not be exceeded. This rating is called the **working voltage direct current** (WVDC) and represents the maximum voltage that can be steadily applied without danger of arc. The working voltage depends on the type of dielectric material used and its thickness (refer to Table 9–3). A high-voltage capacitor requires a thicker dielectric than a low-voltage capacitor, for a given type of dielectric, in order to withstand the higher voltage. Since larger size increases the spacing between the plates, the capacitance is reduced. If a larger capacitance is required, the plate area must be increased to compensate for the greater spacing between plates.

A capacitor that is designed to operate at 500 V dc cannot be safely operated on an ac voltage with an effective value of 500 V. Such a voltage has a peak value of 707 V (see Chapter 15) and may cause the capacitor to short. It is good practice to select a capacitor with a WVDC that is at least 50% greater than the highest voltage to be applied.

11–9
Types of Capacitors

Variable Capacitors

Variable capacitors are so named because their capacitance can be varied. There are two basic types of variable capacitors: rotor-stator and trimmer capacitors.

The *rotor-stator capacitor* consists of two sets of metal plates. The fixed plates are connected together and form the *stator*. The movable plates are connected together on the shaft and form the *rotor*. Capacitance is varied by rotating the shaft in such a way that the rotor plates mesh with the stator plates. There is no electric contact between the two sets of plates; air is the dielectric. When the plates are fully meshed, the effective plate area is greatest, and maximum capacitance results. Conversely, when the plates are out of mesh, the effective plate area is greatly reduced, and minimum capacitance results. Theoretically, an infinite number of continuously variable capacitances can be realized between the limits of maximum and minimum.

Two rotor-stator capacitors are shown in Figure 11–8 along with their symbols. The split-stator capacitor in Figure 11–8*a* is similar to the capacitor used for tuning most older types of radio receivers. The midget single capacitor in Figure 11–8*b* is used for tuning high-fre-

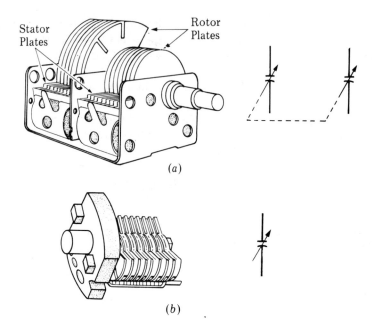

Figure 11–8

Variable capacitors: (a) Split-stator. (b) Midget single.

quency circuits. The dotted line joining the symbols in Figure 11–8a indicates that the capacitors are *ganged together*; that is, both rotors move simultaneously because they are attached to a common shaft.

    Trimmer capacitors are small variable units consisting of two metal plates usually separated by a thin piece of mica. The capacitance is changed by varying the distance between the plates by means of a small screw that forces the plates closer together. Trimmer capacitors are used in receivers and other equipment to permit fine adjustment of capacity of the tuned circuit for accurate alignment.

Fixed Capacitors

Fixed capacitors are manufactured to specific values and cannot be changed. Most capacitors used in industry are of this type. They are made in a number of different shapes and sizes, depending on the required capacity, voltage rating, mounting requirements, and permissible leakage. Customarily, they are named according to the type of dielectric used. The more common kinds are described next, along with several of their salient characteristics. The eight types of fixed capacitors are:

Paper capacitors,
Plastic film extended-foil capacitors,
Mica capacitors,
Ceramic capacitors,
Monolithic ceramic-chip capacitors,
Temperature-compensating ceramic capacitors,
Aluminum electrolytic capacitors,
Tantalum electrolytic capacitors.

Paper capacitors are the least expensive and therefore most widely used fixed capacitors. The dielectric is kraft paper, a relatively heavy, high-strength sulfate paper, between thin aluminum foil plates, rolled together and impregnated with resin. The unit is encapsulated in a tubular plastic or metal sheath to make it impervious to moisture and contaminants. Axial leads are usually brought out of each end. Typical capacities run from about 0.0001 to 2 μF, with voltage ratings from 200 to 600 V.

Plastic films such as polystyrene or Mylar have largely replaced paper as a dielectric for general-purpose capacitors. Plastic is much denser than paper, and contaminating foreign particles are virtually nonexistent. Plastics can withstand higher temperatures and are considerably more stable than paper. Plastic film and paper capacitors are constructed in essentially the same manner. Details of construction are shown in Figure 11–9; this capacitor is known as an *extended-foil* capacitor. It is essentially noninductive, since all layers for one plate are joined at one end.

Mica capacitors are used when capacitors with high voltage ratings are needed. Mica is one of the best insulators and has low loss. Radio transmitters use these capacitors since the voltage rating and current may go as high as 30 kV and 100 A! These units are physically quite large. Capacities in excess of 0.05 μF are uncommon. A variation of this kind of capacitor is the *silver-mica capacitor*. A thin layer of silver is deposited on the surface of the mica, and the resulting capacitor has excellent stability and tolerance. These capacitors are used principally for temperature compensation in radio frequency–tuned circuits owing to their predictable, linear temperature-capacitance (*TC*) variations.

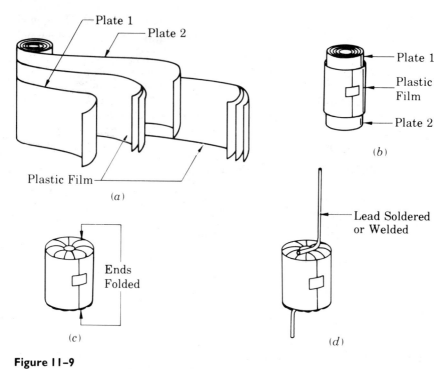

Figure 11-9

Construction of a plastic-film, extended-foil capacitor: (a) Assembly of plates. (b) Plates wound. (c) Ends folded. (d) Leads soldered.

Ceramic capacitors consist of a ceramic disc with silver electrodes (plates) attached to each flat surface. Leads are bonded to these electrodes to provide connection to the unit. Figure 11-10 shows the details of a monolithic ceramic capacitor. Ceramic capacitors are used for applications ranging from low frequency and audio to high, very high, and ultrahigh frequency, through 1000 MHz. (See Chapter 15, Table 15-1, for classifications of various audio and radio frequencies.) Dielectric materials for ceramic capacitors are made from mixtures of barium and strontium titanates, blended with rare earth and other additives for electrical-characteristic enhancement.

Monolithic ceramic-chip capacitors have become very popular because they save space and achieve capacitance values that are difficult to attain by either thick- or thin-film capacitors. Capacitance values in excess of 100,000 pF are easily achievable with ceramic multilayer chips that measure 100 by 180 mils and less. These capacitors are used extensively in all kinds of analog and digital integrated circuits. In off-chip bypass applications, multilayer chips are used be-

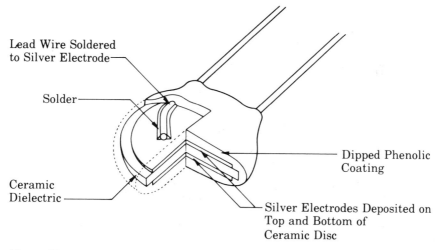

Lead Wire Soldered
to Silver Electrode

Solder

Ceramic
Dielectric

Dipped Phenolic
Coating

Silver Electrodes Deposited on
Top and Bottom of
Ceramic Disc

Figure 11-10

Construction details of a typical monolithic ceramic capacitor (Courtesy Sprague Electric Co.)

cause they have higher volumetric efficiency than ceramic discs. Also, they can be obtained in single and dual in-line configurations.

Multilayer construction techniques produce a very high ratio of capacitance to volume with a minimum of self-inductance. The capacitor sections are made by alternately depositing very thin layers of ceramic dielectric material and metallic electrodes until the desired capacitance is achieved. The resultant capacitors are then fired into an all but indestructible solid block. Construction details of a typical multiple-layer ceramic capacitor are shown in Figure 11-11.

Figure 11-12 shows how the capacitance changes in percentage with various temperature changes. This performance curve is for typical ceramic capacitors in the range of 0.01 to 4.7 μF.

Example 11-6

A typical ceramic capacitor has a capacitance of 3.9 μF at 25 °C. What will its capacitance be at -25 °C?

Solution

Refer to Figure 11-12. The percent change will be approximately -30%, or 0.3×3.9 μF $= 1.17$ μF. The capacitance at -25 °C will then be 3.9 μF $- 1.17$ μF $= 2.73$ μF.

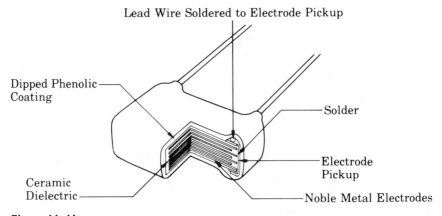

Figure II-II

Construction details of multilayer ceramic capacitor

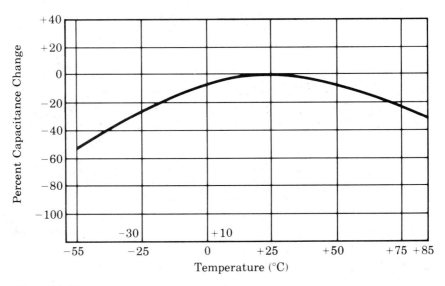

Figure II-12

Performance curve for typical ceramic capacitors in the 0.01 to 4.7 μF range

Temperature-compensating ceramic capacitors exhibit controlled, predictable changes in capacitance with temperature changes. If the capacitance value increases with an increase in temperature, the capacitor has a *positive temperature coefficient* and is classified with the letter P, as in P150. This type of capacitor is rarely used.

Ceramic capacitors that remain stable with changes in temperature are called NP0 (negative-positive-zero) capacitors. These capacitors are even more stable than silver-mica capacitors. They are used in many kinds of receivers and generally have values between 1.0 pF and 0.033 μF.

Ceramic capacitors that have a *negative temperature coefficient* are widely used in TV and quality radio receivers to stabilize tuned circuits that otherwise would change frequency with any change in temperature. Because a radio frequency coil or an intermediate-frequency transformer has a positive temperature coefficient of inductance, its inductance will increase with any increase in temperature, and the tuned frequency will therefore decrease. If the tuned-circuit capacitance change is selected to be exactly equal but opposite to the inductance change, their product will remain constant, and the resonant frequency will not vary with temperature changes. (See Chapter 19 for discussion of tuned circuits.)

These effects are shown graphically in Figure 11-13. A capacitor with a negative temperature coefficient will decrease capacitance with an increase in temperature.

Example I I-7

A certain 300 pF ceramic capacitor has a temperature coefficient of -250 ppm/$^\circ$C. What is its capacitance if the temperature increases 25°C?

Solution

The capacitance change is as follows:

$$300 \text{ pF} \times 10^{-6} \times (-250)\, 25 = -1.875$$
Therefore $\qquad\qquad 300 - 1.875 = 298.125 \text{ pF}$

A commonly used temperature coefficient for temperature-compensating ceramic capacitors is N750. The 750 means that the decrease in capacitance will be 750 parts per million for each degree Celsius of temperature rise. In other words, the capacitance value will decrease 0.075% for a 1°C temperature increase or 1.5% for a 20°C temperature increase. The N750 ceramic capacitors are available from about 4.0 to about 680 pF and are usually rated at 500 WVDC.

Caution: When replacing a defective capacitor, use a capacitor with a higher voltage rating, a tighter capacitance tolerance, a smaller size, or improved electrical characteristics. Even capacitance value can

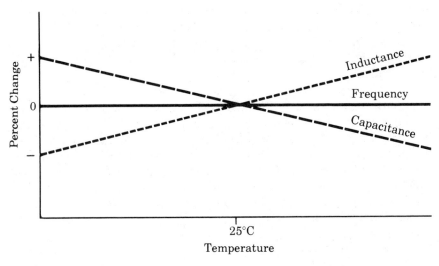

Figure II-I3

Percent change in inductance and capacitance with temperature change

often be increased. *However, if a temperature-compensating capacitor is being replaced, only another temperature-compensating capacitor of the same capacitance value and the same temperature characteristic should be used.*

Example II-8

A certain temperature-compensating capacitor has an N250 co-efficient. What will be the percentage of capacitance increase for a temperature rise of $20°C$?

Solution

For a $1°C$ temperature change:

$$\frac{250}{1 \times 10^6} \times 100 = 2.5 \times 10^{-2} = 0.025\%$$

For $20°C$, $0.025 \times 20 = 0.5\%$.

Aluminum electrolytic capacitors are used when large amounts of capacitance are needed and leakage current is not critical. Their ca-pacities range from a few microfarads to hundreds of thousands of farads. Generally, they are not rated higher than 450 V in values up to several hundred microfarads. The voltage ratings tend to decrease as the capacity increases because thinner dielectrics are used with the larger capacities and therefore lower voltage ratings result (see Equation 11–3).

In its basic form, the electrolytic capacitor's construction is similar to the construction of the paper capacitor. The positive terminal is made of aluminum foil, and one side is covered with an extremely thin layer of oxide, which is the dielectric. In contact with this layer is an electrolyte (from which the capacitor derives its name) in the form of a paste, which serves as the negative plate of the capacitor. A second strip of aluminum foil is placed against the paste to make electric contact. It is this connection that becomes the negative terminal of the capacitor. Large capacities are possible because the metal foil is *chemically etched* to expose the grain structure of the metal, which greatly increases the effective area exposed to the electrolyte (see Equation 11–3).

By an *anodizing process*, a barrier layer of oxide is formed on the foil surface. Its thickness is proportional to the voltage, approximately 14 Å/V; 1 angstrom (Å) $\cong 1 \times 10^{-10}$ m. The combination of an etched foil and an extremely thin dielectric results in very large capacitances. Current flow (leakage) through the capacitor is required to maintain this thin film. Construction details for an aluminum foil capacitor (basically a foil sandwich) are shown in Figure 11–14. The can-type (see Figure 11–14) electrolytics come with a silicone rubber vent plug with a calibrated pressure relief diaphragm that will blow if excessive pressure builds up within the unit. Excessive pressure can result if voltages beyond the rating are applied or if the polarity of the applied voltage is reversed.

Electrolytic capacitors are *polarized*. Should the positive plate be accidentally connected to a negative voltage source, the thin oxide film (dielectric) will disappear and the capacitor will short. Because they are polarity-sensitive, electrolytic capacitors are normally restricted to use with dc circuits or circuits in which a small ac voltage is superimposed on a dc voltage.

There is some leakage current with electrolytics. Usually, it increases with age, especially if the capacitor has not been in use. The amount of leakage is a fair indication of quality, with larger units having more than smaller ones. Figure 11–15a shows several types and styles of electrolytic capacitors, and Figure 11–15b shows the schematic symbol.

All capacitors have characteristics besides capacitance to be considered. These characteristics include the inductance L of the leads and

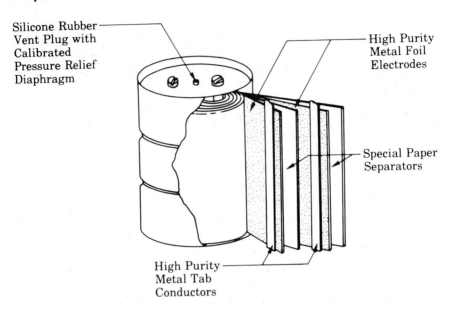

Silicone Rubber Vent Plug with Calibrated Pressure Relief Diaphragm

High Purity Metal Foil Electrodes

Special Paper Separators

High Purity Metal Tab Conductors

Figure 11-14

Construction details of an aluminum foil electrolytic capacitor

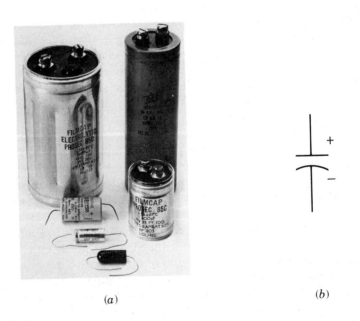

(a)

(b)

Figure 11-15

Electrolytic capacitors: (a) Some common types. (b) Schematic symbol.

plates, as shown in Figure 11–16, the resistance R_2 of the leads and plate material, and the resistance R_1 of the dielectric. Resistance R_2 is the effective series resistance (ESR) of the capacitor. Of course, C represents the capacitor. At low frequencies, the inductance L of the capacitor is insignificant; but at higher frequencies, it becomes important.

Tantalum electrolytic capacitors are the most stable electrolytic capacitors known. Pure tantalum, one of the rarer elements in the earth's crust, is silver-gray metal. The metal was named tantalum because it is so "tantalizingly" unreactive. It remains unaltered when immersed in most acids. The element is exceptionally resistant to attack by most chemicals below 350°F (175°C).

A thin film of tantalum oxide is extremely stable and has excellent dielectric properties. Consequently, tantalum is ideal for use in electrolytic capacitors. Tantalum electrolytics have become the preferred type to use where high reliability and long service life are paramount considerations.

A tantalum electrolytic capacitor consists of two conducting surfaces, or plates, which are separated by an insulating material. This insulator, or dielectric, is tantalum pentoxide, and it is common to all tantalum electrolytic capacitors. (The wet-foil capacitor has a porous paper separator between the foil plates, but this paper serves merely to hold the electrolytic solution and to keep the foils from touching each other.) Tantalum pentoxide is a compound possessing high dielectric strength and a high dielectric constant. A film of this compound is formed on the electrodes of the tantalum capacitor by an electrolytic process. Tantalum pentoxide has a dielectric constant of 26, which is some three times greater than that of aluminum oxide. This high constant, plus the fact that extremely thin films can be deposited in the electrolytic process previously mentioned, makes the tantalum capacitor very efficient with respect to the number of microfarads available

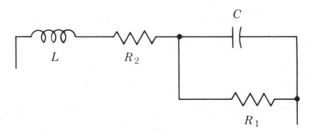

Figure 11–16

Equivalent circuit of a typical capacitor

per unit volume. *Foil tantalum capacitors* are roughly one-third the size of comparable aluminum electrolytic capacitors.

In solid-electrolyte tantalum capacitors, the electrolyte is a dry material, manganese dioxide. This conductive, solid material forms the cathode (negative) capacitor plate.

Foil tantalum capacitors are made by rolling two strips of thin foil, separated by a paper saturated with electrolyte, into a convolute roll. The tantalum foil, which is to be the anode, is chemically etched to increase its effective surface area, thus providing more capacitance in a given volume. This etching step is followed by anodization in a chemical solution under direct voltage. This anodization produces the dielectric tantalum pentoxide film on the foil surface. Foil tantalum capacitors are generally designed for operation over the temperature range of $-55\,°C$ to $+125\,°C$ and are found primarily in industrial and military electronic equipment.

Sintered-anode, wet-electrolyte tantalum capacitors and *sintered-anode, solid-electrolyte capacitors* have in common a pellet of sintered tantalum powder to which a lead has been attached. This anode has an enormous surface area for its size because of the way it is made. Tantalum powder of suitable fineness, sometimes mixed with binding agents, is machine-pressed into pellets. The next step is a sintering operation in which binders, impurities, and contaminants are vaporized and the tantalum particles are sintered (welded) into a porous mass with a very large internal surface area. Following the sintering and before formation of the dielectric film on the pellet, a tantalum lead wire is attached by welding the wire to the pellet.

A film of tantalum pentoxide is electrochemically formed on the surface areas of the fused tantalum particles. Provided sufficient time and current is available, the oxide will grow to a thickness determined by the applied voltage. The pellet is then inserted into a tantalum or silver can that contains an electrolytic solution. Most liquid electrolytes are gelled to prevent the free movement of the solution inside the container and to keep the electrolyte in intimate contact with the capacitor cathode. A suitable end-seal arrangement prevents the loss of the electrolyte. A cutaway view of a sintered-anode tantalum capacitor is shown in Figure 11–17.

Distributed Capacity

Capacitance exists between any two objects separated by a dielectric. Examples are the conductors of an electric cable, the elements of a vacuum tube, an antenna and ground, and the windings of a transformer.

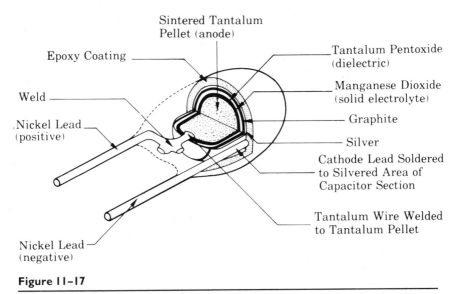

Figure 11-17

Cutaway view of a sintered-anode tantalum capacitor (Courtesy Sprague Electric Co.)

Normally, these capacitances are very small, but when the cables are long or the frequency extremely high, they become significant and can impair operation. In the case of transistors, the capacitance between elements is considerably greater because of the high dielectric constant of the semiconductor material. Good design is required to keep these capacitances to a minimum.

11-10
Capacitors in Parallel and in Series

Parallel Capacitors

According to Equation 11-3, capacitance is a direct function of plate area. Connecting capacitors in parallel effectively increases plate area and therefore capacitance. To calculate the total capacitance of parallel-connected capacitors, we use the following formula:

$$C_T = C_1 + C_2 + C_3 + \cdots + C_n \qquad \textbf{11-6}$$

This equation indicates that the total capacitance is the sum of the individual capacitances.

Example 11–9

Suppose you are given the following capacitors: 0.03 μF, 2 μF, and 0.25 μF. What is their total parallel capacitance?

Solution

Use Equation 11–6:

$$C_T = 0.03 \ \mu\mathrm{F} + 2 \ \mu\mathrm{F} + 0.25 \ \mu\mathrm{F} = 2.28 \ \mu\mathrm{F}$$

If the voltage ratings of the capacitors are different, the maximum potential that can be applied is determined by the voltage rating of the lowest.

Capacitors in Series

When capacitors are connected in series, the effect is as though the thickness of the dielectric were increased. The result is a reduction of capacity. The general equation for series capacitors is as follows:

$$\frac{1}{C_T} = \frac{1}{C_1} + \frac{1}{C_2} + \frac{1}{C_3} + \cdots + \frac{1}{C_n} \qquad \textbf{11–7}$$

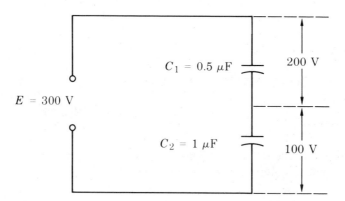

Figure 11–18

Series capacitors in which voltage varies inversely with capacitance

The *equivalent capacitance* will always be *lower than the smallest capacitor* in the series. Total voltage, however, will always be greater than any individual ratings, of which it is the sum. Thus, if three 0.1 μF capacitors were series-connected and each had a 600 V rating, a total of 1800 V could be impressed across the string. Each capacitor receives one-third the applied voltage.

Example 11–10 ━━━━━━━━━━━━━━━━━━━━━━━━━━━━━━━

What is the equivalent capacity of the following three capacitors if they are connected in series: 0.05 μF, 0.15 μF, and 0.025 μF?

Solution

$$\frac{1}{C_T} = \frac{1}{0.05} + \frac{1}{0.15} + \frac{1}{0.025} = 66.67$$

$$C_T = 0.015 \ \mu\text{F}$$

If the circuit contains only two series capacitors, the product-over-sum formula can be used:

$$C_T = \frac{C_1 C_2}{C_1 + C_2} \qquad\qquad \text{11–8}$$

Series capacitors present an interesting problem when their capacitances are different. The voltage across each capacitor is inversely proportional to its capacitance, as shown in Figure 11–18. The reason for this inverse relationship is as follows. Series capacitors have the *same charge* because they are in the same current path. With equal charge, the larger capacitor has the *smaller* voltage. Suppose the charging current in the circuit of Figure 11–18 is 100 μA for 1 s. Then, the charge is as follows:

$$Q = C_{eq} E = \frac{0.5 \times 1}{0.5 + 1} \times 300 = 0.333 \times 300 = 100 \ \mu\text{C}$$

Both C_1 and C_2 have a charge Q of 100 μC. Now, the voltage across C_1 and C_2 will not be the same because their capacitances are different:

$$V_{C_1} = \frac{100 \ \mu\text{C}}{0.5 \ \mu\text{F}} = 200 \text{ V}$$

$$V_{C_2} = \frac{100 \ \mu\text{C}}{1 \ \mu\text{F}} = 100 \text{ V}$$

In some high-voltage power supplies, capacitors are connected in series when the output voltage is greater than the voltage rating of the individual capacitors. The capacitors must have equal capacitances and voltage ratings to be connected in series in this application.

11-11
Losses in a Capacitor

Dielectric Leakage

No dielectric is perfect. Even if its resistance is of the order of 10^8 Ω or more, as it is for a quality capacitor (excluding electrolytics), there will be an extremely small leakage current when a voltage is applied. This leakage represents a power loss and must be supplied by the circuit to which it is connected.

Resistance Losses

The plates of a capacitor have a resistance that, though very small, can cause the charging and discharging currents to reach large maximum values, resulting in a series resistance I^2R loss. At high frequencies, this loss cannot be neglected, particularly since currents tend to travel near the surface of a conductor at these frequencies.

As mentioned earlier, capacitors are normally encapsulated with an insulating material. Even quality products offer high resistance across terminals, which results in some leakage that acts like a shunt across the capacitor.

Dielectric Hysteresis

When an ac voltage is connected across a capacitor, the electrostatic field between the plates reverses polarity with each half cycle. The electrons in the dieletric are strained first toward one plate, then the other, distorting their orbital motion. This distortion consumes some energy and, at high frequencies, can cause heating of the dielectric. The tendency of the electrons to resist this motion is called *hysteresis*. Dielectric materials must be carefully selected for high-frequency applications.

Dissipation Factor

All of the losses in a capacitor cause heating and a reduction in efficiency. A measure of the quality of a capacitor is expressed by the term *dissipation factor* (*D*), which is the ratio of capacitive reactance to equivalent shunt resistance (capacitive reactance is covered in Chapter 16). For commercial capacitors, *D* values may range from 0.025 to 0.0001.

11–12
Frequency and Capacitance Characteristics

A chart showing the frequency characteristics of different types of capacitors appears in Figure 11-19. Large-capacitance aluminum and tantalum electrolytics are useful on dc and relatively low frequencies. The dashed lines indicate frequencies where operation is possible or is being explored.

The data in Figure 11-20 provides an overview of the different capacitor types and their basic capacitance ranges. As expected, the electrolytics offer the greatest capacitance.

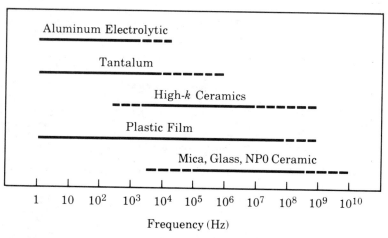

Figure 11-19

Basic frequency ranges for different types of capacitors

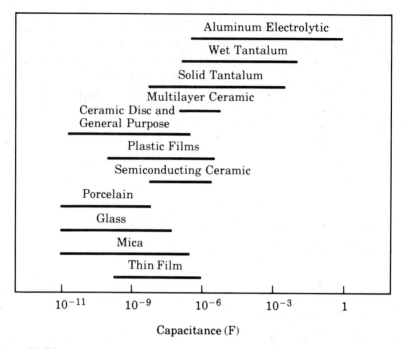

Figure 11–20

Basic capacitance ranges for different types of capacitors

11–13
Capacitor Coding Systems

Capacitance values follow the same standard used with resistors. The following data shows the arrangement from 1.0 to 8.2 μF. Capacitors are available in multiples and submultiples of 10 of the values indicated.

Basic Capacitance Values

1.0	2.2	4.7
1.2	2.7	5.6
1.5	3.3	6.8
1.8	3.9	8.2

Table 11-2 shows the EIA (Electronics Industries Association) codes for capacitor tolerance and temperature compensation. General-purpose ceramic capacitors are normally coded in *picofarads* below 100,000 pF and in *microfarads* above. A typical code printed on a capacitor might be 104M. What capacitance value does this code indicate? The first two digits are significant numbers, and the last digit indicates the number of zeros. The letter following the number is the tolerance (see Table 11-2). Therefore, the capacitor is a 100,000 pF ± 20%.

Example 11-11

What is the value of a ceramic capacitor marked 223J?

Solution

The significant numbers are 22, and there are three zeros. From Table 11-2, the J indicates 5% tolerance. Thus, the value is 22,000 pF ± 5%.

For more details about the EIA numbering system for ceramic capacitors, refer to the Appendix. Details of the EIA numbering system for aluminum and tantalum electrolytic capacitors can also be found in the Appendix.

Table 11-2 EIA Codes for Capacitor Tolerance and Temperature Compensation

Capacitance Tolerance Codes		Temperature Compensation Codes	
		Industry Use	EIA Code
C	±1/4 pF	NP0	C0G
D	±1/2 pF	N033	S1G
F	±1 pF or ±1%	N075	U1G
G	±2 pF or ±2%	N150	P2G
J	±5%	N220	R2G
K	±10%	N330	S2H
L	±15%	N470	T2H
M	±20%	N750	U2J
N	±30%	N1500	P3K
P	−0%, +100%	N2200	R3L
W	−20%, +40%		
Y	−20%, +50%		
Z	−20%, +80%		

┌─Summary

Capacitance is the property of an electric circuit that opposes any change of voltage.

The unit of capacitance is the farad. Because of the extremely large size of this unit, the microfarad or picofarad is used.

The charge on a capacitor may be expressed by the following formula:

$$Q = CE$$

No current flows through a capacitor during its charging or discharging cycle.

The voltage across a fully charged capacitor is equal to the source voltage.

A dielectric is the insulating material separating the plates.

Capacitance of a capacitor is determined by the following formulas:

$$C = 0.08842 \frac{kA}{d} \quad \text{or} \quad C = 0.224 \frac{kA}{d}$$

Temperature-compensating ceramic capacitors exhibit controlled, predictable changes in capacitance with temperature changes.

Plastic films have largely replaced paper as a dielectric.

Electrolytic capacitors are characterized by high capacitance per unit size.

Electrolytic capacitors are polarized and have relatively low leakage resistance.

The total capacitance of series-connected capacitors is found by the following formula:

$$\frac{1}{C_T} = \frac{1}{C_1} + \frac{1}{C_2} + \cdots + \frac{1}{C_n}$$

The total capacitance of series-connected capacitors is always less than the lowest value of the capacitor in the series.

The total capacitance of parallel-connected capacitors is equal to the sum of the total capacitors and is calculated by the following formula:

$$C_T = C_1 + C_2 + C_3 + \cdots + C_n$$

The losses in a capacitor are a result of dielectric leakage, resistance, dielectric hysteresis, and the dissipation factor.

Capacitance values follow the same standard used with resistors.

┌──Progress Test

The bracketed number after each question indicates the section of this chapter where the answer can be found.

　　1.　Capacitance is the property of an electric circuit that: (a) opposes any change of I in the circuit. (b) opposes any change of E in the circuit. (c) is not affected by a change of E. (d) aids any change of I in the circuit. [11–2]

　　2.　Which of the following statements is true regarding the dielectric constant? (a) It is directly related to breakdown voltage. (b) It increases with thickness. (c) It directly affects capacitance. (d) It increases with temperature. [11–6]

　　3.　The capacitive current flowing in a condenser: (a) is a displacement current only. (b) flows through the capacitor. (c) is greatest when the emf is greatest. (d) increases linearly under the applied voltage. [11–2]

　　4.　The energy stored in a capacitor is: (a) proportional to its terminal voltage. (b) measured in amperes. (c) directly proportional to capacitance. (d) inversely proportional to the square of the voltage. [11–3]

　　5.　The charge on a capacitor is: (a) measured in volts. (b) the product of voltage and capacitance. (c) measured in amperes. (d) C times E^2. [11–4]

　　6.　When capacitors are connected in parallel, the resulting capacitance is the: (a) reciprocal of the sum of the capacitances. (b) sum of the individual reciprocals of each capacitor. (c) product of their capacitances divided by their sum. (d) sum of the individual capacitances. [11–10]

　　7.　Which of the following is characteristic of a series-capacitive circuit? (a) The charging current is the same in each capacitor. (b) The applied voltage divides equally across each capacitor. (c) Total capacitance is the sum of the individual capacitances. (d) Capacitance equals the reciprocal of the sum of the capacitances. [11–10]

8. The total capacitance of two 100 μF capacitors connected in series is: (a) 25 μF. (b) 50 μF. (c) 100 μF. (d) 200 μF. [11-10]

9. What is the total capacitance of two 200 μF capacitors connected in series with each other, with the combination in parallel with a 50 μF capacitor? (a) 50 μF (b) 250 μF (c) 100 μF (d) 150 μF [11-10]

10. Factors that determine the capacitance of a parallel-plate capacitor are: (a) the area of the plates and the type of dielectric. (b) the thickness of the plates and the type and thickness of the dielectric. (c) the dielectric constant and the applied voltage. (d) the thickness of the dielectric, the area of the plates, and the polarity of the current flow. [11-7]

11. What is the capacitance of a capacitor that receives a charge of 0.01 C when 25 V is applied? (a) 4 × 10^{-2} F (b) 4 × 10^{-3} F (c) 4 × 10^{-4} F (d) 40 × 10^{-4} F [11-5]

Problems

Answers to odd-numbered problems are at the back of the book.

1. Calculate the energy stored in a 0.022 μF capacitor that has 400 V across it.

2. What is the energy stored in a computer capacitor of 6300 μF when 50 V appears across its charged terminals?

3. How many coulombs are stored in a 6 μF capacitor that has been charged with a 250 V potential?

4. A computer-grade 22,000 μF capacitor has 15 V across its terminals. What is the charge in coulombs?

5. What is the capacity of an electrolytic capacitor having 150 V dc across its terminals and a charge of 0.35 C?

6. A charge of 0.075 C exists on a 33 μF capacitor. What is the voltage across the capacitor?

7. A capacitor is marked 0.00036 μF. What is its value in picofarads?

8. What is 0.39 nF expressed in picofarads? In microfarads?

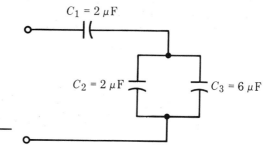

Figure 11–21

Circuit for Problem 18

9. Calculate the capacitance of a mica capacitor with a k of 6.5, a plate area of 55 cm^2, and a spacing between plates of 0.5 cm.

10. Determine the spacing between the parallel plates of a condenser with the following values: $A = 213$ cm^2, $C = 220$ pF, and $k = 3.5$.

11. A certain ceramic capacitor has the following parameters: $C = 810$ pF, parallel-plate separation of 0.15 cm, and $A = 4.7$ cm^2. What is its dielectric constant?

12. Calculate the effective plate area of a capacitor having the following values: plate spacing $= 0.06$ cm, $C = 180$ pF, $k = 2.1$.

13. A 200 pF ceramic capacitor has a temperature coefficient of -350 ppm/°C. If the temperature increases 35°C, what is its new capacitance?

14. Determine the total capacitance of the following capacitors when parallel-connected: 0.033 μF, 1.2 μF, and 0.68 μF.

15. The following capacitors are parallel-connected: 610 nF, 2700 pF, and 0.0015 μF. Calculate C_T.

16. What is the effective capacity of a 4 μF and a 6 μF series-connected combination?

17. Calculate the equivalent capacitance of the following capacitors when they are series-connected: 0.008 μF, 0.0025 μF, and 0.01 μF.

18. Calculate C_T for the circuit in Figure 11–21.

19. Suppose 200 V is connected across a series combination of 3 μF and 5 μF. What voltage appears across each capacitor?

20. Calculate the total capacitance for the circuit in Figure 11–22.

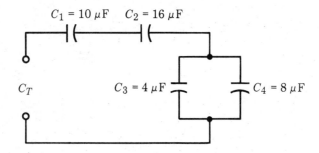

Figure II–22

Circuit for Problem 20

21. What is the capacitance value indicated by the code 103J?

22. A capacitor is coded 471K. What is its capacitance?

23. The code 393G is printed on a capacitor. What is its capacitance?

24. What capacitance is represented by the code 822L?

12

Magnetic Circuits and Electromagnetism

Objectives

After studying this chapter, you will be able to:

Discuss the factors that control the strength of magnetic fields as used in relays, motors and generators, and transformers.

Acquire the prerequisite information necessary for an understanding of inductance.

Introduction

Magnetic phenomena are observed in nature as well as in industrial and scientific applications. The magnetic properties of the earth permit the use of the compass and other more sophisticated instruments for navigational purposes. Many other uses of magnetism can be found in everyday life. Specifically, many electronic circuits employ devices that are magnetic in nature; any study of electric circuits of necessity must include a study of magnetic forces. This chapter will provide the background for an understanding of inductance, one of the basic elements in electronic circuits.

Applications

Although both magnets and electricity have been known and used in their primitive forms for centuries, the discovery made by Hans Christian Oersted in 1820 that the two were related and that their relationship could be combined and studied in *electromagnetism* has resulted in the quality of civilization we experience today.

The first practical application of the electromagnet (which is simply a coil of wire around an iron core) was the telegraph. Later followed telephone, radio, television, stereo loudspeakers, and, most significant of all in terms of raising the quality of civilization, electric power generators. The generator is responsible for creating, running, and lighting our big cities, for energizing underground transit systems and mine hoists, and for equipping the world with cheap power for factories. The generator also makes it possible for people to enjoy other electromagnetic products simultaneously, such as listening to the radio while using a sewing machine, or using a soldering iron to replace a defective component in a piece of electronic equipment while the rest of the family watches TV in an adjoining room. Electric power generators have changed the world from a more bucolic but less comfortable simplicity to the intense and convenience-oriented existence we experience today.

12-1
Permanent Magnets

Basic Concepts

The attraction of a magnet to iron or steel is familiar to almost everyone. But what is this attraction? Why can some materials be magnetized and others not? The answers to these questions can be found in the atomic structure of matter.

A spinning electron has a *magnetic field*. In each shell or energy band, some of the electrons revolve clockwise and others counterclockwise around the nucleus. These groups of electrons each produce magnetic fields. The strength of these fields and their directions (polarity) depend on the number of electrons and their direction of rotation.

Since iron is a common magnetic material, let us examine its atomic structure and see if we can determine why it has magnetic properties. There are a total of 26 electrons in the iron (Fe) atom. The two in the first shell neutralize each other because of opposite rotations. The same is true for the second shell – four revolve in one direction and four in the other. The third shell has 14 electrons – nine revolve in one direction and five in the other. The outermost shell of the iron atom has only two electrons, which revolve in opposite directions. The magnetic properties of iron are determined by the four unbalanced electrons in the third shell.

Nickel and cobalt have fewer magnetic properties than iron. For example, nickel has two unbalanced electrons, and cobalt has three. When a material has the *same number of electrons revolving in each direction*, their magnetic fields *cancel* and the material is said to be *nonmagnetic*. Examples of these materials are paper, wood, air, glass, aluminum, and copper.

Temporary and Permanent Magnets

Residual magnetism   The magnetism remaining after the external magnetizing force is removed.

Retentivity   The ability of a magnetic material to retain its magnetism.

When a piece of hard steel or alloy such as ALNICO (contraction of aluminum, nickel, and cobalt) is subjected to a strong magnetic field, its atoms align themselves with the external field. Once the metal is removed from the external field, nearly all of the atoms remain in their magnetized position. The magnetism remaining after the external magnetizing field is removed is called **residual magnetism**. The ability of a magnetic material to retain its magnetism is called **retentivity**. Permanent magnets have high retentivity and therefore a large amount of residual magnetism. Soft iron has very little retentivity and can be only temporarily magnetized.

Not all the magnetism induced into a piece of hard steel will remain. Over a long period of time, some of the atoms lose their alignment and move toward their former positions. After this *aging period*, the strength of the magnet remains constant, unless abused. Manufacturers of permanent magnets artificially age them by means of a heating process that lasts for several hours. During this period, they lose the same amount of magnetism they would have lost in natural aging.

> **Magnetizing by induction**   The magnetization introduced into a material by bringing it into close proximity, or contact, with a magnet or a magnetic field.

Soft iron becomes temporarily magnetized when brought into contact with, or close to, a permanent magnet or other magnetic field. This temporary process is called **magnetizing by induction**. For example, small nails become temporarily magnetized when attracted to the ends of a magnet. Each nail becomes a small magnet by induction. When removed from the magnet, they lose their magnetism very rapidly.

12–2
Magnetic Fields

Energy Field

> **Lines of force**   The visual representation of the strength and direction of the energy field surrounding a magnet.

Surrounding every magnet is an *energy field*, the strength and direction of which can be represented by **lines of force**. It is customary to represent these magnetic lines as shown in Figure 12–1. Every magnet has two poles, north and south. The lines of force leave the north pole and enter the south pole, as indicated by the arrowheads. Each line is a continuous loop. If a small pocket compass is positioned in the magnetic field, it will orient itself as shown in Figure 12–1, with its north pole pointing toward the magnet's south pole.

Horseshoe-shaped permanent magnets are most frequently used. The pattern formed by their magnetic lines is indicated in Figure 12–2, which shows the concentration between poles. These lines of

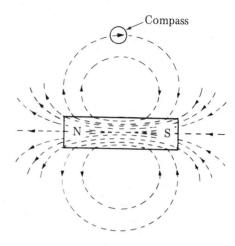

Figure 12–1

Lines of force representing a magnetic field

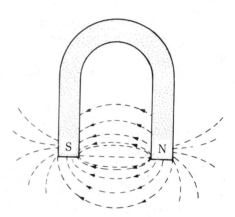

Figure 12–2

Lines of force surrounding a horseshoe magnet

force around a magnet never intersect and are only representative of its field. The actual field is invisible. A more accurate representation of a magnet's field can be obtained by placing a piece of paper over its poles and sprinkling iron filings on the paper. A detailed pattern appears, indicating the shape of the energy, or the magnetic flux field.

Attraction and Repulsion

When *unlike* poles of magnets are placed in close proximity, their lines of force react as indicated in Figure 12–3. Note the attraction between the poles. The opposite effect occurs when *like* poles are placed close

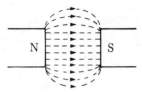

Figure 12–3

Magnetic field between unlike poles

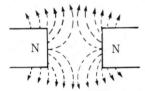

Figure 12–4

Magnetic field between like poles

together, as shown in Figure 12–4. From these observations, we may conclude that *unlike poles attract, and like poles repel.*

The attraction or repulsion between magnets varies *directly* with the product of their strengths and *inversely* with the square of the distance between them.

12–3
Magnetic Domains

Substances made of iron are easily magnetized and are referred to as *ferromagnetic materials.* Iron, nickel, cobalt, and manganese belong in this category. *Paramagnetic materials* are those that can be only feebly magnetized. Some examples are chromium, platinum, titanium, and palladium. *Diamagnetic materials* have the opposite effect of paramagnetic materials. Lines of force tend to go around rather than through them. Examples of this kind of material are copper, brass, antimony, and bismuth.

> **Domains**   Acicular (needle-shaped), microscopic crystals containing a very large number of atoms whose spinning electrons result in a net magnetic effect in a particular direction.

Studies have shown that a ferromagnetic substance contains small regions called **domains**. Domains are acicular, microscopic crystals that contain from 10^{12} to 10^{15} atoms. Within each domain, the magnetic movements of all the spinning electrons result in a net magnetic effect in a particular direction. The net effect of the spinning electrons in the atoms of adjacent domains is likely to be in other random directions. Each domain is spontaneously magnetized to saturation in some direction. Therefore, the material as a whole appears unmagnetized.

Several distinct changes take place in the domains when the material is subjected to a magnetic field. In weak fields, the domains lying in the crystal axis most nearly parallel to the magnetizing force rotate until they are parallel to the external field. This process is accompanied by an outward movement of the domain walls at the expense of other domains, with more diverse directions of magnetization. Since these domains have enlarged slightly, they become predominant, and the material is slightly magnetized. Removal of the magnetizing force allows the domain walls to return to their original positions, and the magnetism disappears.

In stronger magnetic fields, whole domains suddenly rotate by 90° or 180° into parallelism with the crystal axis that most nearly coincides with the direction of the magnetizing force. Finally, when the external field is sufficiently strong, all domains rotate so that they are parallel to the external field. The whole specimen is then in saturation and under some conditions of strain. This strain is immediately relieved as soon as the magnetizing force is reduced. The amount of magnetism remaining is a measure of the material's retentivity.

12–4
Magnetism Produced by Current

Danish physicist Hans Christian Oersted, in the year 1820, discovered that a current-carrying conductor produces a magnetic field. He noticed that a magnetic compass needle placed parallel to the conductor always came to rest at right angles to the conductor. Reversing the current caused the other end of the needle to point to the conductor. From these experiments, he proved that *a conductor carrying an electric current has a magnetic field surrounding it similar to the field around a permanent magnet.*

Oersted's discovery can be further verified by passing a current-carrying conductor through a small hole in a piece of cardboard held horizontally, as represented in Figure 12–5. If iron filings are sprinkled on the cardboard, a series of concentric lines will appear, concentrated near the conductor and diminishing in intensity as the distance from the conductor increases. Note the arrowheads on the lines of force. They would be reversed if the current were in the opposite direction.

Actually, the *magnetic flux* (defined in Section 12–7) around a current-carrying conductor extends its full length, as shown in Figure 12–6*b*. Remember that while the lines of force and their direction are shown, they serve only to indicate the presence of a magnetic field. Figure 12–6*a* represents the end view of the conductor and its field when the current is moving away from the observer. The symbol ⊕ represents the tail of an arrow. Figure 12–6*c* shows the opposite condition, with the current moving toward the observer. The symbol ⊙ represents the tip of the current arrow.

Left-Hand Rule

The direction of the magnetic field surrounding a conductor can be determined by what is known as the *left-hand rule*. It may be stated as follows: If a conductor is grasped in the left hand with the thumb ex-

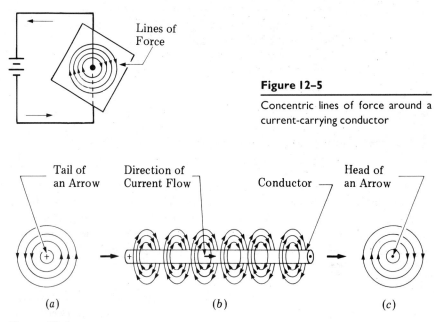

Figure 12–5

Concentric lines of force around a current-carrying conductor

Figure 12–6

Magnetic field surrounding a conductor: (*a*) Cross section of conductor and magnetic field, showing current flowing away from observer. (*b*) Field around a conductor. (*c*) Cross section of conductor and magnetic field, showing current flowing toward observer.

tended in the direction of current, the fingers will point in the direction of the magnetic field. An examination of Figures 12-5 and 12-6 will verify this rule.

Magnetic Field between Current-Carrying Conductors

The magnetic fields produced by currents flowing in parallel conductors either attract or repel the conductors, depending on the direction of current. If the currents are flowing in the same direction, the field between the conductors is weakened because the lines of force oppose each other. In the area outside the conductors, the field is strengthened. Because of the weakened field, the lines of force tend to encircle both conductors somewhat like rubber bands, and they are drawn toward each other.

Parallel conductors having currents in opposite directions produce fields that tend to repel the conductors. Application of the left-hand rule indicates that the lines of force aid each other in the region between the conductors. This strengthened field tends to push the conductors apart.

Motor action   The tendency of a current-carrying conductor to move from a strong to a weak magnetic field, producing a torque.

Summarizing, a conductor carrying an electric current will always tend to move from a strong to a weak magnetic field. This motion resulting from the forces of magnetic fields is called **motor action**.

12–5
Concentration of Magnetic Field by a Solenoid

If a conductor is bent into the form of a loop, as in Figure 12-7, the lines of force surrounding it are as shown. Consequently, they reinforce each other, providing a much stronger field inside the loop.

Current →

Figure 12–7

Concentration of magnetic field inside a loop

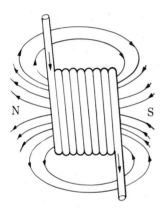

N   S

Figure 12–8

Magnetic field surrounding a solenoid

Solenoids

When a coil is made by wrapping many turns of wire over a relatively long form, a *solenoid* results. If a current is passed through the coil, the field around each turn reinforces the field of adjacent turns. These lines of force produce a greatly strengthened magnetic field, as shown in Figure 12–8. One end of the solenoid has an N polarity (attracts the north-seeking end of a magnetic compass pointer), while the other has an S polarity (attracts the south-seeking end of a magnetic compass needle). Those lines of force passing through the center of the solenoid return to the other end by completing their loops outside the coil.

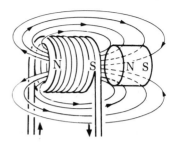

Figure 12–9

Solenoid with iron core

Electromagnets

An *electromagnet* is composed of a coil of wire wound around a core of soft iron. If dc is passed through the coil, it becomes magnetized with the same polarity that it would have without the core. Reversing the current causes the polarity of both coil and core to reverse.

The addition of the iron core accomplishes two things. First, the magnetic flux is greatly increased because the soft iron, in contrast to air, allows much more magnetic flux to flow through it. Air is a poor conductor of magnetic flux, and soft iron has much greater *permeability* than air. Second, the flux is more highly concentrated than it would be if only air were used.

From experience, you know that a piece of soft iron is attracted to either pole of a permanent magnet. A soft iron rod is likewise attracted by a solenoid conducting an electric current, as illustrated in Figure 12–9. From the illustration, you can see that the lines of force extend through the soft iron and magnetize it by induction. Because unlike poles attract, the iron rod is pulled toward the coil. If it is free to move, it will be drawn into the solenoid to a position near the center where the field is strongest. If a spring were attached to the iron rod, it would return to its original position when the current through the coil was interrupted. Solenoids of the basic type shown in Figure 12–9 find widespread application in electromechanical systems.

12–6
Magnetomotive Force

The force produced by current through a coil of wire is called *magnetomotive force (mmf)*. It is the force by which a magnetic field is produced. Just as an increase in electromotive force produces more current in a circuit, so an increase in the mmf increases the strength of the magnetic field. Magnetomotive force, represented by $\mathfrak{F}$, is the product of ampere-turns (A-t) of the coil. *One ampere-turn is the amount of magnetic force or flux produced when one ampere flows through a single turn of an electrical conductor.* This definition is expressed by the following relationship:

$$\mathfrak{F} = \text{A-t} \qquad\qquad \text{12-1}$$

A coil having 10 turns and 1 A of current produces the same magnetic flux as one having 5 turns and 2 A, because the ampere-turns are the same in each case.

Example 12-1 ————————————————————————————

What is the magnetizing force of a coil having 4000 turns and 10 mA flowing through it?

Solution

$$\mathfrak{F} = 4000 \times 0.01 = 40 \text{ A-t}$$

Thus, if a relatively large magnetizing force is needed and only a small amount of current is available, it is necessary to use a great number of turns on the coil.

12-7
Magnetic Flux

Our discussion thus far has made reference to the magnetic flux surrounding a permanent magnet or current-carrying conductor. The symbol for magnetic flux is Φ (Greek capital letter *phi*), and the basic unit for magnetic flux in the SI system is the *weber* (abbreviated Wb). Now, the weber is a very large unit and equals 10^8 lines of force, or *maxwells* (abbreviated Mx) in the CGS (centimeter, gram, second) system referred to by many texts. It is frequently more convenient to use the microweber.

$$1 \ \mu\text{Wb} = 1 \times 10^{-6} \text{ Wb} = 1 \times 10^{-6} \times 10^8 \text{ lines} \qquad \text{12-2}$$
$$= 100 \text{ lines or Mx}$$

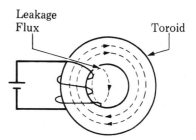

Figure 12-10

Toroidal magnetic circuit

The path of magnetic lines of force may be considered as a magnetic circuit. If the circuit consists of soft iron in the form of a *toroid*, as shown in Figure 12-10, the amount of flux will be large compared with what it would be if the flux had to flow through air. A few flux lines escape from the core and travel through an air path, as illustrated by the *leakage flux* in Figure 12-10. The magnetic circuit is not exactly analogous to the electric circuit, because flux does not flow, as do electrical charges.

Example 12-2

A certain magnet has a flux $\Phi = 1.5$ Wb. How many lines of force does flux represent?

Solution

By definition, 1 Wb $= 10^8$ lines. So,

$$1.5 \text{ Wb} = 1.5 \times 10^8 \text{ lines}$$

Example 12-3

The windings of a particular toroid produce 3000 lines through the core when a current is passed through it. How much flux in webers do these lines represent?

Solution

$$\Phi = \frac{3 \times 10^3}{1 \times 10^8} = 3 \times 10^{-5} \text{ Wb}$$

12-8
Flux Density

Flux density   A measure of the number of magnetic lines of force
per unit of cross-sectional area.

While the total amount of flux produced by a magnet is important, we are more interested in how dense, or concentrated, the flux is
per unit of cross-sectional area. Flux per unit of cross-sectional area is
called **flux density**. Its letter symbol is B, and the SI unit for flux density is the tesla (T).

The relationship between total flux and flux density is given by
the following equation:

$$B = \frac{\Phi}{A} \qquad\qquad 12\text{-}3$$

where B = flux density in teslas
Φ = total magnetic flux of circuit in webers
A = cross-sectional area of magnetic circuit
in square meters

From Equation 12-3, the tesla is automatically defined as being equal
to one weber per square meter:

$$\text{T} = \text{Wb/m}^2 \qquad\qquad 12\text{-}4$$

Example 12-4

Calculate the flux density in a piece of ferromagnetic material
with a cross-sectional area of 0.01 m² containing 100 lines.

Solution

We know that 100 lines equals 1 μWb. Use Equation 12-3:

$$B = \frac{\Phi}{A} = \frac{1\ \mu\text{Wb}}{1 \times 10^{-2}\ \text{m}^2} = \frac{1 \times 10^{-6}\ \text{Wb}}{1 \times 10^{-2}\ \text{m}^2} = 1 \times 10^{-4}\ \text{T}$$

Example 12-5

Determine the cross-sectional area of a toroid that has a flux of
0.5 Wb and a flux density of 25 T.

Solution

$$B = \frac{\Phi}{A}$$

$$A = \frac{\Phi}{B} = \frac{0.5 \text{ Wb}}{25 \text{ Wb/m}^2} = \frac{5 \times 10^{-1} \text{ Wb}}{2.5 \times 10^1 \text{ Wb/m}^2} = 2 \times 10^{-2} \text{ m}^2$$

Example 12–6

An air core coil has 0.65 μWb of flux in its core. Calculate the flux density if the core diameter is 4 cm.

Solution

We must first calculate the area of the core:

$$A = \pi r^2 = 3.14 \times (0.02 \text{ m})^2 = 0.001256 = 1.256 \times 10^{-3} \text{ m}^2$$

$$B = \frac{\Phi}{A} = \frac{0.65 \times 10^{-6} \text{ Wb}}{1.256 \times 10^{-3} \text{ m}^2} = 5.175 \times 10^{-4} \text{ T}$$

12–9
Magnetic Field Intensity

In Section 12–6, we learned that mmf ($\mathfrak{F}$) is the product of ampere-turns. This force may be called the *magnetomotive force*, and it is important because it gives a measure of the magnetic stress imposed on a given core material for its total length. In the MKS (meter, kilogram, second) system, magnetizing force is in ampere-turns per meter (A-t/m). Of primary concern, however, is the magnetomotive force needed to establish a certain flux density in a unit length of the magnetic circuit. *The magnetomotive force per unit length is called the magnetic field intensity.* The letter symbol for magnetic field intensity is H. The following relationship defines H:

$$H = \frac{\mathfrak{F}}{l} \qquad \qquad \text{12–5}$$

where $\mathfrak{F}$ = applied mmf in ampere-turns

   l = average length of magnetic path in meters h in meters

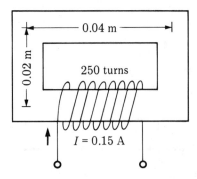

Figure 12–11

Magnetic circuit for Example 12–7

Example 12–7

Find the magnetic field intensity in the magnetic circuit shown in Figure 12–11.

Solution

$$H = \frac{\mathfrak{F}}{l} = \frac{(2.5 \times 10^2)\,(1.5 \times 10^{-1})}{1.2 \times 10^{-1}}$$

$$= \frac{3.75 \times 10^1}{1.2 \times 10^{-1}} = 3.125 \times 10^2 \text{ A·t/m}$$

If the dimensions of the magnetic path were changed, the value of H would also change. For example, if the total length of the magnetic path doubled, we should expect the value of H to decrease to one-half its previous amount. The next example supports this reasoning.

Example 12–8

Calculate the value of H in the magnetic circuit of Figure 12–11 if the physical dimensions are doubled.

Solution

The total length of the magnetic path becomes 0.24 m.

$$H = \frac{\mathfrak{F}}{l} = \frac{3.75 \times 10^1}{2.4 \times 10^{-1}} = 1.5625 \times 10^2 \text{ A·t/m}$$

Clearly, from the example, we can see that, for a given number of ampere-turns, the magnetizing force varies inversely per unit length of the magnetic path.

12–10
Permeability

> **Permeability** A measure of the ease with which magnetic lines of force pass through a given material.

The ability of a material to concentrate magnetic flux is called **permeability**, and its symbol is the Greek lowercase letter *mu*, μ. Any material that is easily magnetized tends to concentrate magnetic flux. Because soft iron is easily magnetized, we say that it has a high permeability. The permeability of a material is a measure of how easy it is for flux lines to pass through it. Numerical values of μ for different materials are assigned by comparing their permeability with the permeability of air or vacuum.

An example of how a soft iron bar can distort the magnetic field around a magnet is illustrated in Figure 12–12. Since soft iron has a high permeability (several hundred times that of air), it is much easier for the magnetic flux to be conducted through it – even though it is a longer path – than to go through an all-air path. It is very difficult for magnetic flux to pass through air.

When an air gap is necessary in a magnetic circuit, it is kept as

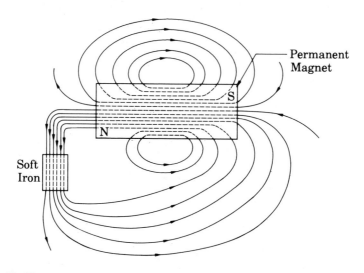

Figure 12–12

Distortion of magnetic field caused by soft iron bar

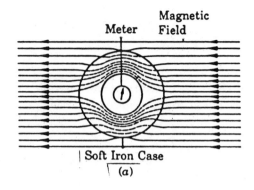

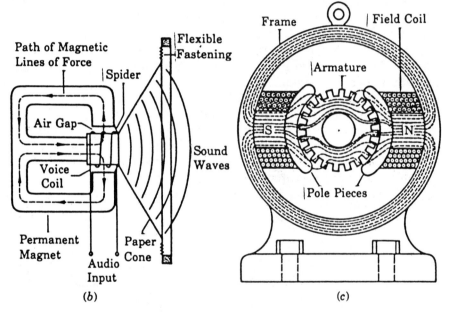

Figure 12-13

Examples of magnetic circuits: (a) Soft-iron case around a meter. (b) Magnetic paths in a typical permanent-magnet loudspeaker. (c) Magnetic paths in the frame, pole pieces, and armature of a simple motor or generator.

small as possible. Figure 12-13 illustrates three different examples of magnetic circuits. If a soft-iron case is placed around a delicate instrument (Figure 12-13a), stray magnetic fields can be shunted around the meter and thereby not cause erroneous readings. The magnetic paths of a typical permanent-magnet (PM) loudspeaker are shown in Figure 12-13b. The gaps between the voice coil (which connects to an audio amplifier) and the permanent-magnet poles must be small to provide a strong magnetic field in the gap but not so close as to cause rubbing between the voice coil and poles. The magnetic paths in a simple motor or generator are shown in Figure 12–13c. It is not necessary to know how these devices operate to understand their magnetic circuits.

Typical values of μ for iron vary from as low as 100 to as high as 5000, depending on the grade used. The permeability of magnetic materials also varies according to the degree of magnetization. Consider the curve for sheet steel shown in Figure 12–14. This curve shows how the permeability of a magnetic material varies with magnetizing force, which is measured in teslas. If too large a magnetizing force is used for a given size and type of magnetic material, the permeability is reduced and the efficiency is impaired.

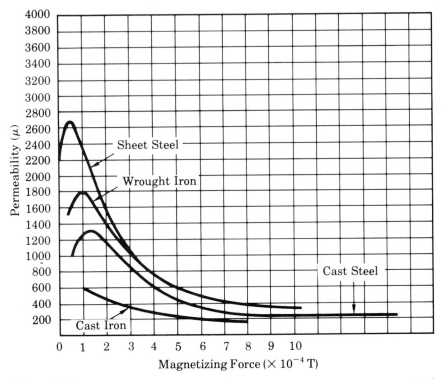

Figure 12–14

Permeability curves for several ferromagnetic materials

Mathematically, μ can be defined as the ratio of flux density to magnetizing force:

$$\mu = \frac{B \text{ (webers per square meter, or teslas)}}{H \text{ (ampere-turns per meter)}} \qquad \text{12–6}$$

In the MKS system, the permeability of air has the symbol μ_0. It has the following value:

$$\mu_0 = 1.26 \times 10^{-6} \text{ T/(A-t/m)} \qquad \text{12–7}$$

The *relative permeability* of a magnetic material, designated μ_r, is the ratio of its absolute permeability μ to that of air. The relationship is expressed as follows:

$$\mu_r = \frac{\mu}{\mu_0} \qquad \text{12–8}$$

Example 12–9

Calculate the absolute value of μ for a magnetic material whose μ_r is 800.

Solution

Use Equation 12–8 and solve for μ:

$$\mu_r = \frac{\mu}{\mu_0}$$

$$\mu = \mu_r \times \mu_0 = 800 \times 1.2 \times 10^{-6} = 960 \times 10^{-6}$$
$$= 0.96 \times 10^{-3} \cong 1 \times 10^{-3} \text{ T/(A-t/m)}$$

The values for μ_r for iron vary from as low as 100 to as high as 5000, depending on the grade used. Special magnetic alloys have been developed, such as supermalloy, which has a μ_r of approximately 800,000. Relative permeability μ_r is a dimensionless quantity since it is the ratio of two numbers.

If the permeability and field intensity of a circuit are known, we can calculate the flux density B by using Equation 12–6 and solving for B:

$$\mu = \frac{B}{H}$$

Therefore
$$B = \mu H \qquad \text{12–9}$$

Example 12–10

Using the value of μ in Example 12–9, calculate the flux density that will result if the field intensity has a value of 500 A-t/m.

Solution

$$B = \mu H = [1 \times 10^{-3}\ \text{T/(A-t/m)}]\ (500\ \text{A-t/m}) = 5 \times 10^{-1}\ \text{T}$$

Observe that the units of ampere-turn per meter cancel, leaving the flux density expressed in teslas.

12–11
B–H Curves

Graphically, we can show the relationship between B and H for a specimen of ferromagnetic material. An example of this relationship is shown in Figure 12–15, which is the *B–H* curve for a particular grade of soft iron. Similar curves are obtainable for all magnetic materials. Note that the overall curve is nonlinear, which indicates that the permeability changes as the magnetizing force changes.

Any magnetic material will go through several stages as it experiences an increasing magnetizing force. Assuming the specimen is free of any magnetism, there is a slight bend in the curve—the *heel*—as it rises from zero. As the magnetizing force increases, the flux den-

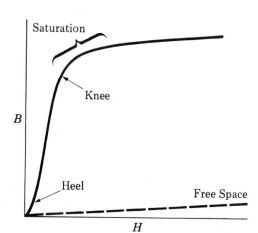

Figure 12–15

B–H magnetization curve for soft iron

sity increases rapidly toward the knee of the curve. Between the heel and knee, the curve is quite linear. The permeability in this region is the highest exhibited by the material.

A further increase in H produces a gradual bend known as the *knee* region of the curve. The slope decreases and the permeability falls off rapidly. Beyond this region, the curve becomes quite flat, and the material enters the *saturation region.* Any further increase in magnetizing force produces very little increase in flux density. In fact, the slope approaches that of the curve for air, as shown by the dotted line labeled "free space" at the bottom of the figure. In the saturation region, the permeability has dropped to a very low value and is nearly flat.

The operation of nearly all electromagnetic devices requires that they be operated in the *linear portion* of their B–H curves. For example, such devices as transformers, loudspeakers, and inductors require that the change in magnetic flux through their magnetic circuits (called the core) conform to the changes of current that take place through the turns of wire around the core (called the winding). Otherwise, the output of the devices (such as a transformer; see Chapter 21) would have a distorted relationship compared with the input.

12–12
Hysteresis Loss

> **Hysteresis** Lagging of the magnetization of a ferromagnetic material behind the magnetizing force.

When the magnetization of a ferromagnetic material lags behind the magnetizing force, **hysteresis** results. In the core of a power transformer, for example, the flux lags behind the increase or decrease of the magnetizing force. This hysteresis results because the magnetic domains are not perfectly elastic or capable of immediately aligning themselves with the changing, external magnetizing force. Also, the domains do not return exactly to their original positions when H is removed.

If the magnetizing force varies at a slow rate (say, up to 60 times per second), the hysteresis loss is small. Beyond this rate, the loss becomes increasingly important. When the magnetizing force reverses thousands of times per second or more, the loss causes a considerable amount of energy to be dissipated in the core. In this case, a large part of H is used in overcoming the internal friction of the domains, and heat is produced. Hysteresis losses vary considerably for different kinds of magnetic materials, being lowest for those having high permeability.

Hysteresis loop   A curve, or loop, plotted on B-H coordinates showing how the magnetization of a ferromagnetic material varies when subjected to a periodically reversing magnetic field.

By using a graph having B-H coordinates, we can plot the hysteresis characteristics of a given ferromagnetic material. Such a curve is plotted in Figure 12–16 and is called a **hysteresis loop**. By periodically

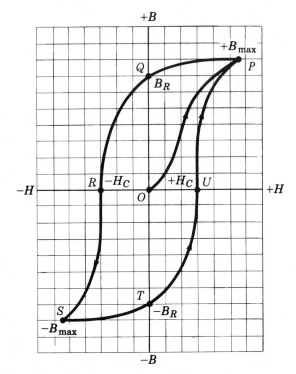

Figure 12–16

Hysteresis loop

reversing the magnetizing force, we can plot the changing values of B within the material.

Actually, a hysteresis loop is a B–H curve under the influence of an ac magnetizing force. Values of flux density B are shown on the vertical axis and are in teslas. Magnetizing force H is plotted on the horizontal axis.

In Figure 12–16, the specimen is assumed to be unmagnetized, and the current is starting from zero in the center of the graph. As H increases positively, B follows the curve OP to saturation, indicated by B_{max}. As H decreases to zero, the flux follows the curve PQ and drops to B_R, which indicates the *retentivity* or *residual induction*. This value represents the amount of flux remaining in the core after the magnetizing force is removed. When H starts in the negative direction, the core will lose its magnetism, as shown by following the curve from Q to R. The amount of magnetizing force required to completely demagnetize the core is called the *coercive force* and is designated as $-H_C$ in Figure 12–16.

As the peak of the negative cycle is approached, the flux follows the portion of the curve labeled RS. Point $-B_{max}$ represents saturation in the opposite direction from $+B_{max}$. From point S, the $-H$ values decrease to point T, which corresponds to a zero magnetizing force. Flux $-B_R$ still remains in the core. A coercive force of $+H_C$ is required to reduce the core magnetization to zero. As the magnetizing force continues to increase in the positive direction, the portion of the loop from U to P is completed. The periodic reversal of the magnetizing force causes the core flux to repeatedly trace out the hysteresis loop.

> **Demagnetization or degaussing** The process by which the magnetization within a ferromagnetic material is reduced to zero by exposing it to a strong alternating magnetic field that is gradually reduced to zero.

Over a period of time, many objects become partially magnetized because of exposure to magnetic fields. These objects include magnetic record/reproduce heads on tape recorders, metal parts of color TV receivers, instruments, and even hand tools. These objects must be occasionally subjected to **demagnetization**, or **degaussing**. To demagnetize any magnetic material, we must reduce its residual magnetism B_R to zero. This task can be done by connecting a suitable coil to a source of alternating current and placing it close to the object to be degaussed. Slowly moving the coil and object away from each other, or reducing the amplitude of the alternating current, causes the hysteresis loop to become progressively smaller. Finally, a point is reached where the loop is reduced to zero and no residual magnetism remains.

12-13
Magnetic Circuits

Reluctance ($\Re$)   The opposition that a magnetic circuit presents to the passage of magnetic lines through it.

Electric and magnetic circuits have some basis for comparison. Magnetic flux Φ corresponds to current and is produced by magnetomotive force mmf. Consequently, mmf ($\mathcal{F}$) corresponds to emf. Just as the electric circuit opposes current, the magnetic circuit opposes flux. This opposition in a magnetic circuit is called **reluctance** ($\Re$). Reluctance varies inversely with permeability.

These three magnetic circuit parameters are related as follows:

$$\Phi = \frac{\text{mmf}}{\Re} \qquad \qquad \textbf{12-10}$$

where $\Re$ is expressed in ampere-turns per weber. This equation expresss *Ohm's law for the magnetic circuit.*

Example 12-11 ───────────────────

An iron core toroid is wound with 200 turns. The core flux is 200 μWb when 0.15 A passes through the coil. What is the reluctance of the core?

Solution

$$\Re = \frac{\text{mmf}}{\Phi} = \frac{(200 \times 0.15)\,\text{A·t}}{200 \times 10^{-6}\,\text{Wb}} = 1.5 \times 10^5\,\text{A·t/Wb}$$

Example 12-12 ───────────────────

The coil in Example 12-11 produces 0.1 μWb when it has an air core. Calculate its reluctance.

Solution

Since 0.1 μWb is 2000 times smaller than 200 μWb, the reluctance must be 2000 times greater. Therefore:

$$\Re = 2000 \times 1.5 \times 10^5 \text{ A·t/Wb} = 3 \times 10^8 \text{ A·t/Wb}$$

Many magnetic circuits, such as loudspeakers and the frame and rotor assembly of motors and generators, contain unavoidable air gaps. Sometimes, small air gaps are deliberately added to the cores of transformers and coils to prevent *core saturation* when large amounts of current flow through the windings. The presence of even a small air gap adds considerable reluctance to the magnetic circuit and thereby reduces the flux and prevents saturation.

An example of a magnetic circuit containing an air gap is the transformer core shown in Figure 12–17. The air gap introduces an additional reluctance to that of the core material. The two reluctances are in series and are treated as series resistances. If we designated the core reluctance as $\Re_c$ and the gap reluctance as $\Re_G$, then the total reluctance is as follows:

$$\Re_T = \Re_C + \Re_G \qquad\qquad \textbf{12-11}$$

Because the gap introduces considerable reluctance to the magnetic circuit, many more ampere-turns are required to produce the same flux than would be the case if no gap were present.

Table 12–1 is a compilation of the major elements of magnetism and their symbols and units. It is presented here for convenient reference.

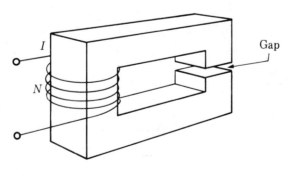

Figure 12–17

Magnetic circuit with air gap

Table 12–1 Symbols and SI Units for Magnetism

Quantity	Symbol	Unit
Flux	Φ	Webers (Wb)
Flux density	$B = \Phi/A$	$Wb/m^2 = teslas$ (T)
Field intensity	$H = \mathcal{F}/l$	Ampere-turns per meter (A-t/m)
Magnetomotive force	$\mathcal{F}$	A-t
Permeability	$\mu_0 = 1.26 \times 10^{-6}$	T/(A-t/m)
Relative permeability	$\mu_r = \mu/\mu_0$	None, pure number
Reluctance	$\mathcal{R} = \mathcal{F}/\Phi$	A-t/Wb

Summary

A magnet can attract ferromagnetic materials.

Retentivity is the ability of a material to retain its magnetism.

Every magnet has a north (N) and a south (S) pole.

The magnetic flux surrounding a magnet is represented by lines of force that never intersect.

The law of magnetic charges states that like poles repel and unlike poles attract.

The attraction or repulsion between magnets varies directly with the product of their strengths and inversely with the square of the distance between them.

Ferromagnetic materials are easily magnetized.

Paramagnetic materials can be only slightly magnetized.

Diamagnetic materials cannot be magnetized; lines of force go around them.

Ferromagnetic materials contain acicular particles called domains.

A current-carrying conductor produces a magnetic field.

The magnetic fields produced by parallel, current-carrying conductors either attract or repel the conductors, depending on the direction of current.

Conductors carrying an electric current always move from a strong to a weak magnetic field.

The addition of a soft-iron core to a solenoid coil greatly increases the magnetic flux.

When current flows through a coil, a magnetomotive force (mmf) is developed, which is the product of ampere-turns.

The symbol for magnetic flux is Φ, and its basic unit is the weber (Wb); $1 \text{ Wb} = 10^8$ lines of force, or Mx.

Flux density is represented by B and can be defined as follows:

$$B = \frac{\Phi \text{ (webers)}}{A \text{ (square meters)}}$$

The SI unit for flux density is the tesla (T).

Magnetic field intensity is defined as follows:

$$H = \frac{\mathcal{F} \text{ (ampere-turns)}}{l \text{ (meter)}}$$

Permeability μ is a measure of how easily flux can flow through a material. It is defined as follows:

$$\mu = \frac{B \text{ (teslas)}}{H \text{ (ampere-turns per meter)}}$$

Relative permeability is defined as follows:

$$\mu_r = \frac{\mu}{\mu_0}$$

Curves for B–H show the relationship between magnetizing force and flux density for a particular type of ferromagnetic material.

Hysteresis loops show the energy losses within a magnetic material.

A magnetic circuit, like the electric circuit, must be complete.

Ohm's law for the magnetic circuit is as follows:

$$\Phi = \frac{\text{mmf}}{\mathcal{R}}$$

Reluctance $\mathcal{R}$ is measured in ampere-turns per weber.

┌── Progress Test

The bracketed number after each question indicates the section of this chapter where the answer can be found.

1.   Magnetic poles are attracted to each other only if: (a) they have the same polarity. (b) they have opposite polarities. (c) there is no air gap between the poles. (d) they are equal.   [12–2]

2.   Magnetic induction results in: (a) induced poles opposite from the original field poles. (b) induced poles the same as the original field. (c) two north poles. (d) an induced magnetic field but no induced poles.   [12–1 and 12–5]

3.   If an iron bar is placed in a magnetic field, the: (a) field lines will be less dense in the iron bar. (b) iron bar has the same effect as air on the field lines. (c) bar will not affect the field lines. (d) iron bar will concentrate the magnetic field lines.   [12–5]

4.   The space in the air gap between two magnetic poles has: (a) no magnetic field. (b) a strong magnetic field. (c) a magnetic field only if the poles touch. (d) no magnetic field unless the poles are far apart.   [12–13]

5.   The domain theory of magnetism assumes that magnetic action is due to the: (a) inability of protons to move within the atom. (b) electron movement and arrangement of the electron orbits. (c) force of the repulsion between electrons accumulated on the dielectric. (d) presence of free electrons in a dielectric.   [12–3]

6.   Which of the following equations properly expresses the relationships in a magnetic circuit? (a) $\Phi = mmf/\mathcal{R}$ (b) $\Phi = \mathcal{R}/\mathcal{F}$ (c) $\mathcal{R} = \Phi/mmf$ (d) $\mu = H/B$ [12–13]

7.   What is the magnetizing force, in ampere-turns, of a solenoid having 2500 turns when 6 mA passes through it? (a) 416.6 (b) 90.7 (c) 15.0 (d) 9.3   [12–6]

8.   A certain magnet has a cross-sectional area of 0.005 m². Calculate the flux density if 250 lines of force are passing through it. (a) 5×10^{-6} T (b) 25×10^{-4} T (c) 5×10^{-4} T (d) 25×10^{-3} T   [12–8]

9.   An inductor produces 1600 lines when 20 mA passes through it. How much flux in webers do these lines represent? (a) 16×10^{-7} (b) 1.6×10^{-5} (c) 1.6×10^{-4} (d) 16×10^{-2} [12–7]

10.   A certain toroid has 1200 turns with 8 mA passing through it. If the mean length of the magnetic path is 4.5×10^{-2} m,

what is the magnetizing force in ampere-turns per meter? (a) 378.1 (b) 319.4 (c) 243.1 (d) 213.3   [12–9]

11.   Determine the flux density in a ferromagnetic material that has an absolute permeability of 4.5×10^{-3} T/(A·t/m) and a field intensity of 300 A·t/m. (a) 1.35 T (b) 13.5×10^{-2} T (c) 1.35×10^2 T (d) 1.35×10^{-3} T   [12–10]

⌐─Problems

Answers to odd-numbered problems are at the back of the book.

1.   Calculate the mmf produced by a solenoid of 1500 turns having 8 mA passing through it.

2.   How much current must pass through a coil of 800 turns to produce a magnetomotive force of 30 A·t?

3.   A toroid has 42,700 lines of force through it. How many microwebers do these lines represent?

4.   A coil wound on a rectangular form (as in Figure 12–11) produces 7.56×10^{-5} Wb when current passes through it. How many microwebers does this force represent?

5.   The iron core of a certain coil has 31.6 μWb through its cross-sectional area. How many webers does this force represent?

6.   Calculate the flux density in a toroid of 15 cm² cross-sectional area with a total flux of 8.5×10^3 Mx.

7.   The core of a relay measures 1.5×1.75 cm through its cross section. What is the flux density if a total of 2.65×10^2 Mx is in the core?

8.   An air coil has 1.63 μWb of flux in its core. What is the flux density if the core has a diameter of 1.6 cm?

9.   The flux density in a certain piece of ferromagnetic material is 7.62×10^{-2} T. What is the total flux if the cross-sectional area is 3.6×10^{-2} m²?

10.   Calculate the magnetic field intensity in the magnetic circuit of Figure 12–18 if the mean diameter is 4 cm.

11.   If the mean diameter of the toroid in the circuit of Figure 12–18 is increased three times, and other conditions remain unchanged, what is the new magnetizing force?

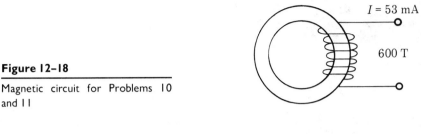

$I = 53$ mA

600 T

Figure 12–18

Magnetic circuit for Problems 10 and 11

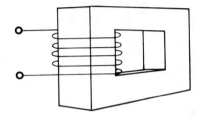

Figure 12–19

Magnetic circuit for Problem 14

12. Calculate the permeability of a magnetic circuit where the flux density is 6.5×10^{-2} T and the magnetic field intensity is 3.5×10^2 A·t/m.

13. A certain ferromagnetic material has an absolute permeability of 4.72×10^{-3} T/(A·t/m). What is its relative permeability?

14. What is the reluctance of the rectangular iron core in Figure 12–19 if the core flux is 425 μWb when 65 mA is passing through the 375 turns of the winding?

15. What ampere-turns are required to produce a flux of 3.68×10^{-5} Wb when the reluctance is 7.14×10^5 A·t/Wb?

16. In Problem 15, if the coil has 1500 turns, what current is required?

17. The flux in a magnetic circuit is 6.35×10^{-4} Wb. Adding a 15-mil air gap reduces the flux 1.5×10^3 times (the ampere-turns remain unchanged). What is the value of the circuit flux?

13

Inductance

Objectives

After studying this chapter, you will be able to:

Describe how solenoids operate.

Discuss the effect that two closely associated coils have on each other.

Recognize how an inductor can store a considerable amount of energy.

Introduction

There are three properties of an electric circuit that can impede the flow of an electric current: resistance (R), capacitance (C), and inductance (L). The first two have been investigated in previous chapters. The characteristics of *inductance* are the subject of this chapter. Since an inductor is basically a magnetic device, a knowledge of electromagnetism is essential to its understanding.

Oersted's discovery of electromagnetism prompted research into methods of producing an electric current by means of a moving field. Several years later, other physicists found that a moving magnetic field forces electrons to move along a conductor. The results of these investigations led to the development of electric generators and other inductive devices.

Applications

The electric power generators we spoke of in Chapter 12 work by means of magnetic inductance. Hundreds or even thousands of turns of copper wire spin rapidly through a strong magnetic field, producing an electromotive force that is used to power external circuits connected to the generator. In large-scale commercial power plants, steam or water pressure is used to provide the energy to spin a large coil of wire inside even more powerful magnetic fields in order to produce power.

Depending on their design, power generators can produce either direct or alternating current. It was found early in the history of commercial electric power that high voltages are required for economical transmission to large areas. Thus, another inductive device is needed to lower the voltage to a level safe for household use. This device is the transformer. Because transformers require alternating current (ac), we do not use direct current (dc) in household applications. Smaller generators, such as those in automobiles or portable lighting equipment, can use direct current, however.

The use of transformers to change one ac voltage level to another is familiar in such applications as power supplies, automobile spark coils, doorbell circuits, and audio amplifier output circuits. Inductors, also called chokes or coils, appear frequently in electronic circuits to perform such functions as tuning, coupling, timing, filtering, and storing energy. They are essential components of most motors, transmitters, and audiovisual equipment.

Inductance is also used in industrial heating processes when the materials to be heated or melted are made of metal, such as ores and alloys. Induction furnaces utilize a crucible that rests inside a coil of wire. When an ac voltage is applied to the coil, electric current (called an eddy current) is induced into the conducting material, and the resistance of the metal causes heat to be produced. Induction heating has a valuable characteristic: The distance below the surface that the eddy currents penetrate depends on the frequency of the applied voltage. Thus, it is possible, through adjustment of the frequency, to treat only the metal's surface. In case hardening of gears, for example, the current is turned off before the resulting heat has had time to penetrate to the gear's interior, and coolant is sprayed on the surface. Only the wear-prone part of the mechanism is hardened, while the inside retains its strength and flexibility.

13-1
Definition of Inductance

Inductance (L)   That property of an electric circuit that opposes any change of current through it.

Inductance (L) is a property of a circuit whereby electrons flow when a nearby source of magnetism moves in relation to the circuit—that is, when the circuit is in the presence of a time-varying magnetic field. Inductance also has the converse property whereby electric current causes a circuit to behave like a magnet.

Oersted's discovery in 1820 that a current-carrying conductor sets up a magnetic field created speculation that the reverse effect could also be produced. In other words, should not a current be induced in a wire by passing it through a magnetic field? Michael Faraday, in 1831 during the course of his experimentation, discovered the principle of developing an electric voltage by passing a conductor through a magnetic field.

Magnetic Induction

Magnetic induction   The process by which an emf is induced in a conductor as it passes through a magnetic field.

The nature of Faraday's discovery can be illustrated by means of the drawing in Figure 13-1. The ends of a conductor are connected to a microammeter with a zero-center scale. When the conductor is passed quickly down through the flux of the magnet, the free electrons in the conductor react with the magnet's field, causing a movement of electrons in the conductor. The meter is momentarily deflected to the right and then returns to zero. If the conductor is now quickly moved out of the field, the electrons in the conductor move in the reverse direction, and the microammeter needle momentarily deflects to the left and then returns to zero. The process shown in Figure 13-1 is called **magnetic induction** because no physical connection exists between conductor and magnet.

If the conductor were moved parallel to the lines of flux, *no voltage* would be induced. If the angle between the flux lines and the path of the conductor were increased, the deflection of the microammeter needle would also increase, and maximum needle deflection would be obtained when the conductor was swept through the field at right angles. If the polarity of the magnetic field were changed, the induced emf would be reversed.

Voltage Induced in a Coil

In the course of Faraday's experiments, he discovered several factors that controlled the magnitude of the induced voltage and resulting current pulse. If the conductor was wound into a coil, as in Figure 13-2, *the current pulse increased directly with the number of turns.* He also observed that the *speed* at which the magnet moved toward or away from the coil determined the strength of the pulse. Also, *stronger magnets produced stronger pulses.* Further, the pulse polarity *changed* when the magnet moved into or out of the coil.

Faraday's conclusions regarding magnetic induction can be summarized as follows. *Whenever there is relative motion between a conductor and a magnetic field, a voltage will be induced in the conductor.*

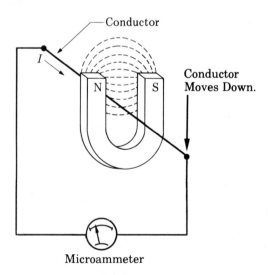

Figure 13-1

Inducing a current in a conductor

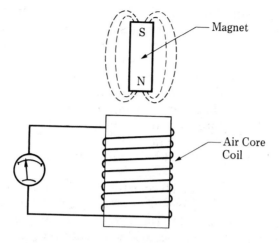

Figure 13–2

Inducing a current by magnetic flux cutting through a coil

13–2
Factors Determining Strength of Induced Voltage

Four basic factors determine the magnitude of an induced emf:

1. The number of turns in the conductor.
2. The strength of the magnetic field.
3. The relative speed between coil and magnetic field.
4. The angle at which the conductor passes through the magnetic field.

Each of these factors will be discussed in the following paragraphs.

The voltage induced into each turn on the coil is in series with the voltages induced into *every other turn*. These voltages are *series-aiding* voltages. The total induced emf is the sum of the individual voltages in each turn.

A strong magnetic field produces a greater reaction with the magnetic fields of the free electrons than a weak one does, causing more electrons to move along the conductor. When the magnet's position is remote from the coil, none of its lines of magnetic flux link with the turns on the coil. As the magnet is brought closer, flux linkages begin to take place, inducing some voltage. Finally, when the magnet is thrust inside the coil, maximum flux linkages occur, and the largest voltage is developed. The magnitude of the emf will become less, and its polarity will reverse, as the magnet is withdrawn from the coil. If an electromagnet were used, the current flowing through it could be varied and the generated, or *induced*, emf thereby regulated.

The rate at which the lines of force, or flux linkages, cut the turns of wire also dictates the amount of induced voltage. If the rate is increased, the voltage will be greater. Therefore, if a large voltage is needed, a high relative speed between conductor and field must be maintained.

The fourth factor involved in the magnitude of voltage has to do with the angle the conductor makes in cutting through the flux. If the conductor is moved parallel to the flux, no emf is produced. The greatest emf is produced when the conductor cuts the flux lines at right angles. Varying the angle between these two values causes the induced emf to fluctuate between zero and maximum.

13–3
Faraday's Law

The statements in Section 13–2 regarding the factors that determine induced voltage outline the experiments performed by Faraday and have become known as *Faraday's law*. These relationships can be expressed mathematically as follows:

$$e_L = N\frac{\Delta\Phi}{\Delta t}$$ 13–1

where e_L = induced voltage
N = number of turns
$\Delta\Phi$ = a small change in flux linkages
Δt = a small change in time

With the ratio $\Delta\Phi/\Delta t$ in webers per second, the induced voltage e_{ind} is in volts. Equation 13–1 is frequently written as follows:

$$e = N\frac{d\Phi}{dt}$$

where d is an abbreviation for the Greek capital letter *delta* Δ. These equations assume that all the flux Φ links with all the turns N.

Example 13–1 ————————————————————————

What voltage is induced in a coil of 500 turns if 0.5 Wb/s pass through it?

Solution

$$e_{\text{ind}} = N\frac{d\Phi}{dt} = 500 \times 0.5 = 250 \text{ V}$$

Example 13–2 ————————————————————————

A magnetic flux of 0.025 Wb cuts across an inductor of 300 turns in 0.05 s. What voltage is induced in the coil?

Solution

$$e_L = 300\,\frac{0.025 \text{ Wb}}{0.05 \text{ s}} = 150 \text{ V}$$

13–4
Nature of Inductance

When a voltage is applied to a coil, the current does not immediately rise to a value equal to E/R. It reaches this level after a brief delay. The extent of the delay is a function of the *number of turns* in the coil and the *permeability* of the core. The delay increases as either of these parameters increases. This characteristic of a coil is called *inductance*. As we will learn in Chapter 14, the exact time required for the inductive current to reach its maximum value depends on the ratio of inductance to series resistance.

By means of a graph, we can illustrate the basic shape of the current wave as it flows through the coil. The inductance of the circuit of Figure 13–3a is indicated by the letter symbol L. The dotted resistor represents the internal resistance of the coil. Let us assume that switch SW is placed in position a at time t_1; refer to the graph of waveforms in Figure 13–3b. Now, if the circuit were purely resistive, the current would rise instantly to a value determined by the ratio E/R and remain there, as represented by the vertical and horizontal dotted lines. The

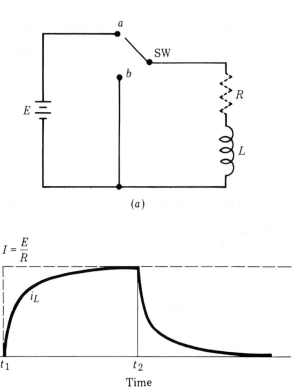

(a)

(b)

Figure 13–3

Current through coil: (a) Circuit. (b) Waveforms for rising and decaying currents.

inductive current i_L is delayed somewhat in reaching its maximum value, as is characteristic of all inductors. (*Note*: An electric value that is changing with time is represented by lowercase letters to distinguish it from dc or steady-state values.)

Next, let us consider what happens when SW is placed in position b at time t_2. The battery is no longer connected across the coil. It would seem logical that the current would immediately drop to zero, as does indeed occur when the circuit is purely resistive (see the vertical line at t_2). However, the characteristic of inductance is such that the inductive current decays according to the shape of the curve to the right of t_2. The reason for the delay in the rise and decay times of i_L will be explored in Section 13–5.

13–5
Counterelectromotive Force

> **Counterelectromotive force (cemf)**   A voltage developed in an inductive circuit by an alternating, or pulsating, current whose polarity at every instant is opposite to the applied voltage.

Let us examine the circuit action that occurs when a voltage is connected across an inductor. The simple circuit of Figure 13–4 will be used in this analysis. Assume that switch SW has just been closed. Current begins to build up. Magnetic flux, starting within the coil, begins to expand, as indicated by the two solid arrows pointing away from the coil. Flux lines created by each turn in the coil encircle adjacent turns, inducing voltages into these turns in a direction opposite to that of the applied emf. The sum of these voltages forms a **counterelectromotive force (cemf)**, sometimes called *back emf*, that very nearly opposes the supply voltage.

The rate of change of current is greatest at t_1, the moment the switch is closed, as shown in Figure 13–3b. As the current continues to rise, its rate of change gradually decreases. Therefore, the cemf that it produces also decreases. As the cemf decreases, it allows the current to increase. The current changes less rapidly as it approaches its ultimate value, and less cemf is produced. Finally, when it reaches E/R, it becomes steady, and no further cemf is produced. The current is now limited only by the resistance of the coil.

Now, let us consider what occurs when the switch in Figure 13–4 is opened. Obviously, the voltage source E will no longer sustain the current. The magnetic flux collapses back into the coil, as indicated by the two dashed arrows. The flux linkages now cut through the coil in

Figure 13–4

Circuit used to explain generation of cemf by expanding and collapsing field

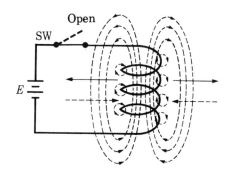

the opposite direction, inducing a voltage into the coil that trys to prevent the current from decreasing. This cemf is opposite to the cemf that opposed the original buildup of current. Because the energy stored in the field is limited, the cemf cannot completely stop the current from decreasing. By referring to Figure 13–3b, we can see the shape of the decaying current waveform as it appears starting at t_2. The magnitude of the cemf diminishes as the rate of change of current (often referred to as roc i) falls off. When the field has totally collapsed, there can be no more cemf, and i_L is reduced to zero.

The above principle was first observed by Russian physicist Heinrich Lenz (1804–1865) and has become known as Lenz's law. Based on the physical concept of the conservation of energy, the law states that *the induced emf in an inductance always opposes the current change that produced it.*

Figure 13–5 is a graph showing the relationships of applied voltage, cemf, and the resulting current flow. This graph should be correlated with Figure 13–3b in the following explanation. Voltage E is applied to the inductive circuit at time t_1. The rate of change of current, roc i, is greatest at this instant and (therefore) the cemf is maximum. The roc i decreases with time, reducing the cemf and allowing i_L to increase. Observe that the curves for i_L and cemf are complements of each other. At t_2, the circuit is opened, and the collapsing field generates a cemf that trys to sustain the current at its maximum value. As the cemf diminishes, i_L decreases until both finally reach zero.

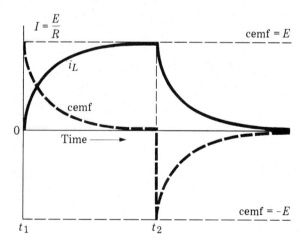

Figure 13–5

Graph showing relationship between inductive current, cemf, and applied E, plotted against time when E is applied, then removed

13–6
Self-Inductance

> ┌─ **Self-inductance**   The ability of an inductor to induce a voltage
> │   into itself.

The property of a circuit that opposes any change of current through it is called *inductance*. The ability of an inductor to induce a voltage into itself is called **self-inductance**. Inductance is present *only* when the current is changing. The magnetic field established around an inductor when current is passing through it represents electromagnetic energy.

The unit of inductance is the henry, abbreviated H, named in honor of the American physicist Joseph Henry (1797–1878), whose research and findings on electromagnetism paralleled those of Michael Faraday. *The henry is the amount of inductance that will produce one volt across it when the current changes at the rate of one ampere per second*. Mathematically expressed, inductance is as follows:

$$L = \frac{e_L}{di/dt}$$

13–2

where
L = inductance of a circuit in henrys
e_L = induced voltage in volts
di/dt = current change in amperes per second

The factor di/dt specifies how fast the magnetic flux is cutting through the inductor to produce e_L. (*Note:* Do not confuse the abbreviation H for henry with the letter symbol H for magnetic field intensity.)

Example 13–3

What is the di/dt rate of an inductive current if it changes from 3 to 6 A in 1 s?

Solution

$$\frac{di}{dt} = \frac{(6 - 3)\ \text{A}}{1\ \text{s}} = 3\ \text{A/s}$$

Example 13-4

If the current through a small inductor changes 7 mA in 1.5 ms, what is its di/dt rate in amperes per second?

Solution

$$\frac{di}{dt} = \frac{7 \times 10^{-3}}{1.5 \times 10^{-3}} = 4.67 \text{ A/s}$$

Example 13-5

Calculate the inductance of a coil with an induced emf equal to 25 V when the rate of change of current is 5 A/s.

Solution

$$L = \frac{e_L}{di/dt} = \frac{25}{5} = 5 \text{ H}$$

Example 13-6

The current through a small inductor changes 30 μA in 5 μs. If 16 mV is produced across the device, what is its inductance?

Solution

$$L = \frac{e_L}{di/dt} = \frac{e_L \times dt}{di} = \frac{(16 \times 10^{-3}) \times (5 \times 10^{-6})}{30 \times 10^{-6}}$$

$$= 2.67 \times 10^{-3} \text{ H} = 2.67 \text{ mH}$$

Example 13-7

What voltage will be induced in a 200 mH inductor if the current changes 50 mA in 1 ms?

Solution

Use Equation 13-2 and solve for e_L:

$$e_L = L\frac{di}{dt} = 0.2 \text{ H} \times \frac{0.05 \text{ A}}{0.001 \text{ s}} = 10 \text{ V}$$

The inductance of a coil depends on the following physical parameters:

1. The number of turns squared.
2. The cross-sectional area of the coil; L varies directly with A.
3. The permeability of the core; L increases as μ_r increases.
4. The length of the coil; L varies inversely with length for the same number of turns, since the flux is less concentrated.

These parameters are related as shown in the following formula, where the length is at least 10 times the diameter:

$$L_{\text{H}} = \mu_r \frac{N^2 A}{l} \times 1.26 \times 10^{-6} \qquad \text{13–3}$$

where
$N =$ number of turns of the coil
$A =$ cross-sectional area in square meters
$l =$ length in meters of the magnetic path
$\mu_r =$ relative permeability

The term 1.26×10^{-6} is the permeability of air, which, when multiplied by μ_r, gives the absolute permeability of the core. (See Equations 12–7 and 12–8.) Therefore:

$$\mu = \mu_r \mu_0 = \mu_r \times 1.26 \times 10^{-6}$$

Example 13–8

Determine the inductance of the coil in Figure 13–6 from the information shown. Assume $\mu_r = 600$.

Solution

First, determine A:

$$A = 0.01 \times 0.01 = 10^{-4}\,\text{m}^2$$

Next, determine the average length of the magnetic path around the core:

$$l = (2 \times 0.06) + (2 \times 0.04) = 0.2\,\text{m}$$

Then:

$$L = 600\,\frac{250^2 \times 10^{-4}}{0.2} \times 1.26 \times 10^{-6}$$

$$= \frac{6 \times 10^2 \times 6.25 \times 10^4 \times 10^{-4}}{0.2} \times 1.26 \times 10^{-6} = 0.0236\,\text{H}$$

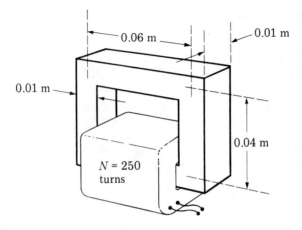

Figure 13–6

Coil for Example 13–8

From Equation 13–3, we can draw the following conclusions:

1. Inductance increases as the square of the number of turns.
2. Inductance increases with the permeability of the magnetic circuit. Where high values of L are needed, coils are wound on iron cores.
3. Inductance increases with the area of the coil.
4. Inductance decreases with the length of the coil.

The inductance of a single-layer, air core coil may also be calculated from the following formula:

$$L_{\mu H} = \frac{r^2 N^2}{9r + 10l} \qquad \text{13–4}$$

where
$r = $ radius in inches
$l = $ coil length in inches
$N = $ number of turns

An alternative formula can be used if the coil dimensions are given in centimeters:

$$L_{\mu H} = \frac{r^2 N^2}{24r + 25l} \qquad \text{13–5}$$

Example 13-9

Determine the inductance of a single-layer, air core coil with the following parameters: $r = 0.375$ in., $l = 1.5$ in., and $N = 48$.

Solution

$$L_{\mu H} = \frac{0.375^2 \times 48^2}{(9 \times 0.375) + (10 \times 1.5)} = 17.63 \ \mu\text{H}$$

To calculate the number of turns of a single-layer, air core coil for a required inductance, use the following formula:

$$N = \sqrt{\frac{L\,(9r + 10l)}{r^2}} \qquad\qquad \text{13-6}$$

Example 13-10

An inductance of 25 μH is needed. The form on which the coil is to be wound has a diameter of 1 in. and is 1.25 in. long. What is the number of turns needed?

Solution

$$N = \sqrt{\frac{25\,(9 \times 0.5 + 10 \times 1.25)}{0.5^2}} = 41.23$$

13-7
Types of Inductors

In spite of the great number of different kinds of inductors, only a few schematic symbols are needed to represent them. For example, a *filter choke* for a power supply is shown in Figure 13-7a, and its symbol is shown in Figure 13-7b. The two parallel lines beside the coil represent an iron core. Inductors with iron cores are suitable for use only with power line and audio frequencies. Above the *audio range* (approximately 20 kHz), the losses in the core become too large to be practical.

Typical *radio frequency (RF) coils* are shown in Figure 13-8a, and the symbol is shown in Figure 13-8b. No iron core is used at these frequencies. Many RF coils have a *ferrite slug* or powdered iron core inside the coil, which can be screwed in or out. The symbol for such a

(a)

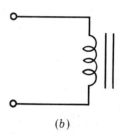

(b)

Figure 13–7

Iron core choke coil: (a) Typical choke. (b) Schematic symbol.

(a)

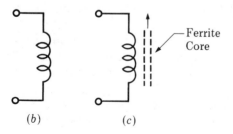

—Ferrite
Core

(b)   (c)

Figure 13–8

Radio frequency coil: (a) Typical RF coils. (b) Schematic symbol. (c) Schematic symbol for coil with adjustable ferrite core.

device is represented in Figure 13–8c. Because of the high permeability of this kind of core, the inductance value can be made to vary over an appreciable range.

Another popular type of inductor is the *transformer*, which can be used with power and audio frequencies as well as radio frequencies. Some typical transformers and their schematic symbols are shown in Figure 13–9. Note the primary and secondary windings for transformers.

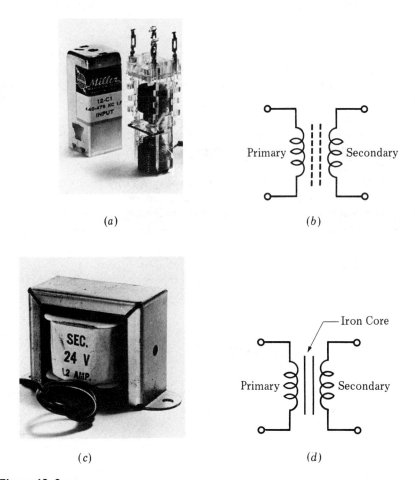

(a)

(b)

(c)

(d)

Figure 13–9

Transformers: (a) RF transformer. (b) Schematic symbol for RF transformer. (c) Power transformer. (d) Schematic symbol for power transformer.

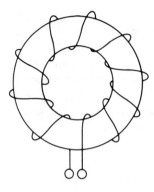

Figure 13–10

Toroid with toroidal winding

A very common type of inductor is the *toroid*, a small dough-nut-shaped coil, as shown in Figure 13-10. The coil may consist of a single-layer winding, a multilayer winding, or even several coils wound on the same toroid. The core is of ferrite material, which has a high permeability. Consequently, the inductance can be quite large for the number of turns. Practically all the flux is confined to the core, and very little shielding is required to prevent interference with adjacent components.

Sometimes, it is necessary to have inductors that can vary the amount of RF energy – or coupling – between primary and secondary windings. Such an inductor is shown schematically in Figure 13-11*a*. In Figure 13-11*b*, a variable or adjustable inductor is represented in schematic form.

13–8
Mutual Inductance

> **Mutual inductance**   The property that exists between two current-carrying conductors when the magnetic field from one links with those of the other.

When two coils are positioned so that flux lines from one cut the turns of another, they are said to exhibit **mutual inductance**. The action of mutual inductance in producing a voltage in another coil is basically

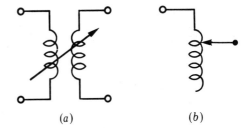

Figure 13–11

Schematic symbols for variable inductors: (a) Variable-coupling RF transformer. (b) Adjustable coil.

(a) (b)

the same as that of self-inductance, which was explained earlier. The main difference between the two is that self-inductance is the property of a single coil, while mutual inductance is the property of two or more coils acting together. Mutual inductance is measured in henrys and is designated by L_M.

Coefficient of Coupling

Coefficient of coupling   The degree of coupling, expressed as a decimal fraction, that exists between two magnetic circuits.

Mutual inductance depends on such factors as the number of turns on each coil, the permeability of the core material, the physical dimensions of the coils, and the **coefficient of coupling**. The coefficient of coupling is a measure of how much flux from one coil cuts the windings of another. If the total flux of one coil links through all the turns of the other, the coefficient of coupling is *unity*. In practice, unity is never attained, since there is always some stray or leakage flux.

Let us investigate mutual inductance by referring to Figure 13–12a. Assume a voltage source is connected to L_1 that produces a current Δi. This changing current creates a fluctuating magnetic field that links with L_2, inducing e_{L_2} into it. If L_1 and L_2 each have the same number of turns, and all the flux of L_1 links with L_2, then the voltage induced into L_2 will equal the voltage induced into L_1.

In Figure 13–12b, the two inductors are wound on a common iron core. The ac generator causes a changing current to pass through L_1, creating a varying flux around it. The induced emf in L_2 causes a current to flow in L_2, whose magnetic field induces a voltage back into L_1. Therefore, mutual inductance exists between the two coils because current flowing in one winding induces voltage into the other. If a cur-

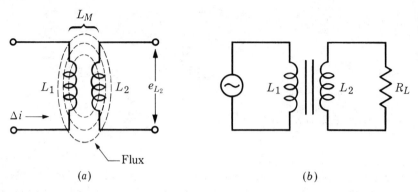

Figure 13–12

Mutual inductance: (a) Between coils L_1 and L_2. (b) Between coils wound on an iron core.

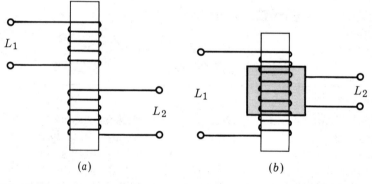

Figure 13–13

Mutual inductance between two coils: (a) Loose coupling. (b) Closer coupling.

rent change of 1 A/s in one coil induces a voltage of 1 V in the second coil, there is 1 H of mutual inductance between them.

Not all the flux of one coil links with the other. The fraction of total flux linking them is referred to as the coefficient of coupling, which is represented by the letter k. If one-half of the flux produced by L_1 in Figure 13–12a links with L_2, $k = 0.5$. In Figure 13–12b, the presence of the iron core suggests that practically all of the flux produced by L_1 links L_2; therefore, $k = 1$, or unity (k is always expressed as a decimal fraction).

In many electronic circuits, k has a fixed value, while others have varying coefficients of coupling. For example, in Figure 13–13a, two coils are wound on a common plastic core and have a fixed k value, typically 0.1, determined by the manufacturer. In Figure 13–13b, one coil is wound on top of the other to achieve a higher k value, typically 0.25 to 0.35. Using a closed core between two inductors increases k to about 1.

Determining Mutual Inductance

Mutual inductance can be calculated by the following formula:

$$L_M = k \sqrt{L_1 L_2} \qquad\qquad \textbf{13–7}$$

where $\quad L_1$ and $L_2 =$ inductance of each coil in henrys

Example 13–11

Calculate the mutual inductance between two coils wound on a closed iron core when $L_1 = 3$ H and $L_2 = 4$ H.

Solution

$$L_M = 1 \sqrt{3 \times 4} = \sqrt{12} = 3.46 \text{ H}$$

This result means that if the current changes at the rate of 1 A/s in either coil, 3.46 V will be induced in the other.

Example 13–12

The characteristics of a certain magnetic circuit are $L_1 = 80$ mH, $L_2 = 40$ mH, and $k = 0.1$. Calculate L_M.

Solution

$$L_M = 0.1 \sqrt{80 \times 10^{-3} \times 40 \times 10^{-3}} = 0.1 \sqrt{3.2 \times 10^{-3}}$$
$$= 0.1 \times 5.66 \times 10^{-2} = 5.66 \text{ mH}$$

Example 13–13

If the two inductors in Example 13–12 had twice the value of L_M, what must k equal?

Solution

$$L_M = k \sqrt{L_1 L_2}$$ 13-8

$$k = \frac{L_M}{\sqrt{L_1 L_2}} = \frac{2 \times 5.66 \times 10^{-3}}{\sqrt{80 \times 10^{-3} \times 40 \times 10^{-3}}}$$

$$= \frac{1.13 \times 10^{-2}}{5.66 \times 10^{-2}} = 0.2$$

Observe that k doubled when L_M doubled.

13-9
Inductors in Series or in Parallel

Calculation of Inductance

When series-connected inductors are far apart or well-shielded, so that no mutual inductance is present, the total inductance is as follows:

$$L_T = L_1 + L_2 + L_3 + \cdots + L_n$$ 13-9

This equation is similar to the equation for resistors in series.

Example 13-14 ──────────────────────

Calculate the total inductance of three series-connected inductors, having no mutual inductance, whose values are $L_1 = 1.5$ H, $L_2 = 0.4$ H, and $L_3 = 2$ H.

Solution

$$L_T = 1.5 + 0.4 + 2 = 3.9 \text{ H}$$

Inductors that are parallel-connected behave like paralleled resistors, assuming no mutual inductance:

$$\frac{1}{L_T} = \frac{1}{L_1} + \frac{1}{L_2} + \frac{1}{L_3} + \cdots + \frac{1}{L_n}$$ 13-10

When only two inductors are involved, the product-over-sum formula may be used:

$$L_T = \frac{L_1 L_2}{L_1 + L_2}$$ 13-11

Example 13-15

Find the equivalent inductance of two parallel inductors, where $L_1 = 2$ H, $L_2 = 4$ H, and $k = 0$.

Solution

$$\frac{1}{L_T} = \frac{1}{2} + \frac{1}{4} = \frac{2}{4} + \frac{1}{4} = \frac{3}{4}$$

$$L_T = 1.33 \text{ H}$$

When two series-connected inductors have mutual inductance between them, the total inductance depends on the amount of coupling and whether the inductors are connected in *series-aiding* or *series-opposing* connections. A series-aiding connection means that the magnetic field of one inductor aids, or adds to, the magnetic field produced by the other inductor. With a series-opposing connection, the reverse is true. A series-aiding connection is shown in Figure 13–14a, and a series-

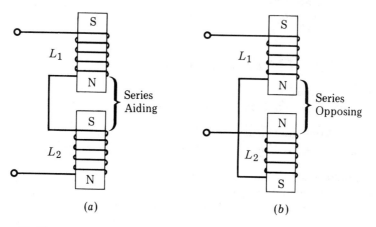

(a)   (b)

Figure 13–14

Two series inductors: (a) L_M is series aiding. (b) L_M is series opposing.

opposing connection is shown in Figure 13–14b. Note the direction of the induced magnetic fields in both cases.

The total inductance of two mutually coupled coils can be determined from the following formula:

$$L_T = L_1 + L_2 \pm 2\,L_M \qquad \text{13–12}$$

Term L_M is positive (plus) when the coils are series aiding and negative (minus) when series opposing.

Example 13–16

Two inductors are series aiding and their values are $L_1 = 20$ mH and $L_2 = 15$ mH. Their mutual inductance is 8 mH. Calculate L_T.

Solution

$$L_T = 20 + 15 + (2 \times 8) = 51\ \text{mH}$$

Example 13–17

Assume the two inductors in Example 13–16 are series opposing and L_M is unchanged. What is L_T?

Solution

$$L_T = 20 + 15 - (2 \times 8) = 19\ \text{mH}$$

Calculation of Mutual Inductance

From Equation 13–11, we can determine the value of mutual inductance between two coils by following two steps:

1. Measure the series-aiding inductance L_a of the two coils.
2. Measure the series-opposing inductance L_o of the coils.

Note: Inductance measurements can be easily made with a special type of instrument called an impedance bridge.

The formula for the mutual inductance is as follows:

$$L_M = \frac{L_a - L_o}{4} \qquad \text{13–13}$$

Example 13–18 ───────────────────────────────────

When two coils have a series-aiding connection, their total inductance is 120 mH. When they have a series-opposing connection, their total inductance measures 40 mH. What is the value of L_M?

Solution

$$L_M = \frac{L_a - L_o}{4} = \frac{120 - 40}{4} = 20 \text{ mH}$$

If we know the mutual inductance, the coefficient of coupling can be determined by the following formula:

$$k = \frac{L_M}{\sqrt{L_1 L_2}} \qquad\qquad 13\text{–}14$$

Example 13–19 ───────────────────────────────────

Two inductors, $L_1 = 200$ mH and $L_2 = 80$ mH, have a mutual inductance of 30 mH. Calculate the value of k.

Solution

$$k = \frac{L_M}{\sqrt{L_1 L_2}} = \frac{30}{\sqrt{200 \times 80}} = 0.24$$

13–10
Energy Stored in an Inductor

When current increases through an inductor, energy is added to the magnetic field. When the field collapses, the energy is returned to the circuit. The electric energy delivered to or consumed by any circuit is the product of $ei \times$ time. When an inductive circuit, such as shown in Figure 13–15a, is first energized, i_L and cemf vary with time. Therefore, the rate at which energy is stored varies with time and is represented in Figure 13–15b by the shaded area under the $W = ei$ curve. If the switch

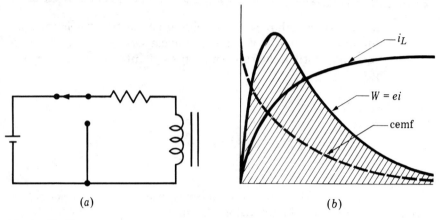

Figure 13-15

Energy in inductance: (a) Inductive circuit. (b) Energy curve.

in the circuit is placed in the alternate position, the energy will be released in the form of heat in the resistance.

   The average value of the current is equal to one-half the steady-
state or maximum value E/R, even though the current does not change
uniformly. Therefore:

$$W = 1/2\,LI^2 \qquad\qquad \textbf{13-15}$$

where $W =$ stored energy in joules
            $L =$ inductance in henrys
            $I =$ steady-state current in amperes $(I = E/R)$

Example 13-20 ──────────────────────────────

   Calculate the energy stored in an inductor of 2 H when 1.5 A is
flowing.

   Solution

$$W = 1/2\,LI^2 = 1/2 \times 2 \times (1.5)^2 = 2.25\ \text{J}$$

Summary

Inductors store electromagnetic energy.

A voltage is induced in a conductor when it passes through a magnetic field.

The polarity of an induced voltage depends on the direction in which it cuts through magnetic lines of force.

The magnitude of an induced voltage depends on the number of turns in the conductor, the strength of the magnetic field, the relative speed between coil and magnetic field, and the angle at which the conductor passes through the field.

Faraday's law is as follows:

$$e_{\text{ind}} = N \frac{\Delta \Phi}{\Delta t}$$

Inductance is that property of an electrical circuit that opposes any change of current.

When a voltage is applied to an inductor, the current does not immediately rise to E/R.

Inductance is measured in henrys (H), and its letter symbol is L.

When the current in an inductive circuit is interrupted, it does not immediately drop to zero.

A cemf is developed across an inductor whenever any change of current occurs through it.

Lenz's law states that the induced emf in an inductance always opposes the current change that produced it.

The rate of change of current is greatest when the inductor is first connected across a voltage source.

One henry is the amount of inductance that will produce one volt across it when the current changes at the rate of one ampere per second.

The value of an inductance varies with the square of the turns, the cross-sectional area of the coil, the permeability of the core, and the length of the coil.

Mutual inductance results from flux lines of one coil linking the turns of another. It is measured in henrys and is designated by L_M.

The coefficient of coupling k represents the fraction of total flux that links two coils.

Inductors in series behave like resistors in series (assuming no mutual inductance).

Inductors in parallel act like resistors in parallel, assuming no L_M.

The energy stored in an inductor is as follows:

$$W = 1/2 \, LI^2$$

┌──Progress Test

The bracketed number after each question indicates the section of this chapter where the answer can be found.

1. Inductance is the property of an electric circuit that: (*a*) opposes any change to the applied voltage. (*b*) opposes any change in the current through that circuit. (*c*) aids any change in the applied voltage through that circuit. (*d*) aids any change in the current through that circuit. [13–6]

2. The unit of inductance is the: (*a*) ohm. (*b*) farad. (*c*) henry. (*d*) coulomb. [13–6]

3. To increase the inductance of a coil: (*a*) increase the permeability of the core material. (*b*) increase the length of the coil. (*c*) increase the magnitude of the current flow through the coil. (*d*) decrease the cross-sectional area of the coil. [13–6]

4. An inductive circuit of 5 H has a rate of change of current of 4 A in 2 s. What value of cemf is developed? (*a*) 10 V (*b*) 100 V (*c*) 25 V (*d*) 250 V [13–6]

5. When current in an inductor increases, Lenz's law states that the self-induced voltage will: (*a*) tend to increase the amount of current. (*b*) aid the applied voltage. (*c*) oppose the current producing it. (*d*) aid the increasing current. [13–5]

6. If the number of turns in an inductor is doubled and other factors remain unchanged, the inductance is: (*a*) quadrupled. (*b*) doubled. (*c*) tripled. (*d*) the same. (*e*) halved. [13–6]

7. Faraday's law states that the amount of induced voltage in an inductor is: (a) proportional to the strength of the flux. (b) dependent on the amount of current in the inductor. (c) directly proportional to the change in time. (d) proportional to the number of turns. [13-3]

8. The emf induced in an inductor of 200 turns when the flux changes 0.02 Wb in 0.1 s is: (a) 50 V. (b) 40 V. (c) 30 V. (d) 20 V. [13-3]

9. Which of the following statements apply to cemf? (a) It is in phase with the applied voltage. (b) It is greatest when I is at a maximum. (c) It varies inversely with roc i. (d) It is greatest when current begins. [13-5]

10. The following measurements were made on four different inductors. Which is the largest inductor? (a) A 1 A change in 0.1 s induces 10 V. (b) A 50 mA change in 1 ms induces 10 V. (c) A 0.3 A change in 0.01 s induces 35 V. (d) A 25 mA change in 50 μs induces 15 V. [13-6]

11. Which of the following statements apply to mutual inductance? (a) It is affected by the permeability of the core. (b) It varies inversely as the number of turns. (c) It increases as the coefficient of coupling decreases. (d) It is measured in ohms. [13-8]

12. A 0.15 H inductor is in a series-aiding connection with a 0.6 H inductor. What is their total inductance if 0.08 H mutual inductance exists between them? (a) 1.42 H (b) 0.91 H (c) 0.86 H (d) 0.76 H [13-9]

⎾—Problems

Answers to odd-numbered problems are at the back of the book.

1. What is the voltage induced in an inductor of 1500 turns when the flux is changing at the rate of 0.035 Wb per 63 ms?

2. A small inductor has 450 turns. If the current through it causes a flux change of 2.15×10^{-4} Wb in 27 ms, what is the induced voltage?

3. How many turns does an inductor have if 90 V is induced in it when the flux is changing at the rate of 7.6×10^{-3} Wb in 0.02 s?

4. Suppose 165 V is induced in an inductor when the current is changing at the rate of 0.42 A per 2.5 ms. Calculate the value of the inductance.

5. What is the value of inductance if 39 V is induced in it when the current is changing 83 mA in 6 ms?

6. What voltage is induced in an inductor of 47 mH when roc i is 1.6 mA in 750 μs?

7. Calculate roc i when the current through an inductor changes 5.6 mA in 2 ms.

8. What is the di/dt rate of an inductive current that changes from 3.8 to 6.2 μA in 83 μs?

9. For the coil of Figure 13–6, determine the inductance of the coil if the mean magnetic path length is doubled and the turns increased to 500. Assume $\mu_r = 600$.

10. Determine the inductance of an iron core coil with the following parameters: $A = 4 \times 10^{-5}$ m^2, $l_{avg} = 6.5 \times 10^{-2}$ m, $\mu_r = 920$, and $N = 375$.

11. A certain inductor has 0.55 H. The circuit parameters are $A = 8.2 \times 10^{-4}$ m^2, $\mu_r = 390$, $l_{avg} = 7.2 \times 10^{-2}$ m. Calculate the number of turns of the inductor.

12. Determine the relative permeability of the core of a small transformer having the following characteristics: $N = 325$, $l = 2.7 \times 10^{-2}$ m, $A = 1.8 \times 10^{-4}$ m^2, and $L = 860$ mH.

13. Calculate the mutual inductance between two inductors wound on a closed iron core when one has an inductance of 2 H and the other has 5 H.

14. What is the mutual inductance between two coils when one has 30 mH and the other has 65 mH and $k = 0.35$?

15. Two inductors of 45 mH and 80 mH have 22 mH of mutual inductance. What is the coefficient of coupling between them?

16. If 0.74 H of mutual inductance exists between two coils, where one has an inductance of 4.5 H, and the coefficient of coupling is 0.95, what is the inductance of the other?

17. What is the total inductance of three well-shielded, series-connected inductors with values of 2.5, 0.2, and 1.6 H?

18. No mutual inductance exists between three series-connected inductors of 39 mH, 6.8×10^{-4} H, and 18×10^{-3} H. What is their total inductance?

19. What is the total inductance of two parallel-connected inductors of 68 mH and 0.12 H when $k = 0$?

20. Calculate the total inductance between two coils of 270 and 180 μH that are connected in series-aiding fashion when 80 μH of mutual inductance exists between them.

21.　If the inductors described in Problem 20 were connected in series-opposing fashion and all other factors remained the same, what would be the total inductance?

22.　When two inductors are connected in series-aiding fashion, their total inductance is 520 mH. When connected in series-opposing fashion, their combined value is 375 mH. What is the value of mutual inductance between them?

23.　Determine the coefficient of coupling between two coils of 128 and 89 μH when 22 μH of mutual inductance is present.

24.　Calculate the amount of energy stored in an inductor of 4.5 H with a steady-state current of 0.81 A.

25.　What is the average value of current through a 0.18 H inductor that has 0.41 J stored in its field?

26.　Calculate the voltage induced in a 120 mH inductor if the current changes 35 mA in 0.5 ms.

27.　An inductor of 80 μH experiences a current change of 2.5 mA in 10 μs. Calculate the voltage induced across the inductor.

28.　Calculate the inductance of a single-layer, air core coil having the following parameters: $r = 0.125$ in., $l = 1.2$ in., and $N = 60$ turns.

29.　An inductor has the following specifications: $N = 110$, $l = 0.75$ in., and $r = 0.1$ in. Calculate its inductance.

30.　Calculate the number of turns required to provide 68 μH on an air core form, where $l = 0.85$ in. and $d = 0.6$ in.

14

Time Constants

Objectives

After studying this chapter, you will be able to:

Determine the charging and discharging rate of a capacitor.

Calculate the rise and fall of current in an RL circuit.

Explore how voltage transients can be produced in RC and RL circuits for triggering various timing circuits.

Describe the effects of short and long time constants on square wave signals.

Introduction

When current in an inductive circuit is switched on or off, a time interval is required for it to reach its maximum or reduce to zero. The time interval depends not only on the amount of inductance present but also on the value of the series resistance. Likewise, if a voltage is suddenly connected across a capacitor, a short period of time is required to allow the capacitor voltage to reach its maximum value. A discharging capacitor also requires a brief period before the capacitive voltage drops to zero. Again, the amount of resistance in series with the capacitor has a significant effect on the length of the charge and discharge cycle. Details on these topics will be given in the following sections.

Applications

The ignition system of an automobile is one of the most familiar uses of the time constants associated with RL and RC circuits. A conventional ignition system creates the voltage to fire the spark plugs by opening and closing the primary circuit of the spark coil, which is simply a transformer. The transient properties of this RL circuit result in a series of high-voltage, short-duration pulses applied to the spark plugs.

Circuits with long time constants are used to filter out the effects of relatively high frequency signals. Rectifiers, which change ac to dc, require such filters.

The laws of physics governing many natural processes lead to equations very similar to those governing transient electric circuits. When these equations become too complicated to solve mathematically, circuits can be constructed that behave according to equations analogous to those governing the physical problem being studied. The answers to the equations can then be obtained by measuring voltages and currents in the circuit. This procedure is used in the analog computer, and the time constant properties of circuits are of central importance in its applications.

Finally, the synthesizer, which produces electronic music with a tremendous range of tonal qualities, depends heavily for its operation on decaying and rising voltages produced by circuits with time constant properties.

14–1
Time Constant of *RC* Circuits

When a discharged capacitor is suddenly connected across a dc supply, such as E_S in Figure 14–1a, a current immediately begins to flow. At time t_1, the moment the circuit is closed, the capacitor acts like a short circuit. The in-rush current i_C is at its maximum value and is limited only by the series resistance R. However, as soon as the capacitor begins to charge, electrons from the supply begin to build up on the

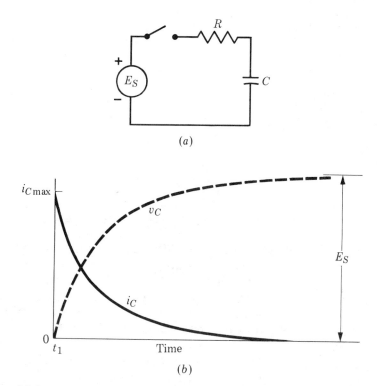

Figure 14-1

Effects in an *RC* circuit: (a) Circuit. (b) Current and voltage waveforms for closed switch.

plate of the capacitor. Simultaneously, electrons from the upper plate are attracted to the positive terminal of the supply. A difference of potential now begins to appear across the capacitor. The polarity, being the same as that of the source, opposes E_S.

> **Exponential curve**   A curve that varies by the square or some other power of a factor instead of linearly.

The net voltage available to charge the capacitor is the difference between the capacitor voltage v_C and E_S. (*Note*: Lowercase letters are used to designate voltage and current values that are changing.) As v_C, the capacitor's emf, increases, this net voltage diminishes. Consequently, the rate of charge slows down. This cumulative effect continues until v_C is approximately equal to E_S, at which time the charging current i_C is reduced to nearly zero. See Figure 14-1b. Theoretically, the capacitor never reaches full charge, but most capacitors can be considered fully charged in several seconds or less. Curves i_C and v_C in Figure 14-1b are examples of **exponential curves**. An exponential curve has the property of dropping or rising very quickly toward a limiting value. The closer it gets to the limit, the more gradual its approach becomes.

14-2
Effect of *R* and *C* on Charging Time

Inasmuch as the charging current must flow through the series resistor, this current inevitably has an effect upon the rate of charge. If the resistance is increased (C unchanged), the IR drop produced by the charging current is greater, and the net voltage charging the capacitor is reduced. Therefore, *the capacitor takes a longer period of time to reach full charge when the series resistance is increased.* Of course, the reverse is true if R is made smaller.

Consider what happens when the series resistance remains constant but the capacitance increases. The initial in-rush of current is the same as was illustrated for Figure 14-1, assuming the same supply voltage. Since the capacitor's charge capacity Q is equal to CE, it is apparent that more time will be required to charge that capacitor because C is larger. The emf across the capacitor builds up more slowly, causing the rate of charge to be slowed down. Hence, the charging time of a capacitor is directly proportional to its capacitance.

14-3
Graphs of Different Charging Times

Figure 14-2 shows two curves that represent different charging times for two different RC circuits. Curve A represents a fast charging time resulting from a small series resistance, small capacitor, or both. Curve B indicates a longer charging cycle resulting from larger capacitors, larger resistors, or both. In both cases, we are to assume that the supply source is the same and that each circuit begins to charge at t_1.

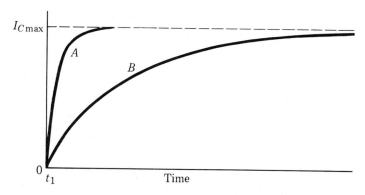

Figure 14–2

Charging curves for capacitors: fast charging (curve A) and slow charging (curve B)

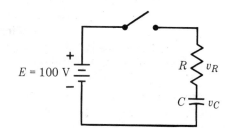

Figure 14–3

Typical RC circuit

Curves A and B are exponential and have the same form as the v_C curve in Figure 14–1b. If the hypothetical circuit assumed for curve A had either its resistance or its capacitance further reduced, the curve would be steeper and would reach the $I_{C\max}$ line faster than before.

Consider the hypothetical circuit represented by curve B. If its series R or C were increased, for example, the steepness of the curve would be reduced. Therefore, more time would be required for the curve to reach $I_{C\max}$.

Thus far, we have given consideration only to the charging current of the capacitor and its emf. What about the voltage across the series resistance? How does it change with respect to v_C? To answer these and other questions, let us examine the RC circuit of Figure 14–3. In Section 14–1, you learned that the in-rush current is maximum when the capacitor is initially connected across the supply, because a discharged capacitor acts like a short circuit for the first instant. Therefore, when the switch is closed in the circuit of Figure 14–3, the initial

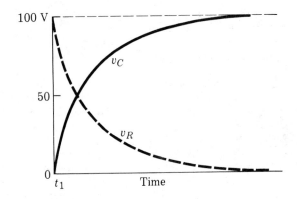

Figure 14–4

Graph of v_C and v_R for RC circuit

charging current produces a voltage drop across R that is approximately equal to that of the supply.

As the charge in the capacitor increases, the charging current drops off exponentially. The voltage drop v_R follows the declining current, as is shown graphically in Figure 14–4. Simultaneously, the capacitive voltage v_C is increasing exponentially. At any instant, the sum of v_R and v_C equals the supply potential and therefore satisfies Kirchhoff's voltage law. We can express this relationship mathematically:

$$E = iR + v_C$$

14–4
Time Constant $\tau = RC$

Whenever a voltage or current constantly changes value, it exhibits *transient effects*. The voltages across the resistance and capacitance in an RC circuit have these characteristics. They are of a transient nature until reaching steady-state values.

Time constant τ   In a capacitor, the time required for a voltage to reach 63.2% of the steady-state or full-charge value. In an inductor, the time required for a current to reach 63.2% of full or steady-state value.

When analyzing the amount of time it takes an RC circuit to reach a steady-state condition, we must deal with a term referred to as the circuit's **time constant**. Expressed mathematically, the time constant τ is as follows:

$$\tau = RC \qquad \qquad 14\text{-}1$$

The time constant τ (Greek lowercase letter *tau*) is expressed in seconds when R is in ohms and C is in farads. That τ is expressed in seconds can be derived as follows:

$$\tau = R \times C = \text{ohms} \times \text{farads} = \frac{\text{coulombs}}{\cancel{\text{volts}}} \times \frac{\cancel{\text{volts}}}{\text{amperes}}$$

$$= \frac{\text{coulombs}}{\text{amperes}} = \frac{\cancel{\text{coulombs}}}{\cancel{\text{coulombs}}/\text{seconds}} = \text{seconds}$$

Now, the circuit's time constant τ *represents the time required for the voltage across the capacitor to reach 63.2% of the steady-state or full-charge value.* It takes four more time constants for v_C to reach a charge value negligibly different from its full-charge value, as demonstrated by the graph in Figure 14–5.

Example 14–1

Calculate the time constant of a series RC circuit where $R = 200$ kΩ and $C = 3$ μF.

Solution

$$\tau = RC = (2 \times 10^5) \times (3 \times 10^{-6}) = 0.6 \text{ s}$$

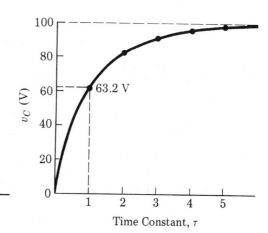

Figure 14–5

Curve of time constants τ in RC circuit

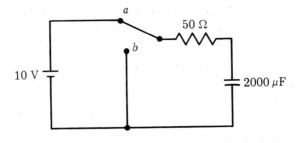

Figure 14–6

RC circuit for Example 14–4

Example 14–2

Determine the time constant of an *RC* circuit when $R = 22 \text{ k}\Omega$ and $C = 0.05 \text{ } \mu\text{F}$.

Solution

$$\tau = (22 \times 10^3) \times (0.05 \times 10^{-6}) = 0.0011 \text{ s} = 1.1 \text{ ms}$$

Example 14–3

What value of resistance must be connected in series with a 20 μF capacitor to provide a τ of 0.1 s?

Solution

$$\tau = RC$$

$$R = \frac{\tau}{C} = \frac{0.1}{20 \times 10^{-6}} = 5 \times 10^3 \text{ } \Omega$$

Example 14–4

Calculate the value of v_R after one time constant has elapsed for the circuit in Figure 14–6. Assume the switch is in position *a*. How long does it take for v_R to reach this value?

Solution

We know that v_C reaches 63.2% of *E* in one time constant. Then:

$$v_C = 0.632 \times 10 \text{ V} = 6.32 \text{ V}$$

Therefore $\quad v_R = E - v_C = 10 - 6.32 = 3.68$ V
$$\tau = 50 \times 2000 \times 10^{-6} = 0.1 \text{ s}$$

14–5
Discharge Current

The time required for a capacitor to discharge is calculated with the same formula that calculates the time required for charging a capacitor. However, when a capacitor is discharging, the time constant represents the time required for v_C to drop 63.2% from its full-charge value. Stated differently, after one time constant, v_C will be equal to 36.8% of full charge. The capacitor current is also discharging at the same exponential rate. If the switch in Figure 14–6 is placed in position b after the capacitor has been fully charged, the rate of current discharge will be as indicated in Figure 14–7.

In some electronics circuits, capacitors are charged relatively slowly through several thousand ohms and then discharged through a low-resistance device such as a strobe or photographic flash. The result is a large, short-duration current pulse. For example, suppose a capacitor is charged to 50 V through a relatively long time constant and then is suddenly connected across a 10 Ω resistive device. The initial current pulse will have an amplitude equal to 50/10, or 5 A!

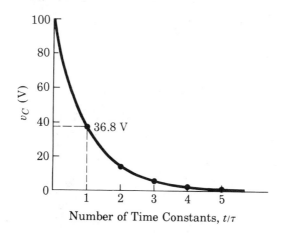

Figure 14–7

Discharge curve for *RC* circuit

14–6
Time Constant of *RL* Circuit

In Chapter 13, you learned that a characteristic of an inductive circuit is to oppose any change of current. When a series *RL* combination is connected across a supply, *voltage and current transients occur until the current attains a steady-state condition.* Consider the circuit of Figure 14–8, where *R* represents the coil's resistance or an external resistance. When the switch is closed, current begins to flow into the inductance. The rate of change of current, roc *i*, will be greatest when the switch is closed. Current and voltage transients will be produced until the current reaches a steady-state level of *E/R*, or 4 A, at which time the coil's effect will be negligible.

The time required for the transient current to reach 63.2% of its maximum value can be calculated by the following equation:

$$\tau = \frac{L}{R} \qquad\qquad \textbf{14–2}$$

where
τ = time constant in seconds
L = inductance in henrys
R = resistance in ohms

The time constant τ also represents the time required for the steady-state current to drop 63.2% when the inductive circuit is opened.

Example 14–5

Determine the time constant of the *RL* circuit in Figure 14–8 when the switch is closed.

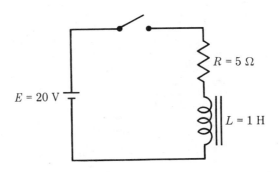

Figure 14–8

Circuit to illustrate *RL* time constant

Solution

$$\tau = \frac{L}{R} = \frac{1}{5} = 0.2 \text{ s}$$

14-7
Voltage Transient Produced in an *RL* Circuit

The time constant of an *RL* circuit changes considerably when it is opened because of the high resistance introduced by the open switch. The open-circuit resistance is not infinite because of leakage resistance around the switch contacts.

Example 14-6 ─────────────────────────────

If the switch is opened in the circuit of Figure 14-8, what is τ if the inductance sees an open-circuit resistance of 5 MΩ?

Solution

The open-circuit resistance is so high that the 5 Ω resistor can be neglected. Therefore:

$$\tau = \frac{L}{R} = \frac{1}{5 \times 10^6} = 0.2 \times 10^{-6} \text{ s}$$

In Example 14-6, notice the extremely short time constant compared with the value when the circuit was closed. Since $v_L = -L(di/dt)$ (from Equation 13-2), the self-induced voltage will be many times larger than the supply potential. The minus sign implies that this cemf is *out of phase* with the supply. This inductive "kick," as it is sometimes called, is often used to *trigger* electronic circuits, as in the firing of spark plugs in the ignition system of automobiles. This high voltage produces arcing across switch contacts and can cause them to become pitted and burned unless an appropriately sized capacitor is placed across these contacts. The capacitor will absorb the charge and then discharge it when the switch is closed again.

14–8
Universal Time Constant Curve

Basic Concepts

Because of the identical nature of the transient responses in RL and RC circuits, a common graph may represent both, as in Figure 14–9. The X axis represents time constants, and the Y axis represents percentage of full current or voltage. Observe that curves A and B have the same shape as the other curves previously shown in this chapter. For RL circuits, curves A and B represent i_L and v_L, respectively. For RC circuits, curves A and B represent v_C and i_C, respectively. For both cases, the rise or fall of the curve changes by 63% in one time constant.

With information obtained from the graph, it is possible to determine the voltage across a capacitor and its charge at any time constant, or fraction thereof, during the charging or discharging cycle. The same statement applies to the inductive circuit.

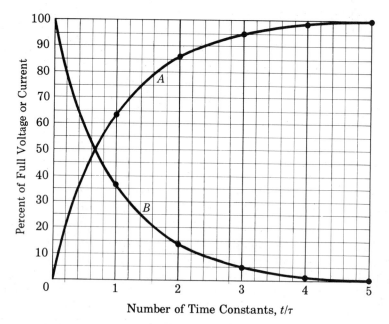

Figure 14–9

Universal time constant graph for RC and RL circuits

Example 14-7 ———————————————————————

If an RC circuit is connected across a 100 V supply, what will v_C be in a time equal to 0.5τ?

Solution

Graphically, this voltage can be determined from curve A of the universal time constant graph (Figure 14-9). Locate 0.5τ on the X axis and project vertically upward until intersecting curve A. Read 40% opposite this point on the Y axis. Therefore:

$$v_C = 40 \text{ V}$$

Example 14-8 ———————————————————————

If an RL circuit has an $I_{max} = 5$ A, what will the current be after 2.5τ when the circuit is opened?

Solution

Locate 2.5τ on the X axis and project upward to curve B. Read approximately 8% opposite this point on the Y axis. Therefore:

$$i_L = 0.08 \times 5 \text{ A} = 0.4 \text{ A}$$

Exponential Nature of Time Constant Graph

The universal time constant graph is based on the following equation, which gives the exponential voltage rise in a capacitive circuit and is derived from the calculus:

$$v_C = E\,(1 - \epsilon^{-t/RC}) \qquad\qquad \textbf{14-3}$$

where E = supply voltage
           ϵ = the base for the Napierian system of logarithms
            and equals approximately 2.7183
    $-t/RC$ = ratio of time to RC time constant

Note: The value of $-t/RC$ is the ratio of actual time of t to the RC time constant.

      The next example illustrates the use of this equation when the time elapsed t is just equal to one time constant – in other words, when $t = RC = \tau$.

Example 14–9 ─────────────────────────────

Determine the rise of voltage across a capacitor in a series RC circuit in one time constant.

Solution

For 1τ, $-t/RC = 1$. Therefore:

$$v_C = E(1 - \epsilon^{-t/RC}) = E(1 - \epsilon^{-1}) = E(1 - 0.3679) = 0.6321E$$

The voltage across the capacitor will rise to 0.6321, or 63.21% of the supply voltage in RC seconds, the value of 1 on the time constant curve. By plotting v_C for different time constants, we obtain the universal curve A of Figure 14–9.

The charging current in a series RC circuit can be calculated for any time constant with the following formula:

$$i = I_{max} \, \epsilon^{-t/RC} \qquad\qquad \textbf{14–4}$$

This equation is the *decreasing* form of the exponential curve (curve B in Figure 14–9).

Example 14–10 ─────────────────────────────

Calculate the value of capacitive current in a series RC circuit in one time constant.

Solution

For 1τ, $-t/RC = 1$. Therefore:

$$i_C = I_{max} \, \epsilon^{-t/RC} = I_{max} \, \epsilon^{-1} = 0.3679 I_{max}$$

The charging current in an RC circuit will have dropped to 0.3679, or approximately 36.8% of its maximum E/R value in one time constant after charging begins. If different time constants are plotted, curve B of Figure 14–9 results.

The exponential curve indicates that a rising current, in approaching a maximum value in a series of time constants, will only increase 63.2% of the remaining value in each successive time constant.

Example 14-11

For the circuit in Figure 14-10, calculate the voltage across the capacitor in 2 s.

Solution

$$v_C = 500\,(1 - \epsilon^{-t/RC}) = 500\,(1 - \epsilon^{-2/0.8}) = 500\,(1 - \epsilon^{-2.5})$$
$$= 500\,(1 - 0.082) = 458.95 \text{ V}$$

For the natural logarithm (base ϵ) of 2.5, the keying sequence on a T1-55-II scientific calculator is as follows:

$$2.5 \boxed{+/-} \boxed{\text{Inv}} \boxed{\text{Inx}} = 0.0820$$

Example 14-12

For the circuit in Figure 14-10, calculate the voltage across the capacitor in 3.5 s.

Solution

$$v_C = 500\,(1 - \epsilon^{-3.5/0.8}) = 500\,(1 - \epsilon^{-4.375})$$
$$= 500\,(1 - 0.01259) = 493.7 \text{ V}$$

If we want to find the voltage across R at the end of 3.5 s, we can use KVL:

$$v_R = 500 - 493.7 = 6.3 \text{ V}$$

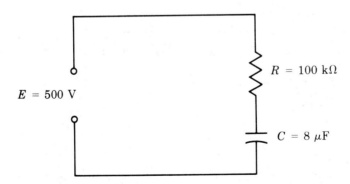

Figure 14-10

Circuit for Examples 14-11 and 14-12

For the series RL circuit, the following formula is used to calculate the inductive current at any instant of time:

$$i = I(1 - \epsilon^{-Rt/L})$$ 14-5

Example 14-13

Refer to Figure 14-11. Calculate i_L at a time 0.2 ms after the circuit is closed.

Solution

First, find Rt/L:

$$\frac{Rt}{L} = \frac{100 \times 0.2 \times 10^{-3}}{20 \times 10^{-3}} = -1$$

Now, calculate i_L:

$$i_L = \frac{10 \text{ V}}{100 \text{ }\Omega}(1 - \epsilon^{-1}) = 0.1\,(1 - \epsilon^{-1}) = 0.1 \times 0.632 = 0.0632 \text{ A}$$

This value represents the current in the circuit after one time constant.

Example 14-14

Refer to Figure 14-12. With S_2 in position a, what will v_C be 3 s after S_1 closes?

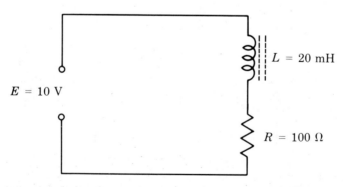

$E = 10$ V

$L = 20$ mH

$R = 100\ \Omega$

Figure 14-11

Series RL circuit for Example 14-13

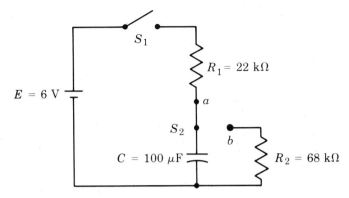

Figure 14–12

Circuit for Examples 14–14 and 14–15

Solution

Here, $RC = 0.022 \ \text{M}\Omega \times 100 \ \mu\text{F} = 2.2$ s. Then:

$$v_C = E \,(1 - \epsilon^{-t/RC}) = 6 \,(1 - \epsilon^{-3/2.2}) = 6 \,(1 - \epsilon^{-1.3636})$$
$$= 6 \,(1 - 0.2557) = 4.466 \text{ V}$$

Example 14–15

Refer to Figure 14–12. If S_2 is suddenly placed to position b, what will the capacitor voltage be 2 s later?

Solution

We must use Equation 14–4 (which is the decreasing form of the exponential curve) but substitute v for i and E_{max} for I_{max}:

$$v_C = E_{\text{max}} \ \epsilon^{-t/RC} \qquad\qquad \textbf{14–6}$$

Now, $E_{\text{max}} = 4.466$ V from Example 14–14 and $RC = 0.068$ $\text{M}\Omega \times 100 \ \mu\text{F} = 6.8$ s. Then:

$$v_C = 4.466\epsilon^{-2/6.8} = 4.466\epsilon^{-0.294} = 4.466 \times 0.745 = 3.328 \text{ V}$$

The percentages of the steady-state values reached after the passage of some of the more commonly used multiples of the time constant are tabulated in Table 14–1. For example, for $\tau = 0.5$, the percentage is 40%.

Table 14–1 τ as Percentage of Ultimate Value

τ	% Ultimate Value	τ	% Ultimate Value
0.1	10	2	86.5
0.5	40	3	95.1
0.7	50	4	98.2
0.9	60	5	99.3
1.0	63.2		

14–9
Short and Long Time Constants

Some characteristics of series RL and RC circuits can be obtained by connecting positive- and negative-going pulses to them. Normally, these pulses would be generated electronically by a *multivibrator* or *square wave generator*. Such a wave can be produced, at slow rates, by simply closing and opening a switch controlling power to a resistive circuit, as in Figure 14–13*a*. The waveform that results is shown in Figure 14–13*b*.

Analysis of Series *RL* Circuit Connected to a Square Wave

The following discussion is an analysis of a series RL circuit connected to a square wave whose period is five times (or more) longer than τ. In the circuit of Figure 14–14*a*, when the switch is in the off position, no

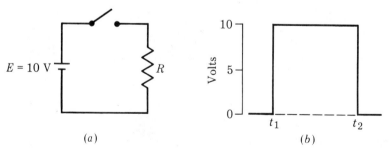

(a) (b)

Figure 14–13

Generating pulses: (a) Circuit. (b) Resulting waveform.

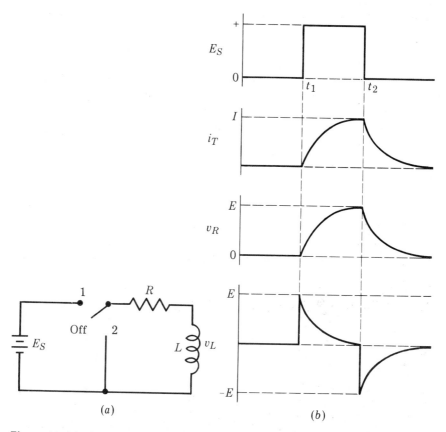

Figure 14–14

RL time constant: (a) Circuit. (b) Waveforms.

voltage is impressed across the circuit and there is no current. Now, assume the switch is moved to position 1 at time t_1. The full supply potential E_S immediately appears across the series RL combination. The current that begins to flow produces the exponential waveform i_T (see Figure 14–14b). This same current produces a voltage drop v_R across R. Observe that i_T and v_R have the same shape. Since the rate of change of current (roc i) is greatest at t_1, the voltage v_L across the inductor L is maximum and *decreases* at an exponential rate. After approximately 5τ, a steady-state condition is reached.

If the switch is suddenly placed in position 2, the supply voltage E_S is now removed, and the voltage across the series RL combination drops to zero. [Placing the switch in position 1 (t_1) and then in position 2 (t_2) effectively places a square wave across RL.] At t_2, the supply current is interrupted and no longer sustains the magnetic field established around the inductor. The field collapses, generating a cemf, as illustrated by v_L, which tries to maintain the original current. The current i_T (and consequently v_R) falls off exponentially. The waveforms indicated in Figure 14–14b are predicated on the switching square wave having $\tau \cong 5$. If the time constant of the circuit were made much smaller relative to the duration of the wave E_S, the pulses would be considerably steeper and of shorter duration. The reverse would occur if the time constant were made five to ten times longer.

Analysis of Series *RC* Circuit Connected to a Square Wave

Suppose a series RC circuit, as shown in Figure 14–15a, is connected to a square wave whose period is five times (or more) longer than τ. The resulting waveforms are shown in Figure 14–15b.

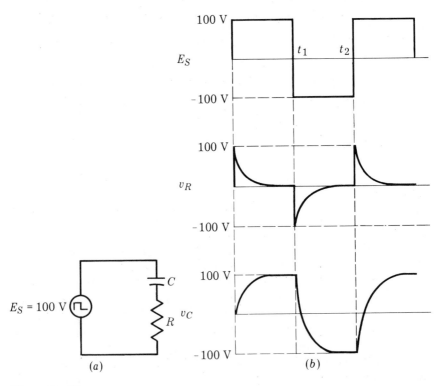

Figure 14–15

RC time constant: (a) Circuit. (b) Waveforms.

Example 14–16

A certain pulse has a repetition rate of 400 pulses per second (pps). What value of time constant would be considered short? What value would be considered long?

Solution

We must first calculate the duration of one pulse:

$$T = \frac{1}{f} = \frac{1}{400} = 0.0025 \text{ s} = 2.5 \text{ ms}$$

Here, f is the frequency in cycles per second (in this case, pps). Therefore:

$$\text{short time constant} = 0.1 \times 2.5 = 0.25 \text{ ms}$$
$$\text{long time constant} = 10 \times 2.5 = 25 \text{ ms}$$

Series RC circuits are more widely used than RL circuits. The pulse, or *spike*, shown in the v_R waveform of Figure 14-15b is frequently used to trigger various kinds of timing circuits. The current transient for this circuit is of the same shape as the v_R waveform, since $v_R = i_C R$.

When a series RC circuit is connected to a square wave whose period is long compared with the circuit's τ (say, ten times τ or more), then the waveforms illustrated in Figure 14-16 are obtained. Note that

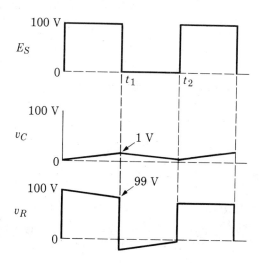

Figure 14–16

Waveshapes for RC circuits with long time constants

v_R has very nearly the same waveshape as the input signal E_S. The circuit current, not shown, would have the identical waveform as v_R, since $v_R = i_C R$. Since the capacitor has very little time to charge before the input pulse goes in the opposite direction, its charge is very small, typically no more than 1 V.

⌐Summary

The in-rush current to a discharged capacitor is limited by the series resistance.

The charging rate of a capacitor is limited by the amount of series resistance.

Capacitors charge at an exponential rate.

The instantaneous sum of the resistor voltage (v_R) and capacitor voltage (v_C) equals the supply voltage.

Time constant τ equals RC.

Time constant τ equals the time required for a capacitor to reach 63.2% of full charge. It also represents the time required for an inductive current to rise to 63.2% of maximum.

Time constant τ also represents the time required for a capacitor to discharge to 36.8% of its maximum value or an inductor to have its current reduced by the same percentage.

The changing capacitive and inductive currents and voltages are referred to as transients.

An inductive circuit produces a very large cemf the instant the circuit is opened.

Universal time constant curves are used to determine the value of i or v in the circuit at any instant during the charge or discharge cycle.

For practical purposes, a capacitor can be said to be fully charged in 5τ.

An inductive current reaches full value in 5τ (practically speaking).

The output waveform of an RC circuit will approximate the input waveform when a long time constant τ is used (voltage taken across R).

Short time constants differentiate the wave and produce short transients or spikes.

A short time constant τ is approximately 0.1 times the duration of the input cycle.

A long time constant τ is approximately 10 times the duration of the input cycle.

⌐─Progress Test

The bracketed number after each question indicates the section of this chapter where the answer can be found.

1. A coil has a resistance of 22 Ω, and in 0.1 s after the switch is closed, the current has reached 63.2% of its final value. The value of inductance is: (a) 0.22 H. (b) 0.44 H. (c) 2.20 H. (d) 4.40 H. [14–6]

2. To increase the time required for the current to reach its steady-state value in a series RL circuit, we must: (a) increase R. (b) decrease R. (c) decrease L. (d) increase the applied voltage. [14–6]

3. The τ of a series RL circuit is: (a) the time required for dc to rise to one-half the maximum value. (b) a method of determining the amount of current after 1 s. (c) the product of the resistance and inductance of a circuit. (d) the time required for the current to reach 63.2% of its steady-state value. [14–6]

4. The difference between E_S and v_C is the: (a) voltage across the series resistor. (b) charge on the capacitor. (c) voltage across the inductor. (d) supply potential. [14–1]

5. The voltage on a 0.82 μF capacitor is 24 V. Its charge is: (a) 0.034×10^{-4} C. (b) $1.97 \times 10^{-5}Q$. (c) $2.926 \times 10^{2}Q$. (d) $3.14 \times 10^{-4}Q$. [14–2]

6. A 0.47 μF capacitor is in series with a 68 kΩ resistor. If 60 V is connected across the combination, the voltage across the capacitor in one time constant is: (a) 1.92 V. (b) 31.96 V. (c) 37.9 V. (d) 40.62 V. [14–4]

7. The time constant of a 0.33 μF capacitor in series with a 27 kΩ resistor is: (a) 8.91 s. (b) 0.89 s. (c) 0.089 s. (d) 8.9 ms. [14–4]

8. A series RC circuit has a time constant of 39 ms. If $C = 0.22$ μF, the resistance is: (a) 177.3 kΩ. (b) 1.77 MΩ. (c) 17.7 kΩ. (d) 0.017 MΩ. [14–4]

9. The time constant of a series RL circuit of 15 Ω and 3.5 H is: (a) 2.33 s. (b) 23.3 ms. (c) 0.233 s. (d) 2.33 ms.   [14–6]

10. A series RL circuit of 2.2 kΩ and 1 H inductance is connected to a 90 V dc source. The current through the inductor in 2τ is: (a) 25.91 mA. (b) 28.46 mA. (c) 35.5 mA. (d) 37.07 mA.   [14–6]

┌──Problems

Answers to odd-numbered problems are at the back of the book.

1. Determine the time constant of a series RC circuit where R = 2.7 kΩ and C = 0.068 μF.

2. How long does it take for the voltage across a capacitor to rise to 63.2% of E_{max} in a series RC circuit where R = 3.9 kΩ and C = 470 pF?

3. How much capacitance must be connected in series with 680 kΩ to provide a 1.5 s time constant?

4. Calculate the required resistance to be used with a 560 pF capacitor to provide a 250 μs time constant.

5. A filter choke has a resistance of 42 Ω. In 0.25 s after the circuit is closed, the current through it reaches 63.2% of its maximum value. What is the value of inductance?

Use the universal time constant graph for Problems 6 through 11.

6. When an emf is applied to an inductor, the current finally stabilizes at 36 mA. Determine the magnitude of the current at τ = 1, 2, 3, 4, and 5 after the voltage was applied.

7. A series RL circuit is connected across a 30 V dc source. Determine the value of the cemf (v_L) at τ = 1, 2, 3, 4, and 5.

8. How many time constants are required for a current to reach 50% of its steady-state value in a series RL circuit after a dc voltage is applied?

9. How many time constants are required for the cemf across an inductor to drop to 25% of its maximum value after the emf is applied?

10. Refer to Figure 14–17. Calculate the number of time constants required for the current to reach 1.7 A after the circuit is closed.

11. How many seconds are required for the current to attain the value indicated in Problem 10?

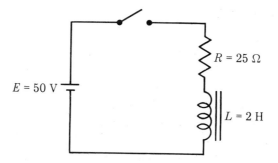

Figure 14–17

Circuit for Problem 10

12. Calculate the instantaneous voltage across a capacitor in a series RC circuit 2τ after the circuit is closed.

13. A series RC circuit is suddenly connected to a 200 V dc source. If $R = 180$ kΩ and $C = 4.7$ μF, what will v_C equal in 1.5 s?

14. A series circuit consists of $R = 220$ kΩ and $C = 0.15$ μF. (a) Calculate v_C 20 ms after a 50 V dc supply is applied to the circuit. (b) What will v_R equal at that instant?

15. In the series RL circuit shown in Figure 14–11, suppose $E = 24$ V, $R = 180$ Ω, and $L = 50$ mH. Calculate i_L 0.1 ms after the circuit is closed.

16. A series circuit consists of $L = 0.25$ H and $R = 10$ Ω. Calculate i_L 50 ms after a 6 V dc supply is connected.

17. Refer to the basic circuit shown in Figure 14–12, with S_2 in position a. If $R_1 = 39$ kΩ, $C = 68$ μF, and $E = 18$ V, what will v_C be 2 s after S_1 closes?

18. Refer to Problem 17 and Figure 14–12. If S_2 is suddenly placed in position b, and if $R_2 = 56$ kΩ, what will v_C equal 3 s later?

19. A series RC circuit is connected across a 200 V dc supply. What will be the value of v_C for $\tau = 0.3$?

20. A certain pulse has a repetition rate of 1200 pps. (a) What would be considered a short time constant for this pulse? (b) What would be a long time constant?

21. A 100 V dc source is connected across a 5.6 MΩ resistor in series with a 2.2 μF capacitor. How long will it take v_C to reach 63 V?

22. A 0.05 μF capacitor has a charge of 132 V. If it is discharged through a 20 kΩ resistor, how much time is required for v_C to reach 66 V?

15

Sinusoidal Alternating Currents

Objectives

After studying this chapter, you will be able to:

Recognize basic ac waveforms.

Determine frequencies of basic ac waveforms.

Calculate the average, effective, and peak values of a sine wave.

Introduction

Our studies have dealt only with direct current (dc) circuits until now. While dc has many applications, the uses of alternating currents (ac) and voltages are just as numerous. In this chapter, we will investigate the characteristics of alternating current waves. While many kinds of ac waves exist (square, triangular, sawtooth, complex), our interests will be directed toward *sine waves*. They are the most frequently encountered type of ac wave, and the mathematics involved in their analysis is relatively straightforward.

Applications

Alternating currents are easily generated and can be efficiently conducted over long distances with small power losses, as demonstrated by the utility companies. They can be converted to dc, where necessary, with ease. On the other hand, conversion from dc to ac is much more difficult.

Many circuits in radios and TVs require the use of ac for their operation, which is generated internally by circuits called oscillators. The whole field of radio communication depends largely on the capability of high-frequency ac to radiate its energy from an antenna into space.

The use of electric currents and voltages that vary in a *sinusoidal* manner is universal. A simple analogy of the repetitive pattern of these ac voltages and currents is the ripples created by dropping a pebble into a pond. More complex waves, such as those used in voice communications or stereo systems, can be mathematically analyzed as being made up of a number of sine waves of differing frequencies and amplitudes.

15–1
Generation of Sine Waves

An ac voltage can be generated by rotating a loop through the flux established between the north and south poles of a magnet. Such an arrangement is shown in Figure 15–1. The ends of the magnet are called the *field poles*. The loop and its supporting structure is designed to rotate freely through the flux and is called the *armature*. Appearing at the ends of the loop will be a voltage that varies in amplitude and periodically reverses itself. This cycle will continue as long as the coil is driven through the flux. A voltage of this type is called an *alternating voltage*.

The loop for an ac voltage consists of many turns of wire so that a large emf can be produced. The free ends of the coil are connected to cylindrical pieces of metal called *slip rings*, which rotate with the loop. Rigidly attached to the frame of this machine, called an *alternator*, are two carbon brushes that ride on the slip rings and provide electric connections to the load.

As the loop rotates on its axis, the magnitude of voltage appearing at the slip rings will vary. In Figure 15–1a, the loop is shown cutting perpendicularly through the maximum lines of flux. Now, the emf induced into one leg (side) of the loop will be opposite in polarity to that induced in the other leg. However, these voltages are *series aiding*, and

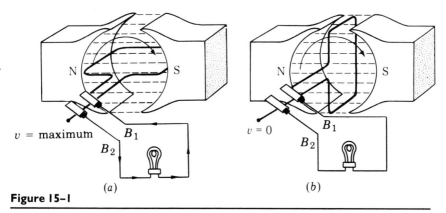

Figure 15–1

Alternating voltage produced by loop rotating in a magnetic field: (a) Rotation perpendicular to flux. (b) Rotation parallel to flux.

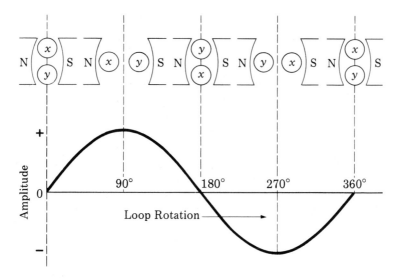

Figure 15–2

Variation of output voltage from simple alternator

therefore, they increase the output voltage. In Figure 15–1b, the loop has rotated 90° and at this instant is moving parallel to the flux lines. No induced voltage is generated in the legs of the loop in this position.

As mentioned in Section 13–2, the amount of voltage induced in a conductor is determined by several factors, including the angle at which it cuts through the flux. Consequently, the induced emf in the loop will vary uniformly between the limits of maximum and zero voltage. Figure 15–2 correlates several positions of the loop and magnetic flux with the amplitude and polarity of the induced voltage. Observe the *reversed polarity* of the ac voltage after the loop completes 180° of rotation. The opposite sides of the loop are designated as x and y so that their angular positions can be easily identified.

15–2
Definition of Cycle

> **Cycle**   The change of an alternating wave from zero to a positive peak to zero to a negative peak and back to zero.

    An ac voltage that starts at zero, rises to a maximum, returns to zero, rises to maximum in the reverse direction, and then returns to zero is said to have completed a **cycle**. This sequence of events is shown in Figure 15–2. In other words, *one cycle represents one complete revolution of the armature* of our simple alternator. One cycle of a voltage, then, is generated every time the armature completes 360°, and the same 0°–360° range is used to describe the successive times along the cycle.

    For convenience, we assumed a cycle starts at 0°. However, we can consider a cycle as starting *anywhere* in its 360° rotation. For example, it may be desirable to consider the cycle as starting at 45° or 90°. In any event, we cannot consider the cycle completed until we

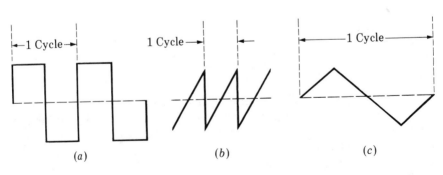

(a)   *(b)*   *(c)*

Figure 15–3

Representation of one cycle for three waveshapes: (a) Square. (b) Sawtooth. (c) Triangular.

continue to the corresponding point on the next cycle. The voltage cycle must go through 360° from whichever starting point is selected.

While our study has focused on the cycle shown in Figure 15-2, we have cycles of other kinds of voltages. For example, Figure 15-3 illustrates several different *waveshapes*, each marked off in cycles. The cycle could be marked off from any two corresponding parts on adjacent waves.

15-3
Radian Measure

> **Radian (rad)**   In a circle, the angle included within an arc equal to the radius of a circle. Numerically, the angle is 57.3°.

Angles are frequently measured in radians, or π measurements, rather than degrees. This natural system of angular measure is used in virtually all of the formulas associated with ac because it is more convenient.

A **radian (rad)** is an angle that, when its vertex is on the center of a circle, intercepts an arc equal to the length of the radius of the circle. This definition is illustrated in Figure 15-4, where the central angle AOB is equal to 1 rad because arc AB is equal to the radius OA. Angle AOB is usually referred to as Θ (Greek capital letter *theta*). The circumference around a circle equals $2\pi r$. Therefore, a circle includes 2π rad since one radian angle equals one length r of the circumference. Hence, one cycle is equal to 2π rad.

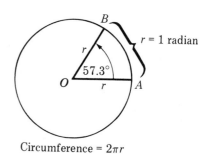

Figure 15-4

Radian measure

Circumference = $2\pi r$

It is frequently convenient to know how many degrees constitute an angle equal to 1 rad. This conversion can be determined as follows:

$$2\pi \text{ rad} = 360°$$

$$1 \text{ rad} = \frac{360°}{2\pi} = \frac{180°}{\pi} = 57.3° \qquad \textbf{15-1}$$

Example 15-1

How many radians are there in two cycles?

Solution

$$1 \text{ cycle} = 2\pi \text{ rad}$$
$$2 \text{ cycles} = 4\pi \text{ rad} = 12.56 \text{ rad}$$

Example 15-2

How many radians are there in 1°?

Solution

$$\frac{\pi}{180°} = 0.01745 \text{ rad}$$

Example 15-3

Convert 60° to radians.

Solution

$$60 \times 0.01745 = 1.047 \text{ rad}$$

15-4
Frequency, Angular Velocity, Period, and Wavelength

Frequency   The number of cycles an ac voltage or current completes each second.

The number of cycles an ac voltage or current completes each second is its **frequency**. The English system used cycles per second (cps) as the unit of frequency measurement. The SI unit is the hertz, abbreviated Hz, named after Heinrich Hertz, a German physicist (1857–1894) who defined the half-wave dipole and built the first radio transmitter. *One hertz equals one cycle per second.* If a voltage or current completes 1000 cps, its frequency is 1000 Hz, or 1 kHz.

Example 15–4

If an alternator completes 1800 revolutions per minute (rpm), what is the frequency of its current waveform?

Solution

$$\frac{1800}{60 \text{ s}} = 30 \text{ Hz}$$

—**Angular velocity**   The rate at which an angle changes, as in the armature of an alternator, expressed in radians per second.

It is apparent from Section 15–2 that frequency is directly related to speed of rotation. Now, the velocity of an object moving uniformly in a straight line is defined as the distance traveled per unit of time. Thus, velocity is measured in feet per second, miles per hour, and so on.

Circular velocity, as in the armature of an alternator, is called **angular velocity** and is symbolized by ω (Greek lowercase letter *omega*). Angular velocity is measured by determining the size of the angle the object generates per unit of time. It is measured in degrees per second or radians per second, the latter being most common.

Angular velocity is obtained by multiplying the number of radians per cycle (2π) by the number of cycles per second (the frequency). Expressed mathematically, angular velocity is as follows:

$$\omega = 2\pi f \qquad\qquad 15\text{--}2$$

where $f =$ frequency in hertz
$2\pi =$ a constant $= 6.28$
$\omega =$ angular velocity in radians per second

Example 15–5 ────────────────────────────────

What is the angular velocity of an ac current having a frequency of 2 kHz?

Solution

$$\omega = 2\pi f = 6.28 \times 2 \times 10^3 = 1.256 \times 10^4 \text{ rad/s}$$

┌─ **Period**   The length of time needed for one cycle.

The length of time needed for one cycle is the **period**, and its symbol is T, for time. The period, or time, of one cycle and its frequency are reciprocals of each other. That is,

$$T = \frac{1}{f} \quad f = \frac{1}{T} \qquad\qquad \text{15–3}$$

Therefore, as the frequency increases, the period becomes smaller.

Example 15–6 ────────────────────────────────

What would be the frequency of the wave in Figure 15–2 if its period were 2 ms?

Solution

$$f = \frac{1}{T} = \frac{1}{2 \times 10^{-3}} = 500 \text{ Hz}$$

Example 15–7 ────────────────────────────────

What is the period of a frequency of 810 kHz?

Solution

$$T = \frac{1}{f} = \frac{1}{8.1 \times 10^5} = 1.23 \times 10^{-6} \text{ s}$$

┌─ **Wavelength**   The spatial length of one complete cycle.

In a periodic wave, the distance between points of corresponding phase of two consecutive cycles is called the **wavelength** and is represented by the Greek lowercase letter *lambda*, λ. In Figure 15–5, a wavelength is shown that has the length of one complete wave or cycle. The wavelength of a wave is related to its velocity of propagation and frequency by the following formula:

$$\lambda = \frac{v}{f}$$ 15–4

where λ = one wavelength in meters
             v = velocity of propagation in meters per second
             f = frequency in hertz

The velocity of electromagnetic *radio waves* through free space is equal to the speed of light, or 3×10^8 m/s. Therefore:

$$\lambda = \frac{3 \times 10^8}{f_{\text{Hz}}} \text{ m}$$ 15–5

Example 15–8

What is the wavelength of a radio wave with a frequency of 44.1 MHz?

Solution

$$\lambda = \frac{3 \times 10^8}{44.1 \times 10^6} = \frac{3 \times 10^2}{44.1} = 6.8 \text{ m}$$

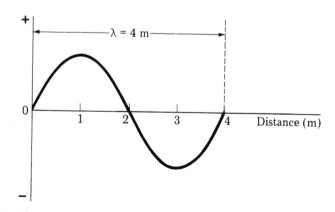

Figure 15–5

Wavelength λ for one cycle

The same basic formula for calculating wavelength (Equation 15–4) can be used to find the wavelength of *sound waves*. Because they are transmitted by mechanical vibrations, their velocity is much slower than that of radio waves. At a temperature of 20°C, their velocity through air is 1130 ft/s. Therefore, the wavelength of a sound wave can be calculated from the following equation:

$$\lambda = \frac{1130}{f_{Hz}} \qquad\qquad \textbf{15–6}$$

Example 15–9

Calculate the wavelength of a 1000 Hz tone.

Solution

$$\lambda = \frac{1130}{1000} = 1.130 \text{ ft}$$

15–5
Frequency Spectrum

Frequency spectrum   A continuous range of electromagnetic radiation from the longest known radio waves to the shortest. For audio, those frequencies between the lowest and highest in the audible range.

Alternating voltages and currents vary over an extremely broad **frequency spectrum**. At the low end are audio frequencies (AF), which range from 20 Hz to 20 kHz. Frequencies above about 300 Hz provide treble notes; below this frequency are the bass tones. The higher the frequency, the higher is the tone or pitch of the sound. Loudness is determined by the amplitude of a wave.

Low-frequency radio waves start above the audio spectrum and continue up toward the frequencies of light. Radio frequencies (RF) can be transmitted by electromagnetic waves and travel 186,000 mi/s, or 3 × 10⁸ m/s. Some of the more common frequency bands are listed in Table 15–1. (*Note:* G = giga = 10^9.)

Table 15–1 Frequency Spectrum

Designations and Abbreviations	Frequency Range
Power frequencies	50 to 400 Hz
Audio frequencies (AF)	20 to 20,000 Hz
Very low frequencies (VLF)	15 to 30 kHz
Low radio frequencies (LF)	30 to 300 kHz
Medium frequencies (MF)	300 to 3000 kHz
High frequencies (HF)	3 to 30 MHz
Very high frequencies (VHF)	30 to 300 MHz
Ultrahigh frequencies (UHF)	300 to 3000 MHz
Superhigh frequencies (SHF)	3 to 30 GHz
Extremely high frequencies (EHF)	30 to 300 GHz

15–6
Sine Wave

Sine Wave   A wave that varies according to the sine function of an angle over a linear period of time.

The wave in Figure 15–5 is characterized by a periodic rising above and falling below a reference line called a *time base*. The rate at which the wave is constantly changing conforms to the sine function of an angle and is therefore called a **sine wave**. (For a brief review of sine, cosine, and tangent functions, refer to the Appendix.)

The graph in Figure 15–6 is that of a sine function and represents one complete cycle, or 360°. Trace the wave starting at 0°. Observe that when it has gone from 0° to 30°, it has already reached 0.5, or half of its ultimate amplitude. At 60°, the wave has an instantaneous value of 0.866 of its peak or maximum value. At 90° and 270°, the wave has reached its *maximum instantaneous value* in the positive and negative directions. Note that at 120° the amplitude is the same as at 60°, and 150° is the same as 30°.

By inspection, it is evident that the negative half of the cycle is inverted and displaced 180° in time (π rad) with respect to the first half cycle. Observe that the amplitude of the sine wave at 210° and 330° is

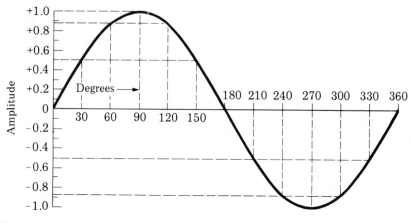

Figure 15–6

Graph of sine function

the same as it is at 30° and 150°, but with reversed polarity. Other corresponding points could easily be established throughout the complete sine wave cycle. By exercising a little care, we can graphically determine the approximate sine of any angle from the graph in Figure 15–6.

Example 15–10

From the sine function graph in Figure 15–6, approximate the sine of 60°.

Solution

By inspection, the sine ≅ 0.87. Using a scientific calculator, you can verify that the sine of 60° is 0.866.

15–7
dc and ac Symbols

In a study of alternating current, it is necessary to establish a nomenclature to identify voltages and currents that have steady values versus those that depend on an exact instant of time. In the preceding chapters, the voltage sources, currents, voltage drops, and power did not depend on any exact instant of time. These steady values were represented by uppercase letter symbols. In ac circuits, these values can vary from one instant to another; their exact value depends on the instant in time that the measurement or calculation is made. Such instantaneous values are represented by lowercase letters. The following statements summarize these concepts:

1. Instantaneous values, which depend on an exact instant of time for their numerical value, are represented by lowercase letters.
2. Steady-state values, which are not dependent on any instant of time for their numerical value, are represented by uppercase letters.

Voltage sources are designated by E, which is the letter symbol for an emf, and represent a potential rise due to the presence of generators or batteries in a circuit. The letter symbol for voltage drop is V. Thus, we can write $E = V_1 + V_2 + V_3 + \cdots + V_n$, indicating that the algebraic sum of the emf's must equal the algebraic sum of the voltage drops around the circuit.

15–8
Peak and Instantaneous Values

> **Peak value**   The amplitude of a wave from the zero or reference axis to the maximum point in either direction.

> **Peak-to-peak value**   The amplitude of an alternating wave measured from positive peak to negative peak.

> **Instantaneous value**   The exact value of a voltage or current at a particular instant in the cycle.

The **peak value** of a sine wave is its amplitude from the zero or reference axis to the maximum point in either direction. Figure 15–7 shows the peak value of a sine wave. The **peak-to-peak value** (abbreviated *pp*) is twice the peak value, because of the symmetrical nature of

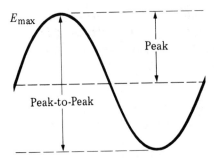

Figure 15–7

Peak and peak-to-peak values

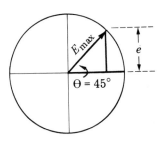

Figure 15–8

Voltage $e = E_{max} \sin \Theta$

each half of the wave. Oscilloscopes are generally used to measure peak-to-peak values, although some ac voltmeters have peak-to-peak scales.

There are many times when it is necessary to know the exact value of a sine voltage at a particular instant in the cycle. An **instantaneous value** can be calculated as follows:

$$e = E_{max} \sin \Theta \qquad \textbf{15–7}$$

where
$$e = \text{instantaneous voltage}$$
$$E_{max} = \text{maximum or peak voltage}$$
$$\sin \Theta = \text{sine of the desired angle}$$

Example 15–11

Calculate the instantaneous value of a sine wave at 45° when its peak value is 100 V. See Figure 15–8.

Solution

$$e = E_{max} \sin \Theta = 100 \times 0.707 = 70.7 \text{ V}$$

The equation for calculating the instantaneous value of a sinusoidal current is as follows:

$$i = I_{max} \sin \Theta \qquad \textbf{15–8}$$

Example 15-12

What is the instantaneous value of a sinusoidal current at $15°$ if $I_{max} = 1.5$ A?

Solution

$$i = I_{max} \sin \Theta = 1.5 \times 0.2588 = 0.3882 \text{ A}$$

In Section 15-4, we saw that the angular velocity of a wave can be expressed as $\omega = 2\pi f$. Since $e = E_{max} \sin \Theta$, the angle Θ must depend on the angular velocity of the wave and the time (t) that it has been in motion. The angle, in radians, is $\Theta = \omega t$. Therefore, we can write the equation for instantaneous sinusoidal voltage as follows:

$$e = E_{max} \sin \omega t \qquad\qquad \text{15-9}$$

Example 15-13

A 60 Hz voltage has a maximum value of 156 V. What is its instantaneous value at the end of 0.002 s? (Assume $t = 0$ when the voltage is zero and increasing in a positive direction.)

Solution

$$\omega = 2\pi 60 = 377 \text{ rad/s}$$
$$e = 156 \sin(377 \times 0.002) = 156 \sin 0.754 \text{ rad V}$$

where 0.754 is the time angle in radians. Since 1 radian $= 57.3°$, e is as follows:

$$e = 156 \sin(0.754 \times 57.3°) = 156 \times 0.685 = 106.8 \text{ V}$$

15-9
Average Value of a Sine Wave

Average value   The sum of the instantaneous voltages in a half-cycle waveshape divided by the number of instantaneous voltages. In a sine wave, it is equal to 0.637 times the peak voltage.

The **average value** of a sinusoid is zero, because the positive and negative alternations have equal areas under their curves. There are certain circuits—power supplies, for example—where it is necessary to compute the average value of one-half of a cycle. Mathematically, this computation can be accomplished by adding a series of instantaneous values of the wave between 0° and 180° and then dividing this sum by the number of values used. This computation gives a figure of 0.637 of the peak value, which indicates the average value of one half cycle. Expressed mathematically, the average voltage is as follows:

$$e_{avg} = 0.637E_{max} \qquad \text{15-10}$$

where
$$e_{avg} = \text{average voltage of one half cycle}$$
$$E_{max} = \text{maximum or peak value}$$

By substituting current for voltage, we can use the formula to indicate average current values:

$$i_{avg} = 0.637I_{max} \qquad \text{15-11}$$

Example 15-14 ────────────────────────────────

Find the average value of a sine wave of current with a maximum value of 85 mA.

Solution

$$i_{avg} = 0.637I_{max} = 0.637 \times 85 = 54.145 \text{ mA}$$

If the average value of a wave is given, its peak value can be easily determined by using the reciprocal value of the averaging factor:

$$\text{maximum value} = \frac{\text{average value}}{0.637}$$

15-10
Effective or rms Value of a Sine Wave

Effective or rms value   That value of alternating current that produces the same heating effect as the corresponding value of direct current. For sine wave currents, the effective value is 0.707 times the peak value.

At times, it is necessary to compare ac and dc voltages or currents. But which ac values should be used? The only rational bases for comparison are the values of ac and dc voltages needed to produce a certain amount of power in a given resistance. For example, 10 V dc connected across a 10 Ω resistor will develop 10 W of power, as shown by the following calculation:

$$P = \frac{E^2}{R} = \frac{10^2}{10} = 10 \text{ W}$$

Now, how much ac voltage is required to produce the same wattage in the 10 Ω resistor? This question can be resolved in the following manner:

1. Find the sine of the angle for every 10° from 0° to 90°.
2. Square each of these values.
3. Find the sum of these values.
4. Determine the mean or average of this sum. (Since there are 10 values, divide by 10.)
5. Calculate the square root, or mean square value, of this average.

For convenience, Table 15-2 includes the steps just listed. From the table, we see that the factor 0.707 is derived as the square root of the average (mean) of all the squares of the sine values. Therefore, whenever

Table 15-2 Determination of rms Value

$\theta°$	Sin θ	Sin$^2\theta$
0	0.0000	0.0000
10	0.1736	0.0302
20	0.3420	0.1170
30	0.5000	0.2500
40	0.6428	0.4132
50	0.7660	0.5868
60	0.8660	0.7500
70	0.9397	0.8830
80	0.9848	0.9698
90	1.0000	1.0000
		Total: 5.0000
		÷ 10: 0.5000
		$\sqrt{0.5000}$: 0.7071

the peak value of an ac wave is multiplied by 0.707, we have its **effective, or rms, value.** Expressed mathematically, the rms, or effective, value is as follows:

$$e_{rms} = e_{eff} = 0.707E_{max} \qquad \qquad 15-12$$

Let us return now to the original question. What value of ac voltage is necessary to produce 10 W in the 10 Ω resistor? Solve for E_{max} in Equation 15–12:

$$E_{max} = \frac{e_{rms}}{0.707} \qquad \qquad 15-13$$

Therefore, the following ac voltage is needed:

$$E_{max} = \frac{10 \text{ V}}{0.707} = 14.14 \text{ V}$$

An ac voltage having a peak value of 14.14 V will produce the same wattage in a 10 Ω resistor (or any resistance) as 10 V dc.

Example 15–15

What is the effective value of an ac voltage with a peak value of 8.91 V?

Solution

$$e_{rms} = 0.707E_{max} = 0.707 \times 8.91 = 6.3 \text{ V}$$

Example 15–16

An ac voltage has an rms value of 120 V. Calculate its peak value.

Solution

$$E_{max} = \frac{e_{rms}}{0.707} = \frac{120}{0.707} = 169.73 \text{ V}$$

Example 15–17

An rms voltmeter indicates a certain circuit voltage is 37 mV. If an oscilloscope were to measure the same voltage, what would it indicate?

Solution

$$e_{pp} = 2E_{max} = 2\,\frac{e_{rms}}{0.707} = 2 \times \frac{37\ mV}{0.707} = 104.67\ mV$$

Note: This much accuracy is rarely obtainable with an oscilloscope. The reading would probably be rounded off to about 105 mV.

15–11
e and *i* Relationships in Resistive Circuits

In-Phase Condition

> **In phase**   The condition that occurs when two waves of the same frequency pass through their maximum and minimum values at the same time and in the same direction.

If a sinusoidal voltage is applied to a resistance, the resulting current will be sinusoidal, as shown in Figure 15-9. Observe that the current waveform is in step with the voltage. When two waves are precisely in step, they are said to be **in phase**. They must go through their maximum and minimum values at the same time and in the same direction.

Out-of-Phase Condition

> **Out of phase**   The condition that occurs when waves do not pass through their maximum and minimum values at the same time.

Frequently, waves are **out of phase** with each other (a condition to be explained in subsequent chapters), as shown in Figure 15–10. The

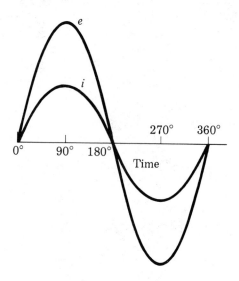

Figure 15–9

Graph showing *e* and *i* waves in phase

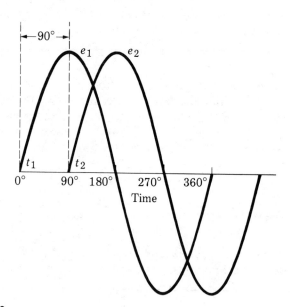

Figure 15–10

Voltage waves 90° out of phase

illustration represents a 90° out-of-phase condition. Waves can also be as much as 180° out of phase, as in Figure 15–11. In fact, any intermediate value is possible between in phase and 180° out of phase.

> **Leading**   A predominately capacitive circuit where the current leads the voltage.

> **Lagging**   A predominately inductive circuit where the current lags the voltage.

To further describe the phase relationship between two waves, the terms **leading** and **lagging** are used. This out-of-phase condition is measured in degrees. Refer to Figure 15–10. Wave e_2 is seen to start 90° later than e_1. Thus, e_2 lags e_1 by 90°. It is just as correct to say that e_1 leads e_2 by 90°.

The 180° out-of-phase condition shown in Figure 15–11 is quite common. Observe that, although the waves pass through their maximum and minimum values at the same time, their instantaneous values are always of opposite polarity. Therefore, they would cancel each other if their amplitudes were the same.

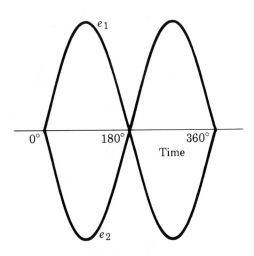

Figure 15–11

Voltage waves 180° out of phase

Addition of Sine Waves

Many electronic circuits combine different ac voltages. The resultant voltage and waveform depend on the amplitudes of the combined voltages and their phase relationship to each other. Consider Figure 15–12, which shows two sinusoids (sine curves) in phase: $e_1 = 6 \sin \omega t$ and $e_2 = 4 \sin \omega t$. The result of these two voltages is $e_R = 10 \sin \omega t$, which could be graphically verified by taking a number of accurate amplitude measurements of e_2 and adding each of them to e_1 at corresponding points in time.

When two 180° out-of-phase voltages with different amplitudes are added, they produce a resultant wave that has the polarity of the larger and an amplitude equal to their differences. Such a situation is indicated in Figure 15–13.

A third situation is represented in Figure 15–14: two sine waves 90° out of phase. Again, the resulting voltage is the algebraic addition of the instantaneous values of each voltage. Observe that, in each of these three figures, the resultant wave has the same frequency as the original waves.

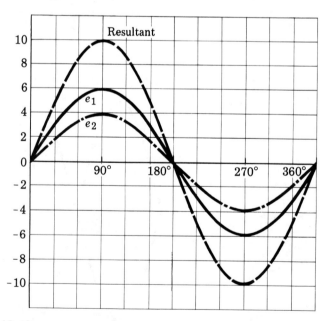

Figure 15–12

Resultant waveform from addition of e_1 and e_2

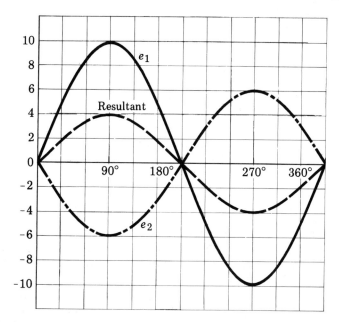

Figure 15–13

Addition of two 180° out-of-phase voltages of unequal amplitude

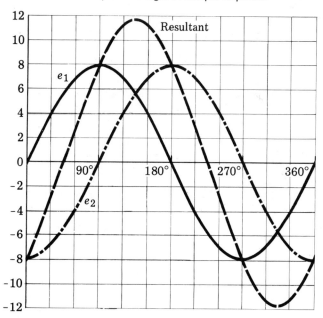

Figure 15–14

Addition of two voltages that are 90° out of phase

15–12
Power in Resistive Circuits

> ── **Instantaneous power**   The product of the instantaneous voltage
> and current in a circuit.

Instantaneous power in a resistive circuit is the product of instantaneous voltage and current and is expressed as follows:

$$P = ei \qquad \text{15–14}$$

The product of e and i over a period of one cycle for a resistive circuit is shown in Figure 15-15. The power curve resulting from the negative values of e and i is shown as positive because the product of two negative numbers is positive. Power is dissipated regardless of the direction in which current flows through the load. The power will be positive as long as e and i are flowing in the same direction. Because the positive and negative halves of the cycle are equal, the average power is the value halfway between the maximum and zero values of power. Therefore, the average power is $0.5P_p$ (P_p is the peak value):

$$P_{avg} = 0.5E_{max}I_{max} = \frac{E_{max}I_{max}}{2} \qquad \text{15–15}$$

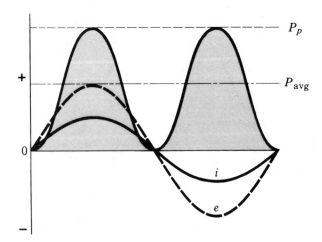

Figure 15–15

Power curve for purely resistive circuit

If the peak values of e and i are converted to rms values, the average power will be obtained (*Note*: $0.707 = 1/\sqrt{2}$):

$$P_{avg} = 0.707 E_{max} \times 0.707 I_{max} \qquad \text{15–16}$$

Stated differently, the average power is as follows:

$$P_{avg} = \frac{E_{max}}{\sqrt{2}} \times \frac{I_{max}}{\sqrt{2}} \qquad \text{15–17}$$

Example 15–18

A 30 V_p sinusoidal voltage is connected across a 10 Ω resistor. Calculate the peak power and the average power.

Solution

First, calculate the peak current by dividing the peak voltage by the resistance:

$$I_p = \frac{30\,V_p}{10\,\Omega} = 3\text{ A}$$

Then
$$P_p = 30\text{ V} \times 3\text{ A} = 90\text{ W}$$

$$P_{avg} = \frac{30\,V_p}{\sqrt{2}} \times \frac{3\text{ A}}{\sqrt{2}} = 45\text{ W}$$

Several conclusions can be drawn regarding power in resistive circuits. They are as follows:

1. Two power curves exist for each cycle.
2. The power curves are always positive.
3. The power curves have a sinusoidal form.

15–13
Vectors and Phasors

> **Vector** A straight line representing the magnitude and direction of a quantity.

Physical forces, frequently shown by arrows that indicate the *magnitude* and *direction* of the forces, are represented by **vectors**. For example, a person pulls on a rope (attached to a post) with a force of 20 kilograms (kg). This pulling force can be represented by a vector, as shown in Figure 15-16a. If a second person attaches another rope and pulls at right angles with a force of 30 kg in the direction illustrated, the vector would be as shown in Figure 15-16b. A resultant force would be established as indicated and at an angle Θ with respect to the first force.

In the past, voltages and currents were commonly represented by vectors in graphical solutions to problems. The reference vector was plotted in the horizontal section of the graph – to the right – and other vectors and their phase angles were measured clockwise or counterclockwise from this reference point.

Phasor   A quantity that has magnitude and direction in the time domain, that is, is constantly changing as instantaneous values of a sine wave.

Strictly speaking, ac voltages and currents are not vectors in that they cannot be defined in terms of force and direction alone. They also have a time relationship to each other. Thus, ac values in the time domain are called **phasors**.

Phasors are of fixed length in any given problem or condition and do not generally represent instantaneous values. Customarily, they represent *effective values*, although they can also represent maximum or average values, or even an instantaneous value for one specific instant of time.

Phasors representing all voltages (or currents) in the same diagram must be drawn to the same scale. If two voltages are in phase and their values are $E_1 = 3$ V and $E_2 = 2$ V, they would be plotted as in Figure 15-17a. If a voltage is not in phase with a current, it will have a time relationship that is either ahead of or behind the current. The voltage can be said to lead or lag the current by some angular amount, usually expressed in degrees. Suppose a current of 3 A leads a voltage of 4 V. This relationship could be plotted as in Figure 15-17b or 15-17c. In either case, the current is leading the voltage. (It would be just as correct to say that the voltage lags the current by 90°.)

With phasors, it is assumed that the values they represent are of the same frequency. If the frequencies were different, the phasor diagram would be valid for only one instant of time.

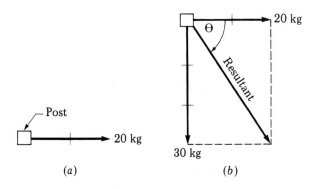

Figure 15–16

Representations of vector forces: (a) A force of 20 kg. (b) A second force of 30 kg at right angles to the first force.

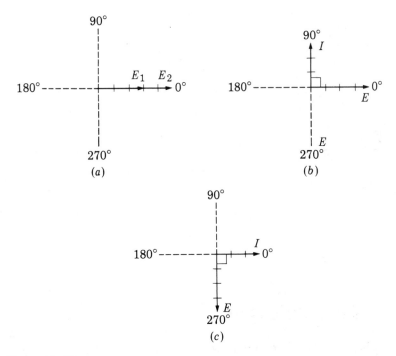

Figure 15–17

Phasors representing sine waves: (a) Two voltages in phase. (b) Current leading voltage. (c) Current leading voltage.

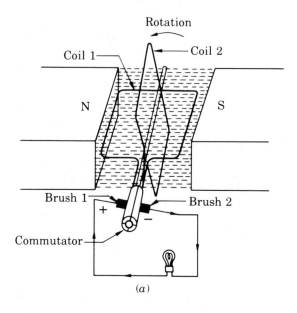

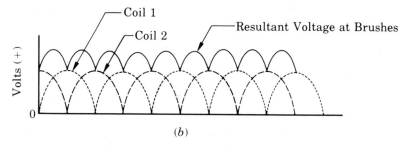

Figure 15–18

Two coil dc generator: (a) Basic design. (b) Voltage output of each coil and the resultant voltage.

15–14
Polyphase Voltages

Polyphase   Having more than one phase.

    Multiphase, or **polyphase**, ac generators have two or more single-phase windings symmetrically spaced around the stator. The rotor is essentially an electromagnet that induces voltages into the stator

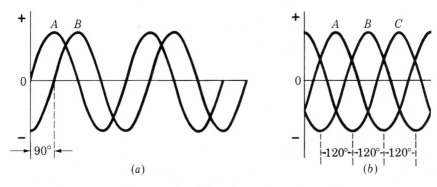

(a)   *(b)*

Figure 15–19

Polyphase voltages: (a) Two phases. (b) Three phases.

windings. In a two-phase alternator (ac generator), there are two single-phase windings physically spaced so that the ac voltages induced into them will be 90° apart. Figure 15-18a shows the basic design of a two-coil dc generator, which is similar to an ac generator except that an alternator uses slip rings in place of a commutator. In Figure 15-18b, the waves (voltages) generated by each coil are shown along with the resultant waveform (voltage).

The voltages generated by a two-phase alternator are represented in Figure 15-19a. Notice that voltage B reaches its peak value 90° later than A. We can say that voltage B lags voltage A by 90°. The voltages represented in Figure 15-19b are generated by a three-phase alternator. Each voltage is out of phase by 120° with the adjacent wave.

Three-phase voltages are very common in industrial applications. The power supplied to radio and TV stations uses this arrangement because of increased efficiency in the rectifying system. Three-phase motors develop more power for a given size and weight than single-phase motors do. Alternators of this type can furnish power to three separate circuits simultaneously.

15–15
Nonsinusoidal Waves

Complex wave   Nonsinusoidal wave consisting of a fundamental frequency plus one or more harmonics of the fundamental.

> ┌─ **Harmonic**   Sinusoidal wave having a frequency that is an inte-
> │   gral multiple of the fundamental frequency.

Complex waves occur in speech, music, TV, rectifier outputs, and many other applications of electronics. If a sine wave is applied to a *nonlinear device* (any device or equipment, such as a diode or an ampli-fier, whose output is not a faithful replica of its input) or a nonlinear circuit, its output will be changed or distorted into a complex wave out-put. Complex waves are made up of the fundamental frequency plus one or more harmonics of the fundamental. **Harmonics** are multiples of the fundamental waveform. For example, the second harmonic of a 1 kHz wave is 2 kHz. Its fifth harmonic is 5 kHz. Since the amplitude of each successive harmonic diminishes, the amount of power each represents (compared with the fundamental frequency) becomes less as the har-monics increase.

Any *nonsinusoidal waveform* that occurs periodically can be constructed by combining a sine wave at the fundamental frequency and appropriate harmonics, plus, if necessary, a dc voltage. The sine waves must have the correct amplitude and phase relationships for the particular waveshape desired.

Figure 15–20 shows a fundamental sine wave with an in-phase second harmonic. By graphically combining the amplitudes of each wave over discrete periods of time throughout the cycle, we obtain a composite wave that is identified as the resultant wave. Notice that this

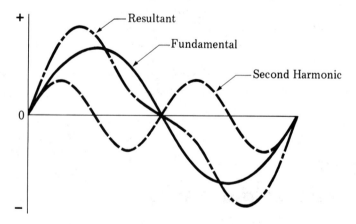

Figure 15–20

Fundamental plus second harmonic in phase

new wave is not sinusoidal in shape, even though it is the result of combining two sine waves.

Suppose the phase relationship between the fundamental and its harmonic is changed. Will the resultant still have the same shape? The answer is no. The resultant wave will undergo changes for each change in phase relation between fundamental and harmonic throughout 360° of the harmonics cycle.

The result of adding a third harmonic to a fundamental frequency is illustrated in Figure 15–21. Notice the symmetrical nature of the resultant wave for each half cycle. If the positive half were shifted 180° to the right, it would be a mirror image of the negative alternation.

If the third harmonic is shifted 180° with respect to its fundamental, the condition shown in Figure 15–22 prevails. There is a profound effect on the shape of the resulting wave. The resulting wave produced by the addition of the third harmonic (Figures 15–21 and 15–22) is not a sinusoid *even though the waves that produced it are.*

When *odd-order harmonics* are combined with the fundamental, at least to the seventh harmonic, a reasonably square wave results. The more odd harmonics that are added, the squarer the wave becomes. A perfect square wave theoretically contains an infinite number of odd harmonics.

A common example of a nonsinusoidal wave is the output of a *half-wave rectifier,* such as is used in the high-voltage supply for TV receivers and oscilloscopes. Figure 15–23a shows this waveform, some-

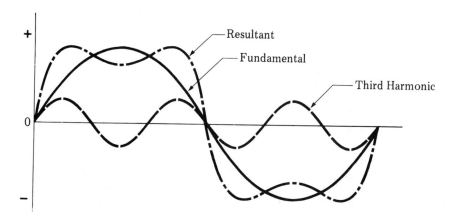

Figure 15–21

Fundamental plus third harmonic in phase

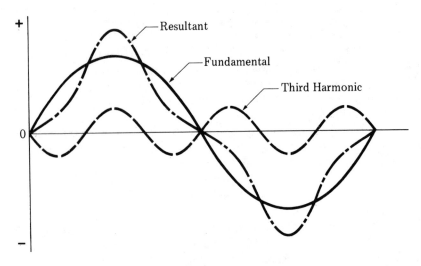

Figure 15–22

Fundamental plus third harmonic 180° out of phase

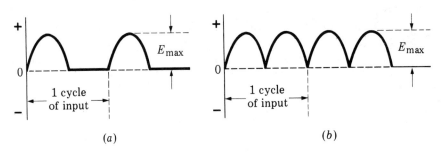

Figure 15–23

Rectifier waveforms: (a) Half wave. (b) Full wave.

times called a *pulsating dc*, although it is actually a complex wave. (The input to the rectifier is a sine wave for most applications.) Because the average value is not zero, it is apparent the wave contains a dc component. The average value of the half wave is $0.637E_{max}$. It follows that since the next half cycle is zero, the average amplitude of one full cycle is E_{max}/π, or $0.318E_{max}$, which represents the dc component of the wave.

Figure 15-23*b* shows the output waveform of a *full-wave recti-fier*, the type most commonly used because of its efficiency. The average value of the output (dc component) over one cycle is $2/\pi$, or 0.637.

Summary

An alternating voltage varies in amplitude and periodically reverses itself.

If the variations of an ac wave conform to the sine function of an angle, it is called a sine wave.

A cycle represents one complete alternation of a wave and is equal to 360°.

The number of cycles a current or voltage completes per second is called its frequency.

Radian measure is commonly used in electronics calculations. One radian equals 57.3°.

Frequency is measured in hertz (Hz).

The period of a wave is the length of time needed for one cycle.

The wavelength of a wave is the spatial length of one complete cycle.

Frequency and period are related by $T = 1/f$.

Angular velocity in radians is expressed as $\omega = 2\pi f$.

The principal measurements of a sine wave's amplitude are peak, peak-to-peak, effective or rms, and average values.

The effective value of an ac voltage is the value that will produce the same power in a given resistance as a corresponding amount of dc voltage.

Both e and i are in phase in a resistive circuit.

When two sine waves are out of phase, they cancel (or partially cancel, depending on their frequencies, amplitudes, and out-of-phase condition), and the resultant wave is not sinusoidal.

Phasors are used to show the time relationship of voltages or currents, usually of the same frequency.

Nonsinusoidal waves (complex waves) result from the addition of two or more waves that may or may not be in phase.

Progress Test

The bracketed number after each question indicates the section of this chapter where the answer can be found.

1. Two radians are equal to: (a) 0.017°. (b) 0.009°. (c) 57.3°. (d) 114.7°. [15–3]

2. The number of radians in 30° is equal to: (a) 0.417. (b) 0.523. (c) 0.581. (d) 0.631. [15–3]

3. If an alternator rotates at 3600 rpm, its frequency is: (a) 120. (b) 90. (c) 60. (d) 30. [15–4]

4. The angular velocity of a voltage having a frequency of 800 Hz is: (a) 5.024×10^3 rad/s. (b) 2.512×10^3 rad/s. (c) 5.024×10^3 degrees. (d) 2.512×10^3 degrees. [15–4]

5. The frequency of a wave having a period of 16.7 ms is approximately: (a) 6 Hz. (b) 60 Hz. (c) 600 Hz. (d) 6 kHz. [15–4]

6. A 10.7 MHz signal has a period of: (a) 0.0934×10^{-6} s. (b) 9.35 μs. (c) 9.34 ms. (d) 0.934×10^{-6} s. [15–4]

7. The wavelength of a radio wave with a frequency of 810 kHz is: (a) 4.216×10^2 m. (b) 3.707 m. (c) 5.14×10^2 m. (d) 3.703×10^2 m. [15–4]

8. The frequency of a wave with a wavelength of 0.78 m is: (a) 2.34×10^8 Hz. (b) 268.2×10^6 Hz. (c) 31.84 MHz. (d) 384.6 MHz. [15–4]

9. A sine wave of 169 V_p has an instantaneous value at 45° of: (a) 119.5 V. (b) 84.5 V. (c) 146.4 V. (d) 238.9 V. [15–8]

10. The rms value of a sinusoid with a peak-to-peak value of 240 V is: (a) 77.82 V. (b) 84.84 V. (c) 89.15 V. (d) 94.68 V. [15–10]

11. An rms voltmeter reads 31.5 mV. The peak-to-peak voltage is: (a) 89.08 mV. (b) 78.23 mV. (c) 68.55 mV. (d) 62.48 mV. [15–10]

12. Measurements made on the collector load resistor of a certain transistor indicate $V_{pp} = 3.82$ V and $I_{pp} = 1.63$ mA. The rms power dissipated in this resistor is: (a) 0.54 mW. (b) 0.63 mW. (c) 0.78 mW. (d) 0.86 mW. [15–12]

┌─── Problems

Answers to odd-numbered problems are at the back of the book.

1. How many radians are there in: (a) 30° (b) 45° (c) 90°?

2. A radar antenna rotates 10 rpm. What is its angular velocity in radians per second?

3. A timing motor operates at 1800 rpm. What is its angular velocity in radians per second?

4. Determine the period of an electromagnetic wave with a frequency of 456 kHz.

5. If a radio wave has a frequency of 10.7 MHz, what is the time of one cycle?

6. A technician measures the time of one cycle, using an oscilloscope, and finds it to be 4.25 μs. What is the frequency of the signal?

7. An oscilloscope displays one cycle in 8.2 ms. Calculate the frequency of the signal.

8. A radio station operates on 146.52 MHz. What is the wavelength of this signal?

9. A broadcast station operates on 610 kHz. What is its wavelength?

10. What is the frequency of a microwave station with a wavelength of 0.12 m?

11. An alternating voltage has a peak value of 69 V. Determine the instantaneous values of the following points in its cycle: (a) 17° (b) 63° (c) 148°.

12. An alternating current has a maximum value of 17.5 mA. Determine its instantaneous values at: (a) 22° (b) 76° (c) 115°.

13. An ac voltage has an instantaneous value of 229 V at 41°. What is its peak value?

14. A 400 Hz signal has a value of E_{max} = 25 mV. What is its instantaneous value 300 μs after the waves cross the zero axis in the positive direction?

15. Determine the average value of one half cycle of an alternating voltage with a maximum value of 85 V.

16. Calculate e_{avg} for an ac voltage whose E_{max} is 169 V.

17. If a sinusoidal current has an average value of 29 mA, what is its peak value?

18. What is the effective value of a voltage with a peak value of 76 V?

19. The maximum value of an alternating current is 3.1 A. What is its effective value?

20. An rms voltmeter reads 120 V. Calculate the maximum value of the voltage.

21. An oscilloscope reading indicates that a certain sinusoid has a 313 mV peak-to-peak value. What is the effective value of this voltage?

22. If an rms voltmeter indicates a circuit voltage to be 19.5 V, what is its peak-to-peak value?

23. Measurements in a certain transistor circuit indicate that the peak values of voltage and current in the load resistor are 2.13 V and 0.89 mA. (a) What is the peak power? (b) What is the rms power dissipated?

24. A transmission line is terminated in a 75 Ω resistance. An oscilloscope reads 1.5 V_{pp} across the resistance. What is the rms power dissipated in the resistor?

25. The output voltage of a simple half-wave rectifier is 34 V_p. What is the value of the dc component of the voltage?

26. Calculate the period of the following frequencies: (a) 60 Hz (b) 20 Hz (c) 1 kHz (d) 560 kHz (e) 27 MHz (f) 250 MHz.

27. Calculate the frequency for each of the following periods of sine waves: (a) 16.67 ms (b) 8.33 ms (c) 20.0 μs (d) 0.3 μs (e) 9.9 ns.

28. A displayed sine wave on an oscilloscope is 3.4 cm long. If the SEC/DIV switch is positioned at 20 μs/DIV, what is the frequency of the displayed wave?

29. Calculate the wavelength in meters of the following sine waves: (a) 60 Hz (b) 550 kHz (c) 27 MHz (d) 98.1 MHz (e) 3.5 GHz.

30. Convert each of the following voltages to the units indicated: (a) 320 V_{pp} to rms (b) 150 V_{pp} to average (c) 84 V_{pp} to effective (d) 23 V_{pp} to rms (e) 15 mV_{pp} to average (f) 65 μV_{pp} to effective.

31. Convert each of the following voltages to the units indicated: (a) 175 V_{rms} to peak (b) 84 V_{rms} to maximum (c) 45 V_{rms} to peak-to-peak (d) 24 mV_{rms} to average (e) 65 μV_{rms} to peak-to-peak.

32. A 60 Hz power line frequency is used to verify the time-base accuracy of a simple oscilloscope. What should the period of the displayed waveform be?

16

Inductive and Capacitive Reactance

Objectives

After studying this chapter, you will be able to:

Explain how reactance opposes alternating currents.

Determine how the reactance in a circuit alters the phase relationship between voltage and current.

Introduction

The effect of resistance in an ac circuit has been explored in the previous chapter. Yet to be considered is the effect that inductance and capacitance, as circuit elements, have on alternating currents. Their individual and combined behavior is investigated in this chapter. Their presence in an ac circuit may upset the in-phase relationship of voltage and current. Sometimes, this disturbance is advantageous; other times, it presents a problem.

Applications

That the effect of inductance and capacitance on alternating current depends on the frequency of the current leads to many important applications. Certain combinations of coils and capacitors known as filters have the property of rejecting all frequencies outside a given range. A *low-pass filter* is used to smooth out high-frequency ripple currents to improve the performance of rectifiers. A *high-pass filter* in a phonograph amplifier removes the low-pitched rumble from the turntable motor. A *bandpass filter* can be used to select a specific frequency band from the many voice channels in a high-frequency communications system.

The term *choke* is often used to describe inductors, because of their property of resisting high-frequency currents. Conversely, capacitors easily allow the passage of high-frequency currents. When the out-of-phase relationships of *e* and *i* become excessive for certain applications, the judicious addition of either inductance or capacitance can provide the necessary compensation.

16–1
Inductive Reactance

> **Inductive reactance (X_L)**   Opposition to the flow of an alternating current by the inductance of a circuit; equal to $2\pi fL$ and measured in ohms.

When a changing current flows through an inductor, a self-induced voltage is developed. Its polarity is such that it opposes the change. The cemf varies directly with the rate of change of current. This opposition is called **inductive reactance**, and it is a property of all inductors. Inductive reactance is measured in ohms and is represented by X_L, where X indicates *reactance* and the subscript L implies that it is *inductive*. If the current change is sinusoidal, the inductive reactance can be calculated from the following equation:

$$X_L = 2\pi fL \qquad \text{16–1}$$

where
2π = one cycle in radian measure
f = frequency in hertz
L = inductance in henrys

Because angular velocity, $2\pi f$, is frequently written as ω, we can write Equation 16–1 as follows:

$$X_L = \omega L \qquad \text{16–2}$$

This formula and all others to follow in subsequent chapters will be predicated on the use of *sine waves*. The calculations of circuit opposition to ac current flow for other waveforms than sine waves involve mathematics at a level beyond the scope of this text. Fortunately, the majority of circuits encountered in our studies involve sinusoidal currents.

An analysis of Equation 16-1 indicates that inductive reactance is directly proportional to frequency and inductance. The next example explores this relationship.

Example 16-1 ─────────────────────────────────

Calculate the inductive reactance of an 8 H filter choke at 60 Hz and at 120 Hz.

Solution

At 60 Hz $X_L = 2\pi fL = 6.28 \times 60 \times 8 \cong 3016\ \Omega$
At 120 Hz $X_L = 6.28 \times 120 \times 8 \cong 6032\ \Omega$

This example clearly illustrates that if the frequency is doubled, the reactance is doubled.

Example 16-2 ─────────────────────────────────

What is X_L for a 25 μH coil at 10.7 MHz?

Solution

$$X_L = 6.28 \times 10.7 \times 10^6 \times 25 \times 10^{-6} \cong 1681\ \Omega$$

If the reactance of a particular inductance is known at a given frequency, its inductance can be determined by rearranging Equation 16-1 as follows:

$$L = \frac{X_L}{2\pi f} \qquad\qquad 16\text{-}3$$

Example 16-3 ─────────────────────────────────

A small inductor used in a TV receiver has a reactance of 5 kΩ at 45 MHz. What is its inductance?

Solution

$$L = \frac{X_L}{2\pi f} = \frac{5 \times 10^3}{6.28 \times 45 \times 10^6} = 1.769 \times 10^{-5} = 17.69 \ \mu\text{H}$$

16-2
Graph of X_L

Equation 16-1 is linear because X_L varies directly as the product of the linear terms f and L. If you plot inductive reactance against frequency, you will have a graph like the one shown by the solid line in Figure 16-1, which happens to be for the 8 H filter choke of Example 16-1. If the frequency were increased beyond 120 Hz, the reactance line would increase beyond the limits shown.

Suppose the inductance used in Example 16-1 were changed to 4 H. How would the reactance vary with frequency? Since there is a direct relationship between inductive reactance and frequency, we can

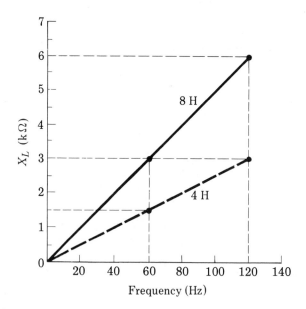

Figure 16-1

Graph of X_L versus frequency for an 8 H and a 4 H inductor

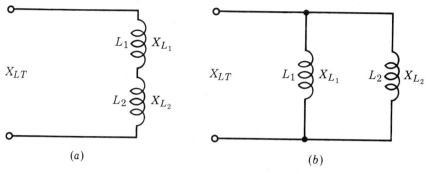

Figure 16–2

Reactances: (a) In series. (b) In parallel.

expect X_L to be reduced to one-half of its value at 60 Hz and 120 Hz. Plotting the reactance of the 4 H inductor on the graph of Figure 16–1 results in the dotted line indicated. Observe that the slope (steepness) is one-half that of the 8 H inductance line. If the value of L were increased to 16 H, we could expect the slope of its line to be twice that of the 8 H line.

16–3
X_L in Series and in Parallel

Inductive reactances in series or in parallel are treated in the same manner as resistances, as is shown in Figure 16–2. Hence, the total inductive reactance of two or more series-connected inductors is as follows:

$$X_{LT} = X_{L_1} + X_{L_2} + \cdots + X_{L_n} \qquad \text{16–4}$$

For two parallel-connected inductive reactances, the formula is as follows:

$$X_{LT} = \frac{X_{L_1} X_{L_2}}{X_{L_1} + X_{L_2}} \qquad \text{16–5}$$

When three or more inductive reactances are in parallel, the reciprocal formula is used:

$$\frac{1}{X_{LT}} = \frac{1}{X_{L_1}} + \frac{1}{X_{L_2}} + \frac{1}{X_{L_3}} + \cdots + \frac{1}{X_{L_n}} \qquad \text{16–6}$$

Voltage drops and current flows are calculated by using Ohm's law, except that X_L is substituted for R.

Example 16–4

Calculate X_{LT}, I_T, V_{L_1}, and V_{L_2} for the series-inductive circuit of Figure 16–3.

Solution

$$X_{LT} = X_{L_1} + X_{L_2} = 500 \ \Omega + 1000 \ \Omega = 1.5 \ \text{k}\Omega$$

$$I_T = \frac{E_S}{X_{LT}} = \frac{30 \ \text{V}}{1.5 \ \text{k}\Omega} = 20 \ \text{mA}$$

The current is an rms value because the supply voltage is assumed to be rms, or *effective*, voltage.

$$V_{L_1} = I_T X_{L_1} = 20 \ \text{mA} \times 500 \ \Omega = 10 \ \text{V}$$
$$V_{L_2} = I_T X_{L_2} = 20 \ \text{mA} \times 1000 \ \Omega = 20 \ \text{V}$$

Since X_{L_2} has twice the reactive value of X_{L_1}, it has double the voltage drop.

Example 16–5

For the circuit of Figure 16–4, calculate X_{LT}, I_{L_1}, I_{L_2}, and I_T.

Solution

$$X_{LT} = \frac{X_{L_1} X_{L_2}}{X_{L_1} + X_{L_2}} = \frac{200 \times 400}{200 + 400} = 133.3 \ \Omega$$

$$I_{L_1} = \frac{10 \ \text{V}}{200 \ \Omega} = 0.05 \ \text{A} = 50 \ \text{mA}$$

$$I_{L_2} = \frac{10 \ \text{V}}{400 \ \Omega} = 0.025 \ \text{A} = 25 \ \text{mA}$$

Thus, we see that the larger reactance has the smaller current flow.

$$I_T = I_{L_1} + I_{L_2} = 50 + 25 = 75 \ \text{mA}$$

As with the resistive circuit, the voltage across each branch is the same.

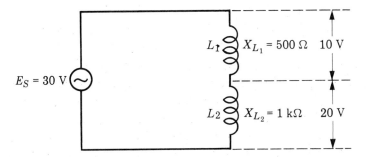

Figure 16–3

Circuit for Example 16–4

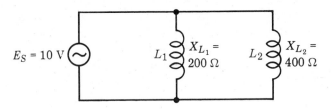

Figure 16–4

Circuit for Example 16–5

16–4
Phase Relationship of e_L and i_L

From Equation 13–2, the inductance of a coil is as follows:

$$L = \frac{e_L}{di/dt}$$

where e_L is the induced voltage across the inductor. Then, the cemf is as follows:

$$e_L = -L\frac{di}{dt} \qquad\qquad \text{16–7}$$

The applied voltage E, being $180°$ out of phase, is as follows:

$$e_L = -\left(-L\frac{di}{dt}\right)$$

This equation gives us the following result:

$$e_L = L \frac{di}{dt}$$ 16-8

This expression reveals that di/dt is positive when the applied voltage is positive, and it is negative when the applied voltage is negative. When di/dt is zero, then e is equal to zero.

Figure 16-5 illustrates the voltage and current relationships for a purely inductive circuit with an applied sine wave. At a time corresponding to point a, the applied e is positive, di/dt is positive, and therefore, *the current is increasing with time.* At point b, the voltage is negative, di/dt is less than zero, and the current is decreasing. For some value of time, then, between points a and b, di/dt is equal to zero; that is, it crosses the zero axis. Thus, at 180° on the time axis, the current curve must flatten out (neither increase nor decrease) as it passes through its maximum value. In like manner, the negative-going portions of the current curve may be drawn. From Figure 16-5, then, it becomes apparent that i is zero when the applied e is maximum, or it is maximum when e is zero. Consequently, the current *lags* the applied voltage by 90°.

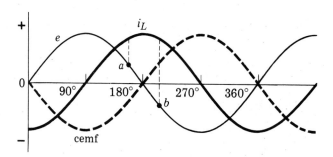

Figure 16-5

Current and voltage relationship in an inductive circuit

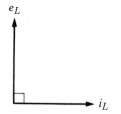

Figure 16-6

Phasor representation of e and i in a purely inductive circuit

Graphically, e_L and i_L can be represented by phasors. Figure 16–6 clearly indicates the 90° lag of current behind the applied voltage e_L for a purely inductive circuit. At any point in time in the ac cycle, this 90° lag of i with respect to e is operating.

16–5
Power in Purely Inductive Circuit

Instantaneous power in a circuit containing pure inductance is the product of the instantaneous voltages and currents. Figure 16–7 shows the plot of the products e and i over a period of one cycle. Observe that *the power curve is a sine wave of twice the frequency of e and i.* Since half of the power curve is below the zero axis and the other half above, the average power over one complete cycle is zero.

Wattless or reactive power   The power that appears to be consumed by a reactive circuit but is returned on the next half cycle.

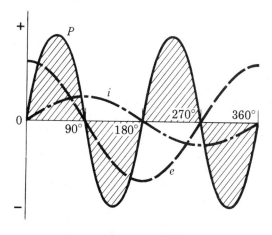

Figure 16–7

Power curve for a purely inductive circuit

During the first quarter cycle, both e and i are positive, as is the power curve. During this interval, the increasing current is storing energy in the magnetic field around the inductor. As the current drops to zero between 90° and 180°, the magnetic field collapses and returns its power back to the line. This portion of the power curve is negative and is the product of $-e$ and $+i$. Between 180° and 270°, both e and i are negative, but their product is positive. The energy stored in the magnetic field is returned during the 270° to 360° time interval. We may correctly conclude that *in a purely inductive circuit, no power is consumed.*

If an ac voltmeter and ammeter were connected into such a circuit, a person could be led to believe that the product of their readings was the actual power being dissipated. Not so! This power is referred to as **wattless** or **reactive power.** The ac voltmeter and ammeter do not indicate the phase relationship between e and i, and therefore, their product does not give a true indication of the actual power consumed. This subject is explained in Section 17–14.

16–6
Inductor Losses

Pure inductances are only hypothetical. All have losses of one kind or another. *Wire losses* are among the most important. These losses include dc voltage drops across the device as a result of any dc component flowing through it. A wire loss can be minimized by using larger wire, but the trade-off is larger size, increased weight, and more expense. At high frequencies, the currents travel near the surface of the conductor (skin effect) and effectively increase the resistance. Special conductors, such as *Litz wire*, keep this loss to a minimum.

Core losses have been partly discussed in Chapters 12 and 13. *Eddy current losses* will be dealt with in Chapter 21 (transformers). Stray-flux linkages can cause another kind of loss by inducing voltages into nearby components, thereby taking power out of the inductor, with a resulting *power loss.*

Inductors are frequently shielded to prevent interaction with adjacent components, but the shields themselves use some energy. At lower frequencies, such as power and audio, inductors are often enclosed in a soft-iron shield. At radio frequencies, an aluminum shield usually encloses the coil to prevent *radiation losses.*

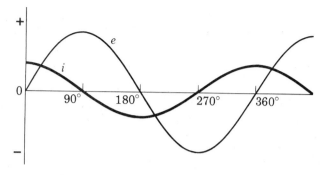

Figure 16-8

Current and voltage in a capacitive circuit

16-7
Capacitive Reactance

If a sinusoidal voltage is applied to a pure capacitance (no series or parallel resistance), the current is maximum when the voltage begins to rise from zero. One-quarter cycle later, the current is zero when the voltage across the capacitor is maximum. This condition, illustrated in Figure 16-8, shows that the current waveform leads the applied emf by 90°.

Capacitive current flow depends on the size of the capacitor and the rate of charge and discharge. At higher frequencies, the rate of charge and discharge increases per unit time. For a purely capacitive circuit, the charging current is as follows:

$$i = \omega CE \qquad \text{16-9}$$

> **Capacitive reactance** (X_C) Opposition to the flow of an alternating current by the capacitance of a circuit; equal to $1/2 \pi fC$ and measured in ohms.

The ratio of effective voltage across the capacitor to the effective current is called the **capacitive reactance** and represents the opposition to current flow. Its symbol is X_C, and is measured in ohms. Mathematically, capactive reactance is expressed as follows:

$$X_C = \frac{1}{2\pi fC}$$ 16–10

where f = frequency in hertz
C = capacitance in farads

Example 16–6

What is the reactance of an 8 μF filter capacitor at 120 Hz?

Solution

$$X_C = \frac{1}{2\pi fC} = \frac{1}{6.28 \times 120 \times 8 \times 10^{-6}} = 165.87 \ \Omega$$

When the capacitance is given in microfarads, the equation can be modified by placing 10^6 in the numerator (10^{12} for C in picofarads). A simplification results by placing the reciprocal of 2π in the numerator. The equation then is as follows:

$$X_C = \frac{0.159 \times 10^6}{fC}$$

16–8
Graph of X_c

Equation 16–10 indicates there is an inverse relationship between capacitance and frequency and a capacitor's reactance. That is, as the former two increase, the latter decreases. See the graph in Figure 16–9. Note that X_C is shown in the fourth quadrant, opposite X_L, which is in the first quadrant. At very low frequencies, approaching dc, the reactance increases at a rapid rate. At dc, the denominator of Equation 16–10 would be zero and the reactance infinite. As the frequency increases, the reactance becomes smaller, approaching 0 Ω at very high frequencies.

As is evident from Figure 16–9, the curve represents the reactance characteristics of a 0.1 μF capacitor. For other capacitor values, the curve assumes slightly different shapes. If the capacitance is re-

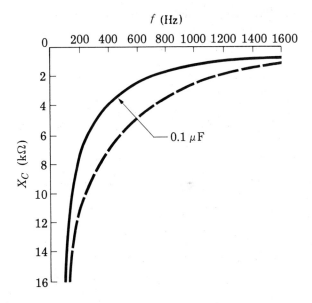

Figure 16-9

Graph of X_C versus frequency

duced, the curve moves toward the right (dotted line); for increased capacities, it moves toward the left. In other words, the solid curve shown represents the varying reactance characteristics of a 0.1 μF capacitor only.

16-9
X_c in Series and in Parallel

Series and parallel combinations of capacitive reactance are treated in the same manner as inductive reactances. Therefore, the total reactance of two or more series-connected capacitors is as follows:

$$X_{CT} = X_{C_1} + X_{C_2} + X_{C_3} + \cdots + X_{C_n} \qquad \textbf{16-11}$$

Two parallel capacitive reactances have the following total reactance:

$$X_{CT} = \frac{X_{C_1}X_{C_2}}{X_{C_1} + X_{C_2}} \qquad \textbf{16-12}$$

which is the familiar product-over-sum formula. If three or more capacitors are parallel-connected, their net reactance is as follows:

$$\frac{1}{X_{CT}} = \frac{1}{X_{C_1}} + \frac{1}{X_{C_2}} + \frac{1}{X_{C_3}} + \cdots + \frac{1}{X_{C_n}}$$ 16–13

Currents and voltages experienced around the capacitive circuit are calculated by using Ohm's law, where X_C is substituted for R. The next two examples illustrate the procedure.

Example 16–7

Using the circuit of Figure 16–10, calculate X_{CT}, I_T, V_{C_1}, and V_{C_2}.

Solution

$$X_{CT} = X_{C_1} + X_{C_2} = 800\ \Omega + 1600\ \Omega = 2400\ \Omega$$

$$I_T = \frac{E}{X_{CT}} = \frac{60\ \text{V}}{2.4\ \text{k}\Omega} = 25\ \text{mA}$$
$$V_{C_1} = I_T X_{C_1} = 25\ \text{mA} \times 0.8\ \text{k}\Omega = 20\ \text{V}$$
$$V_{C_2} = I_T X_{C_2} = 25\ \text{mA} \times 1.6\ \text{k}\Omega = 40\ \text{V}$$

Clearly, the voltage drops are proportional to the series reactances.

Example 16–8

Calculate the following values for the circuit of Figure 16–11: X_{CT}, I_{C_1}, I_{C_2}, and I_T.

Solution

$$X_{CT} = \frac{X_{C_1} X_{C_2}}{X_{C_1} + X_{C_2}} = \frac{500 \times 250}{500 + 250} = 167\ \Omega$$

$$I_{C_1} = \frac{25\ \text{V}}{500\ \Omega} = 50\ \text{mA}$$

$$I_{C_2} = \frac{25\ \text{V}}{250\ \Omega} = 100\ \text{mA}$$

$$I_T = I_{C_1} + I_{C_2} = 50 + 100 = 150\ \text{mA}$$

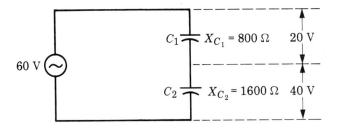

Figure 16-10

Circuit for Example 16-7

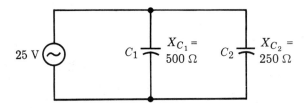

Figure 16-11

Circuit for Example 16-8

16-10
Phase Relationship of e_c and i_c

The voltage-current relationships in a purely capacitive circuit are the exact opposite of those relationships for the purely inductive circuit. That is, the current leads the applied voltage by $90°$, as shown in Figures 16-12 and 16-8. Let us assume that the sinusoidal voltage applied to the capacitor begins its cycle at $0°$; its amplitude, therefore, is zero. The current waveform is at its positive peak and is leading the applied voltage by $90°$ because of the following conditions.

The capacitor current is proportional to the rate at which the applied voltage is changing. Hence:

$$i_C = C\frac{de_C}{dt} \qquad \text{16-14}$$

The applied capacitor voltage is experiencing its greatest rate of change, de_C/dt, as it starts up from $0°$. At this instant, the capacitor is

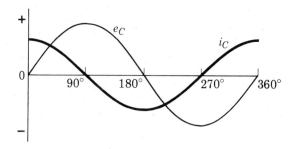

Figure 16–12

Current and voltage relationship in a capacitive circuit

charging at its maximum rate. Therefore, the capacitive current is shown at its maximum positive value. The slope of the charging voltage curve diminishes, until at 90°, $de_c/dt = 0$. During this interval, the charging current is tapering off, until at 90°, the current drops to zero, because of the following condition:

$$i_C = C\frac{de_C}{dt} = C \times 0$$

Past 90°, the voltage drops off in the negative direction, with the slope of the curve becoming steeper as it approaches 180°:

$$-e = \frac{de_C}{dt}$$

During this period, the capacitor is discharging, reaching its maximum rate at 180°.

As the negative half cycle of voltage begins (180°), it becomes more negative but at a decreasing rate. Finally, at 270°, $de_c/dt = 0$ again. During this period, the capacitive discharge current drops to zero from its maximum negative value.

As the last quarter cycle begins, the voltage becomes less and less negative. Its rate of change increases, until at 360°, de_c/dt reaches its maximum value. From Figure 16–12, we see that the charging current is positive and peaks out at 360°. At this point, one cycle has been completed and the next is ready to commence. A close inspection reveals that the sinusoidal current waveform reaches maximum value 90° ahead of the impressed voltage. Since this condition will always prevail in any purely capacitive circuit, regardless of frequency, we may conclude that the current leads the impressed voltage by 90° in a *purely capacitive circuit*.

Graphically, e_C and i_C can be represented by phasors, which

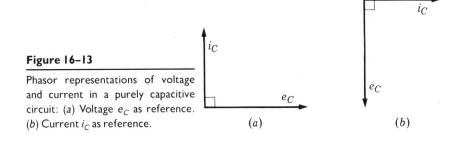

Figure 16–13

Phasor representations of voltage and current in a purely capacitive circuit: (a) Voltage e_C as reference. (b) Current i_C as reference.

(a) (b)

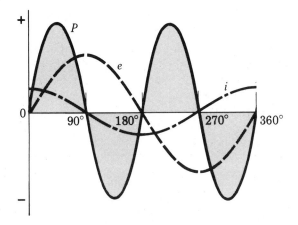

Figure 16–14

Power curve for purely capacitive circuit

show the relative magnitudes and directions of electrical quantities, as in Figure 16–13. In Figure 16–13a, the voltage phasor e_C is considered as the reference, and i_C is shown as by leading 90°. Figure 16–13b essentially shows the same condition, except that i_C is shown in the horizontal and represents the *reference phasor* from which the angle of lead or lag of the quantity being considered is measured. Voltage e_C is shown in a 90° clockwise position, which signifies 90° lag with respect to i_C.

16–11
Power in Purely Capacitive Circuit

The power curve for a purely capacitive circuit has the same form as the purely inductive circuit. Figure 16–14 shows this curve, which is a plot of the instantaneous products of e and i over a period of one cycle. Note that the power curve is a *sine wave of twice the frequency* of e and i. Average power is zero over one cycle, since the positive and negative curves are of equal area.

During the first quarter cycle, e and i are both positive. Current is flowing into the capacitor, and the energy is stored in the dielectric field. The power curve is positive during this interval.

Between 90° and 180°, the supply voltage is dropping, and the charge in the dielectric field is being returned to the line. Also, e is still positive while i is negative; their product is the negative power curve shown.

Between 180° and 270°, e is increasing to its maximum negative value, and the current is flowing back into the capacitor. Another positive power curve results during this time interval.

From 270° to 360°, the applied voltage is dropping back to zero. Thus, the energy in the dielectric field is returned to the source. Also, because e is negative and i is positive, their product produces the negative power curve shown.

At any instant, the power can be computed from the following equation:

$$p = ei$$

The next example illustrates the use of this equation.

Example 16–9

Determine the instantaneous power in a purely capacitive circuit at the moment when $e = 23$ mV and $i = 8$ mA.

Solution

$$p = ei = 23 \times 10^{-3} \times 8 \times 10^{-3} = 184 \times 10^{-6} \text{ W} = 184 \ \mu\text{W}$$

Summary

The opposition that an inductor offers to the passage of ac is called inductive reactance. Its symbol is X_L, and it is measured in ohms.

Inductive reactance is calculated from $X_L = 2\pi f L$.

Inductive reactances in series or in parallel are treated in the same manner as resistances.

In a purely inductive circuit, with a sine wave ac applied, the current lags the impressed voltage by 90°.

Phasors are used to show the time relationships between voltage and current in inductive and capacitive circuits.

The power consumed in a purely inductive circuit is zero, because the energy put into the magnetic field on one half cycle is returned on the other.

The power curve for a purely inductive or capacitive circuit is a sine wave of twice the line frequency (assuming sinusoidal input).

Losses in an inductor include wire (resistance), skin effect, core, and radiation losses.

The opposition that a capacitor offers to a changing voltage is called capacitive reactance. Its symbol is X_C, and it is measured in ohms.

Capacitive reactance is calculated from $X_C = 1/2\pi fC$.

At very low frequencies, X_C approaches infinity; at very high frequencies, it approaches zero.

Capacitive reactances in series or in parallel are treated in the same manner as resistance and X_L.

In a purely capacitive circuit, the current leads the line voltage by 90°.

The power consumed in a purely capacitive circuit is zero, because the energy put into the dielectric field on one half cycle is returned on the other.

The power curve for a purely capacitive circuit is a sine wave of twice the line frequency (assuming sinusoidal input).

Progress Test

The bracketed number after each question indicates the section of this chapter where the answer can be found.

1. Inductive reactance: (a) is a characteristic of all ac circuits. (b) varies directly with frequency and inductance. (c) varies inversely with frequency but directly with inductance. (d) can be calculated with any kind of ac in an inductive circuit. [16–1]

2. Which of the following combinations of L and f exhibit the greatest inductive reactance? (a) 2 H at 60 Hz (b) 0.2 H at 400 Hz (c) 50 mH at 1 kHz (d) 22 μH at 10 MHz   [16-1]

3. A 33 mH inductor has a reactance of 2500 Ω at a particular frequency. If it is series-connected (no mutual inductance) with an identical unit, the total reactance is: (a) 5 kΩ. (b) 2.5 kΩ. (c) 1.25 kΩ. (d) 0.625 kΩ.   [16-3]

4. Two 1.2 kΩ inductive reactances are parallel-connected (no L_M). The combined reactance is: (a) 2.4 kΩ. (b) 1.2 kΩ. (c) 600 Ω. (d) 450 Ω.   [16-3]

5. A certain inductor has a reactance of 4 kΩ at 5 kHz. Its reactance at 15 kHz is: (a) 8 kΩ. (b) 10 kΩ. (c) 12 kΩ. (d) 20 kΩ.   [16-1]

6. In a purely inductive circuit, if the frequency is doubled and the applied voltage reduced to one-half, the resulting current becomes: (a) 1/5. (b) 1/4. (c) 1/3. (d) 1/2.   [16-3]

7. Capacitive reactance: (a) varies inversely with frequency and capacitance. (b) is a property of all ac circuits. (c) can be calculated with any kind of applied ac. (d) is proportional to capacitance and inversely proportional to frequency.   [16-7]

8. Which of the following combinations of C and f present the greatest capacitive reactance? (a) 1 μF at 1 kHz (b) 0.005 μF at 50 kHz (c) 250 pF at 10 MHz (d) 0.01 μF at 5 kHz   [16-7]

9. A certain capacitor has a reactance of 8 kΩ at 10 kHz. Its reactance at 2.5 kHz is: (a) 16 kΩ. (b) 24 kΩ. (c) 32 kΩ. (d) 64 kΩ.   [16-7]

10. Two capacitive reactances of 1000 Ω each, when parallel-connected, have a total reactance of: (a) 250 Ω. (b) 500 Ω. (c) 707 Ω. (d) 2000 Ω.   [16-9]

⌐──Problems

Answers to odd-numbered problems are at the back of the book.

1. Calculate X_L for an inductor of 2 H at 60, 120, and 1000 Hz.

2. What is the reactance of a tuning inductor of 56 μH at 550 kHz? At 1650 kHz?

3. What is the value of inductance with a reactance of 1250 Ω at 1 MHz?

4. An inductor with a reactance of 1500 Ω is connected across a 5 V ac source. (a) Calculate the current through the coil. (b) What is the emf across the coil?

5. At what frequency does a 250 μH inductor have a reactance of 1 kΩ?

6. Refer to Figure 16–3. What is the value of L_1 and of L_2 if the frequency of the source is 450 kHz?

7. What is the equivalent inductance of the two reactances shown in Figure 16–4 if the frequency of the source is 20 kHz?

8. Two inductors, $L_1 = 6$ μH and $L_2 = 9$ μH, are series-connected to a 45 MHz source. (a) Calculate their total reactance. (b) What will the voltage drop across each be if a 15 V_{rms} supply is used?

9. Calculate the reactances of a 240 pF capacitor at the following frequencies: (a) 600 kHz (b) 1500 kHz.

10. What value of capacitance is required to provide a reactance of 920 Ω at 10.7 MHz?

11. Calculate the needed capacitance to give a reactance of 2200 Ω at 27.5 MHz.

12. At what frequency will a 39 pF capacitor provide a reactance of 450 Ω?

13. A certain capacitor has a reactance of 1000 Ω at 500 kHz. What is its reactance at 1 MHz? At 5 MHz?

14. Refer to Figure 16–10. What is the circuit's equivalent capacitance at a frequency of 400 Hz?

15. Three capacitors are connected in series across a 90 V ac source. Their reactances are $X_{C_1} = 200$ Ω, $X_{C_2} = 400$ Ω, and $X_{C_3} = 600$ Ω. (a) Draw the schematic diagram. (b) Calculate their total reactance. (c) Calculate I_T. (d) Determine the voltage across each capacitor.

16. Three capacitive reactances of 900 Ω each are parallel-connected. (a) What is the equivalent reactance? (b) If they are connected to a 600 kHz source, what is their equivalent capacitance?

17. The input resistance of a transistorized audio amplifier is 5 kΩ. Calculate the value of coupling capacitor necessary if its reactance must equal the input resistance at 20 Hz.

18. A bypass capacitor passes 0.025 A of current when connected across a 208 V, 60 Hz line. Calculate the reactance and value of the capacitor. To double the current, what value of C do you need?

19. Determine the frequencies that will cause the following capacitors to have an X_C of 5 kΩ: (a) 220 pF (b) 0.068 μF (c) 4 μF.

20. Calculate the inductance of a coil that has 500 Ω inductive reactance at 10.7 MHz.

21. At what frequency will an inductor of 27 mH have $X_L = 1.3$ kΩ?

22. A bypass capacitor in a transistor amplifier has a value of 100 μF. What current will pass through it at 20 Hz when $V_C = 2.5$ V_{rms}?

23. A certain inductor has a reactance of 1000 Ω at 610 kHz. A capacitor connected across it must have the same reactance at the same frequency. What value of capacitance is required?

24. The output of a transistor is capacitively coupled to the input of another that has an input resistance of 7.5 kΩ. Calculate the required value of capacitance if its reactance must be 0.1 times the input resistance of the second stage at 16 Hz.

17

Series and Parallel ac Circuits

474

Objectives

After studying this chapter, you will be able to:

Calculate the value of current, voltage, and power in ac circuits.

Explain the use of rectangular and polar coordinates in solving complex circuits.

Introduction

Purely inductive and capacitive circuits have more academic interest than practical value. While it is important to know the voltage and current relationships in these ideal devices, it is equally important to understand their practical aspects when series or parallel resistance is present, because the amount of resistance, either in series or in parallel, or both, has a significant effect on actual performance. It can vary the device's characteristics between the extremes of a pure inductance or capacitance to one that is nearly all resistance.

Applications

An understanding of the behavior and uses of the most complicated networks of resistors, inductors, and capacitors begins with the analysis of the simpler combinations discussed in this chapter. In earlier application sections, devices such as filters and resonant circuits have been mentioned. All of the ac properties of these circuits depend on the phase-shifting behavior of capacitors and inductors and can be directly visualized by using an oscilloscope.

An understanding of the performance of series and parallel ac circuits is imperative for the study of typical AM and FM receivers, since these contain a number of series and parallel combinations of L, C, and R. The same can be said of transmitting circuits, such as in the broadcast and TV industries. Also, the popular CB units and cassette recorder/reproducers are further examples of these basic circuits.

17–1
The *j* Operator

j **operator**   A term used to indicate that an electrical quantity has gone through a 90° rotation; that is, j refers to 90° in a counterclockwise (ccw) direction, and $-j$ refers to 90° in a clockwise (cw) direction.

A positive number may be defined as any number greater than zero. It is plotted to the right of zero on the x axis, as shown in Figure 17–1. A negative number is some quantity away from a reference point (to the left of zero along the x axis in Figure 17–1). In electronics and electrical operations, it is frequently necessary to show quantities that are rotated exactly 90° in either a clockwise (cw) or counterclockwise (ccw) direction from the x axis. Engineers have adopted the *j* oper-

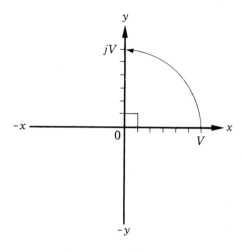

Figure 17–1

Representation of voltage V affected
by the operator j

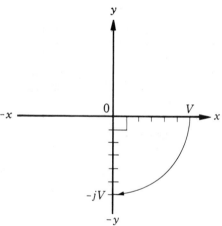

Figure 17–2

Representation of voltage V affected
by the operator $-j$

ator to designate this operation. When the letter j is placed in front of an electrical quantity, the implication is that it has gone through a 90° rotation.

In Figure 17–1, the voltage V, lying on the x axis, can be operated on by the *operator j* to become jV. The quantity still has the same magnitude but has been rotated ccw exactly 90° and lies on the y axis. We are *not* to assume that jV means j multiplied by V. Placing j in front of V implies that it is operating on the quantity V and that it has been rotated 90° in a ccw direction.

Any quantity operated on by j will rotate 90° in a ccw direction. Any quantity operated on by $-j$ will rotate 90° in a cw direction. See Figure 17–2.

An inductively reactive voltage of 20 V would be written as $j\,20$ V, implying that the voltage is leading the current by 90°. If the voltage were capacitively reactive, it would be written as $-j\,20$ V. Similarly, the j operator can be used to indicate leading or lagging currents or inductive and capacitive reactances.

17–2
Series *RL* Circuits

When resistance is added in series with a pure inductance, the circuit's characteristics change. The current no longer lags exactly 90° behind the applied voltage, but it varies somewhere between 90° and 0°, depending on the relative values of R and X_L. In the circuit of Figure 17–3, the line current I_S is the same through each component, a basic characteristic of series circuits. Since R and X_L each equal 100 Ω, we can assume that the IR and IX_L drops will be the same. However, the arithmetic sum of V_R and V_L does not equal the supply, as is the case with purely resistive circuits, but the vector sum does.

Previously, we saw that the reactive voltage across an induc-

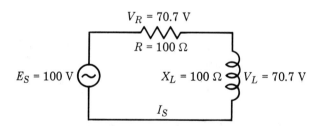

Figure 17–3

Series *RL* circuit

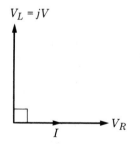

Figure 17–4

Phasor diagram of series *RL* circuit

tor leads the current through it by 90°. Also, we know that the voltage drop across a resistor is in phase with the current through it. Therefore, voltages V_R and V_L are 90° out of phase, as represented by the phasor diagram in Figure 17–4. Observe that the supply current (line current) is in phase with V_R and that V_L is 90° counterclockwise from this reference and is written as jV.

Waveforms

Any graph showing several waveforms synchronized to a common time base(s) can be called a *synchrogram*. The synchrogram in Figure 17–5 shows the time relationship of the voltages and current for the series RL circuit of Figure 17–3. The supply current sinusoid I_S is the *reference waveform* and produces a corresponding waveform across the resistor V_R, which is in phase with it. Observe that the reactive

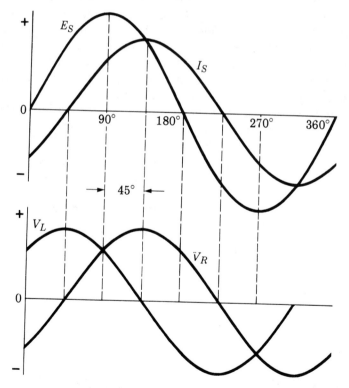

Figure 17–5

Synchrogram of current and voltage waveforms for the circuit in Figure 17–3 voltage V_L leads V_R by 90°. Each of these waveforms has the same amplitude because R and X_L are each 100 Ω.

Now, look at the waveforms of the supply voltage E_S and the supply, or line, current I_S. Notice that they are only 45° apart, with E_S leading. In other words, a circuit of this kind, having equal values of R and X_L, will have the line current lagging 45° behind the supply voltage, not lagging 90°, as in a purely inductive circuit. If the resistance of the circuit in Figure 17–3 were increased and X_L left unchanged, the phase angle between E_S and I_S would *decrease*. Conversely, if R were made smaller than X_L, the phase angle would *increase*.

Combining V_R and V_L

Since the voltages across R and L in the circuit of Figure 17–3 are 90° out of phase (see Figure 17–4), they cannot simply be added to find the total circuit voltage or supply. By adding the phasors in Figure 17–4, we can readily determine the magnitude of the supply potential. This phasor addition is done by adding the tail of the V_L phasor to the arrowhead of the V_R phasor, at an angle of 90°. The resulting phasor diagram is shown in Figure 17–6, where the hypotenuse of the right triangle represents E_S. Since V_R is in phase with the line current, the angle Θ (Greek capital letter *theta*) represents the number of degrees that the line current lags the supply voltage (in this case, 45°).

Figure 17–6 is a graphical, or phasor, representation of the circuit in Figure 17–3. Mathematically, this relationship can be expressed as follows:

$$E_S = \sqrt{V_R^2 + V_L^2} \qquad\qquad \textbf{17–1}$$

Example 17–1

From the voltages across R and L in the circuit of Figure 17–3, verify that $E_S = 100$ V.

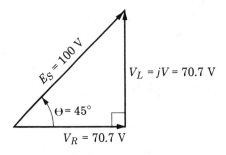

$V_L = jV = 70.7$ V

Θ = 45°

$V_R = 70.7$ V

Figure 17–6

Addition of phasors

Solution

$$E_S = \sqrt{V_R^2 + V_L^2} = \sqrt{70.7^2 + 70.7^2}$$
$$= \sqrt{5000 + 5000} = \sqrt{10,000} = 100 \text{ V}$$

Example 17-2

In a series RL circuit, $V_R = 50$ V and $V_L = 40$ V. What is the supply voltage?

Solution

$$E_S = \sqrt{50^2 + 40^2} = \sqrt{2500 + 1600} = \sqrt{4100} = 64 \text{ V}$$

By transposing Equation 17-1, we can find V_L when E_S and V_R are known. The next example shows the procedure.

Example 17-3

The following voltages are observed in a series RL circuit: $E_S = 50$ V and $V_R = 40$ V. What is the value of V_L?

Solution

Remove the radical sign in Equation 17-1:

$$E_S^2 = V_R^2 + V_L^2$$
$$V_L^2 = E_S^2 - V_R^2$$
$$V_L = \sqrt{E_S^2 - V_R^2} = \sqrt{50^2 - 40^2} = \sqrt{2500 - 1600} = \sqrt{900}$$
$$= 30 \text{ V}$$

17-3
Phase Angles

Phase angle The angle a phasor makes with respect to the zero or horizontal axis; the lead or lag of voltages (or currents) with respect to each other expressed in degrees or radians.

In Figures 17–5 and 17–6, we observed that the current was lagging behind the supply voltage by 45°. This lag is referred to as the **phase angle** between E_S and I_S. By trigonometry, we can calculate the phase angle Θ from its sine, cosine, or tangent functions. In this case, we will use the cosine function:

$$\cos \Theta = \frac{V_R}{E_S}$$

The next example illustrates this calculation.

Example 17–4

Verify that the phase angle between V_R and E_S for the circuit in Figure 17–3 is 45°.

Solution

$$\cos \Theta = \frac{V_R}{E_S} = \frac{70.7}{100} = 0.707$$

$$\cos^{-1} 0.707 = 45°$$

Therefore $\qquad\qquad \Theta = 45°$

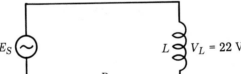

Figure 17–7

Series *RL* circuit for Example 17–5

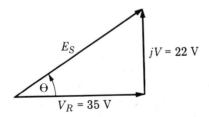

Figure 17–8

Phasor diagram for Example 17–5

Because the circuit contains inductance, the angle represents a lagging condition of I_S with respect to E_S.

Example 17–5

From the information given in the circuit of Figure 17–7, determine the phase angle between the supply voltage and the line current, and calculate E_S.

Solution

Since the supply voltage is not given, we must use the tangent function to find Θ:

$$\tan \Theta = \frac{V_L}{V_R} = \frac{22 \text{ V}}{35 \text{ V}} = 0.63$$

$$\tan^{-1} 0.63 = 32.15°$$

We can calculate the magnitude of E_S by either the sine or the cosine function. In this case, we will use the sine. Refer to the phasor diagram in Figure 17–8.

$$\sin \Theta = \frac{V_L}{E_S}$$

Transpose
$$E_S = \frac{V_L}{\sin \Theta} = \frac{22 \text{ V}}{0.532} = 41.34 \text{ V}$$

17–4
Series *RC* Circuits

The procedures outlined in the previous sections for the series RL circuit can be adapted to the series RC circuit. The only difference is that the phase angle between supply voltage and line current will be in the fourth quadrant, thus indicating a leading phase angle.

The series RC circuit of Figure 17–9 will be used to explain how unknown voltages and phase angles can be calculated. Several examples will also demonstrate these procedures.

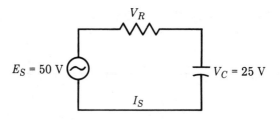

Figure 17–9

Series *RC* circuit

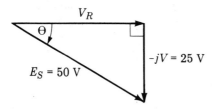

Figure 17–10

Phasor diagram for the *RC* circuit in Figure 17–9

Example 17–6

Calculate the value of V_R for the circuit in Figure 17–9 from the information given.

Solution

$$E_S = \sqrt{V_R^2 + V_C^2} \qquad \text{17–2}$$

This equation is the same as Equation 17–1, except that V_C has been substituted for V_L. Solve for V_R:

$$V_R^2 = E_S^2 - V_C^2$$

$$V_R = \sqrt{E_S^2 - V_C^2} = \sqrt{50^2 - 25^2} = \sqrt{2500 - 625} = \sqrt{1875}$$

$$= 43.3 \text{ V}$$

In an *RC* circuit, a leading phase angle exists between the supply voltage and the line current. The phasor diagram of Figure 17–10 represents the circuit voltages shown in Figure 17–9. Angle Θ is the angle by which I_S leads E_S.

Example 17-7

By what angle does the current lead the supply voltage in the circuit of Figure 17-9?

Solution

Use the phasor diagram:

$$\sin \Theta = \frac{V_C}{E_S} = \frac{25 \text{ V}}{50 \text{ V}} = 0.5$$

$$\sin^{-1} 0.5 = 30°$$

Example 17-8

In an RC circuit, the following voltage readings were obtained: $V_R = 45$ mV and $V_C = 86$ mV. By what angle does the current lead the voltage?

Solution

We must use the tangent function for this solution:

$$\tan \Theta = \frac{V_C}{V_R} = \frac{86 \text{ mV}}{45 \text{ mV}} = 1.91$$

$$\tan^{-1} 1.91 = 62.38°$$

Example 17-9

Determine the magnitude of V_R trigonometrically from the data given in Example 17-7.

Solution

Any of the three basic trigonometric functions can be used in this solution. We will use the cosine:

$$\cos \Theta = \frac{V_R}{E_S}$$

$$V_R = E_S \cos \Theta = 50 \text{ V} \times 0.866 = 43.3 \text{ V}$$

(Angle $\Theta = 30°$ from Example 17-7.)

17–5
Impedance

> ┌─ **Impedance (Z)** The total opposition (that is, resistance and reactance) a circuit offers to the flow of an alternating current; measured in ohms.

Alternating current in a circuit containing resistance and reactance is not limited by the sum of R and X. These quantities must be *combined* vectorially in the series circuit in order to determine the total opposition. The vector sum of R and X is called **impedance**, and its symbol is Z. Impedance is measured in ohms.

Impedance, then, is *the total opposition that a circuit offers to an alternating current.* For a circuit containing only R and X_L, its impedance may be calculated from the following formula:

$$Z = \sqrt{R^2 + X_L^2} \qquad \text{17–3}$$

Equation 17–3 is similar to Equations 17–1 and 17–2, which are used in the solution of right-angle triangles. For example, if a series circuit has $R = 15\ \Omega$ and $X_L = 20\ \Omega$ $(j = 20\ \Omega)$, it can be represented by the vector diagram in Figure 17–11. The impedance of the circuit can be calculated by one of two methods, as shown in the following two examples.

Example 17–10 ────────────────────────────

Determine the value of Z for the circuit represented by the vector diagram in Figure 17–11. Use the right-triangle method.

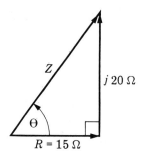

$j\ 20\ \Omega$

θ

$R = 15\ \Omega$

Figure 17–11

Vector diagram of series *RL* circuit

Solution

$$Z = \sqrt{R^2 + X_L^2} = \sqrt{15^2 + 20^2} = \sqrt{225 + 400} = \sqrt{625} = 25 \ \Omega$$

Example 17-11

Calculate the impedance of the circuit represented by the vector diagram in Figure 17-11. Use a trigonometric function.

Solution

The first step is to determine the phase angle Θ:

$$\tan \Theta = \frac{X_L}{R} = \frac{20}{15} = 1.33$$

$$\tan^{-1} 1.33 = 53.13°$$

Now, to calculate Z, we can use either the sine or cosine function. In this case, we use the cosine:

$$\cos \Theta = \frac{R}{Z}$$

$$Z = \frac{R}{\cos \Theta} = \frac{15}{\cos 53.13°} = \frac{15}{0.6} = 25 \ \Omega$$

This result agrees with the result obtained in Example 17-10.

To calculate the impedance of a series RC circuit, use the following formula:

$$Z = \sqrt{R^2 + X_C^2} \qquad\qquad \text{17-4}$$

17-6
Z in Rectangular Notation

Rectangular notation A convenient method of expressing an impedance by means of x (resistance) and y (reactance) values.

The convenient method of expressing an impedance by means of x and y values is known as **rectangular notation**. In a series circuit, R

is always assumed to be on the *real* or *x* axis, and the reactive components are on the *y* axis. Therefore, the general expression for impedance
may be written as follows:

$$Z = R \pm jX \qquad\qquad \textbf{17-5}$$

where $+j$ represents inductive reactance and $-j$ represents capacitive
reactance.

In the circuit of Figure 17-12a, the circuit impedance is written
as $Z = 8 + j\,6\ \Omega$. This equation accurately describes the circuit as consisting of a resistance of 8 Ω in series with an inductive reactance of 6 Ω. In
Figure 17-12b, the circuit impedance is $Z = 10 - j\,15\ \Omega$. The first term of
the impedance is always the resistance, and the second is always the reactance. If a circuit has a pure capacitive reactance of 120 Ω, its impedance
is written as $Z = 0 - j\,120\ \Omega$. Since no resistance is present, we must
write a zero; it is not correct to simply omit the term.

When opposing reactances are in series, as in Figure 17-13,
the impedance can be written as follows:

$$Z = R + j\,(X_L - X_C) \qquad\qquad \textbf{17-6}$$

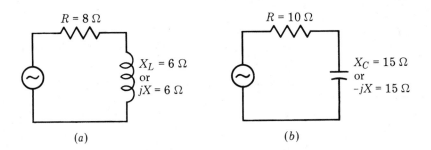

Figure 17-12

Series circuits: (a) Impedance $Z = 8 + j\,6\ \Omega$. (b) Impedance $Z = 10 - j\,15\ \Omega$.

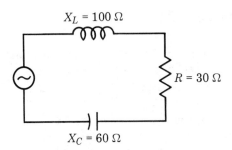

Figure 17-13

Series *RCL* circuit

Example 17–12 ━━━━━━━━━━━━━━━━━━━━━━━━━━━━━━━

Write the impedance of the circuit in Figure 17–13 in rectangular form.

Solution

$$Z = R + j(X_L - X_C) = 30 + j(100 - 60) = 30 + j\,40\ \Omega$$

17–7
Voltage in Rectangular Notation

The total voltage of a series circuit containing resistive and reactive voltages can be expressed in rectangular form by modifying Equation 17–5 to state the voltages in the circuit. Thus:

$$E_T = V_R \pm jV \qquad \textbf{17–7}$$

The next example illustrates this equation.

Example 17–13 ━━━━━━━━━━━━━━━━━━━━━━━━━━━━━━━

Express the total voltage of the circuit in Figure 17–14a in rectangular notation.

Solution

$$E_T = V_R - jV = 60 - j\,80\ \text{V}$$

The phasor diagram of this circuit is shown in Figure 17–14b.

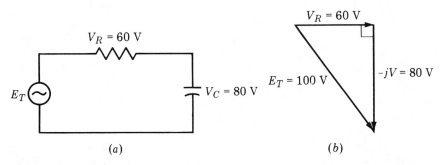

(a)   (b)

Figure 17–14

Series circuit for Example 17–13: (a) Circuit. (b) Phasor diagram.

17–8
Z in Polar Notation

> ─ **Polar notation**   A convenient method of expressing an impedance in terms of a phasor and an angular quantity.

We have shown that an impedance can be described by means of its rectangular coordinates. There is another, equally convenient way to express an impedance, by **polar notation**. For example, the phasor diagram in Figure 17–15 represents a series RL circuit with a rectangular impedance of Z, which is equal to $8 + j\,6\,\Omega$. This impedance can also be described as a phasor with a magnitude of $10\,\Omega$ at an angle of $36.87°$. This description is polar notation and is written as follows:

$$Z = 10 \angle 36.87° \; \Omega \qquad\qquad \textbf{17–8}$$

If the circuit had $6\,\Omega$ of capacitive reactance instead of $6\,\Omega$ of inductive reactance, the impedance in polar form would be as follows:

$$Z = 10 \angle -36.87° \; \Omega$$

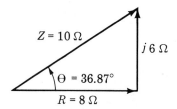

Figure 17–15

Phasor diagram of series RL circuit with polar notation

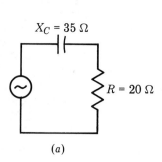

(a)

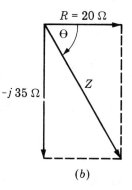

(b)

Figure 17–16

Series RC circuit for Example 17–14: (a) Circuit. (b) Phasor diagram.

Observe that the sign of the phase angle is the same as that of j in rectangular form. Neither the rectangular form nor the polar form is a method for solving series circuit problems. They are simply convenient forms of notation that describe circuit conditions from mathematical and electrical viewpoints.

Example 17–14

Determine the impedance of the circuit in Figure 17–16a in polar form.

Solution

Given
$$Z = R - jX = 20 - j\,35\ \Omega$$

So
$$\tan \Theta = \frac{X}{R} = \frac{-35}{20} = -1.75$$
$$\tan^{-1} 1.75 = -60.26°$$

$$Z = \frac{X}{\sin \Theta} = \frac{35}{\sin 60.26} = 40.31\ \Omega$$

or
$$Z = \frac{R}{\cos \Theta} = \frac{20}{\cos 60.26} = 40.31\ \Omega$$

Therefore
$$Z = 40.31\ \angle{-60.26°}\ \Omega$$

The phasor diagram is shown in Figure 17–16b.

Example 17–15

Determine the impedance of the series RCL circuit in Figure 17–17a in polar form.

Solution

Given
$$Z = R + jX - jX = 40 + j\,60 - j\,30 = 40 + j\,30\ \Omega$$

See the phasor diagram in Figure 17–17b.
$$\tan \Theta = \frac{X}{R} = \frac{30}{40} = 0.75$$
$$\tan^{-1} 0.75 = 36.87°$$

$$Z = \frac{R}{\cos \Theta} = \frac{40}{\cos 36.87°} = \frac{40}{0.8} \doteq 50\ \Omega$$

Therefore
$$Z = 50\ \angle 36.87°\ \Omega$$

The total voltage of a series circuit containing resistive and reactive voltages can also be expressed in polar form.

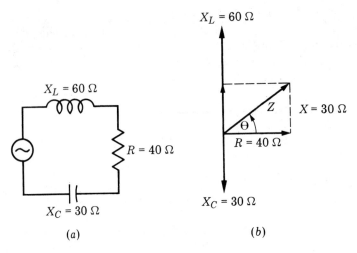

Figure 17-17

Series *RCL* circuit for Example 17-15: (a) Circuit. (b) Phasor diagram.

17-9
Ohm's Law for ac Circuits

Ohm's law can be readily applied to an ac circuit by merely substituting Z for R. Thus:

$$I = \frac{E}{Z} \quad \text{or} \quad I = \frac{V}{Z} \qquad \text{17-9}$$

Example 17-16

A series circuit consists of 125 Ω of resistance and 180 Ω of inductive reactance. Determine the current if 12 V, 60 Hz is applied.

Solution

Determine the circuit impedance:

$$Z = \sqrt{125^2 + 180^2} = \sqrt{48,025} = 219.15 \ \Omega$$

Determine the current:

$$I = \frac{E}{Z} = \frac{12 \ \text{V}}{219.15 \ \Omega} = 0.055 \ \text{A} = 55 \ \text{mA}$$

Example 17-17 ─────────────────────────────────

Calculate the impedance of a circuit if 85 mA flows when 22 V ac is applied.

Solution

$$I = \frac{E}{Z}$$

$$Z = \frac{E}{I} = \frac{22 \text{ V}}{85 \text{ mA}} = 0.259 \text{ k}\Omega = 259 \text{ }\Omega$$

Example 17-18 ─────────────────────────────────

What voltage will appear across an impedance of 650 Ω if 75 mA flows through it?

Solution

$$V = IZ = 0.075 \text{ A} \times 650 \text{ }\Omega = 48.75 \text{ V}$$

17-10
Series *RCL* Circuits

When a series circuit contains resistance plus capacitive and inductive reactance, the reactances *oppose* each other. The circuit exhibits a net reactance *equal to the difference* between them and *in favor of the largest reactance*. For example, assume a circuit has the following characteristics: $R = 15 \text{ }\Omega$, $X_L = 35 \text{ }\Omega$, and $X_C = 25 \text{ }\Omega$. A vector diagram of this circuit is shown in Figure 17-18. The opposing effects of X_L and X_C are plainly manifest. Note that the resulting, or net, reactance is in the direction of the larger reactance and is j 10 Ω. The impedance triangle of the circuit (drawn in heavy lines in the figure) is $R = 15 \text{ }\Omega$, $jX = 10 \text{ }\Omega$, and an impedance Z equal to the circuit's impedance.

The ohmic value of Z can be determined by the following equation:

$$Z = \sqrt{R^2 + (X_L - X_C)^2} \qquad \textbf{17-10}$$

Observe that the terms X_L and X_C (in the parentheses) are subtracted, as they were in the vector diagram of Figure 17-18.

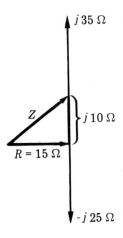

$j\,35\;\Omega$

Z

$j\,10\;\Omega$

$R = 15\;\Omega$

$-j\,25\;\Omega$

Figure 17–18

Net reactance for a series *RCL* circuit

Example 17–19

From the circuit parameters indicated in Figure 17–18, determine the value of Z.

Solution

$$Z = \sqrt{R^2 + (X_L - X_C)^2} = \sqrt{15^2 + (35 - 25)^2}$$
$$= \sqrt{225 + 100} = \sqrt{325} = 18.03\;\Omega$$

Of the two reactances, X_L predominates. Therefore, the current will lag the supply voltage.

Example 17–20

Determine the phase angle between supply voltage and line current in the circuit represented by the vector diagram of Figure 17–18.

Solution

$$\cos\Theta = \frac{R}{Z} = \frac{15}{18.03} = 0.832$$

$$\cos^{-1}0.832 = 33.69° \quad \text{(lagging)}$$

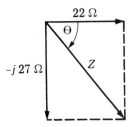

Figure 17–19

Impedance vector for Example 17–21

Example 17–21

An RCL circuit has the following parameters: $R = 22\ \Omega$, $X_L = 12\ \Omega$, and $X_C = 39\ \Omega$. Calculate Z in rectangular and polar form.

Solution

$$Z = 22 + j\,12 - j\,39\ \Omega = 22 - j\,27\ \Omega$$

This solution is represented by the vector diagram in Figure 17–19.

For the polar form, determine the phase angle Θ:

$$\Theta = \tan^{-1}\frac{-27}{22} = -50.83°$$

$$Z = \frac{22}{\cos\Theta} = \frac{22}{0.632} = 34.83\ \Omega$$

Therefore $Z = 34.83\ \angle{-50.83°}\ \Omega$

Circuits may consist of more than one element of R, C, and L. A multielement passive circuit is shown in Figure 17–20. While this circuit is more complicated than those previously considered, it can still be solved by using the same procedures of simpler circuits. Begin by grouping all like parameters into Equation 17–10. Then, form the calculations as required. The next example indicates the solution.

Example 17–22

For the circuit of Figure 17–20, calculate the circuit impedance, calculate the line current, and determine the voltage drop across X_{L_2}.

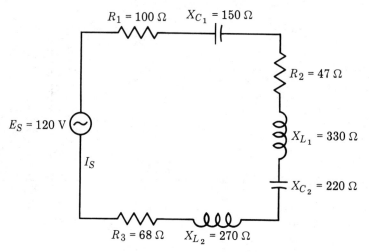

Figure 17–20

Multielement series *RCL* circuit

Solution

Determine circuit impedance:

$$Z = \sqrt{(R_1 + R_2 + R_3)^2 + [\,(X_{L_1} + X_{L_2}) - (X_{C_1} + X_{C_2})\,]^2}$$

$$= \sqrt{(100 + 47 + 68)^2 + [\,(330 + 270) - (150 + 220)\,]^2}$$

Combine terms:

$$Z = \sqrt{215^2 + (600 - 370)^2} = \sqrt{215^2 + 230^2}$$

$$= \sqrt{46{,}225 + 52{,}900} = \sqrt{99{,}125} = 314.84 \ \Omega$$

Determine line current:

$$I_S = \frac{E_S}{Z} = \frac{120 \text{ V}}{314.84 \ \Omega} = 0.381 \text{ A} = 381 \text{ mA}$$

Determine the voltage drop across X_{L_2}:

$$V_{L_2} = I_S X_{L_2} = 0.381 \times 270 = 102.91 \text{ V}$$

17–11
Parallel *RL* Circuits

The impedance of a parallel *RL* circuit *cannot* be determined in the same manner as the impedance of a series circuit. In series circuits, current or resistance serves as the reference vector. In a parallel circuit, the applied voltage is the reference vector, because it is common to all elements. Because the currents in each branch are out of phase, their phasors must be added. Because phasors require adding, the impedance of a parallel circuit cannot be determined as though we were dealing with two parallel resistors. The next example illustrates the correct procedure.

Example 17–23 ────────────────────────────

For the circuit of Figure 17–21, calculate the circuit impedance.

Solution

Calculate each branch current:

$$I_R = \frac{E_S}{R} = \frac{100 \text{ V}}{100 \text{ } \Omega} = 1 \text{ A}$$

$$I_L = \frac{E_S}{X_L} = \frac{100 \text{ V}}{100 \text{ } \Omega} = 1 \text{ A}$$

Because X_L represents a pure inductance, its current will be 90° out of phase with I_R. Therefore:

$$I_S = \sqrt{I_R^2 + I_L^2} = \sqrt{1^2 + 1^2} = \sqrt{2} = 1.41 \text{ A} \qquad \textbf{17–11}$$

Therefore $Z = \dfrac{E_S}{I_S} = \dfrac{100}{1.41} = 70.7 \text{ } \Omega$

A phasor diagram illustrating the currents in the circuit of Figure 17–21 is shown in Figure 17–22. Note that the inductive current phasor is in the fourth quadrant and is lagging behind the supply by angle Θ. Since the resistive current I_R is in phase with the supply, it serves as the reference.

If no voltage is indicated in a parallel circuit, any convenient value can be *assumed* when determining the circuit's impedance, because the voltage cancels out in the final calculation. A circuit's impedance is independent of the supply voltage. The next example illustrates the solution technique.

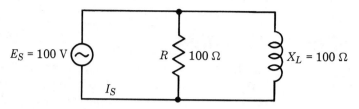

Figure 17-21

Parallel *RL* circuit for Example 17-23

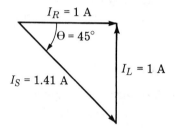

Figure 17-22

Phasor diagram for *RL* circuit in Figure 17-21

Example 17-24

Determine the impedance of the circuit in Figure 17-23.

Solution

Since no emf is indicated, we will assume one – let's say 100 V.

$$I_L = \frac{100 \text{ V}}{1.5 \text{ k}\Omega} = 67 \text{ mA}$$

$$I_R = \frac{100 \text{ V}}{1.2 \text{ k}\Omega} = 83 \text{ mA}$$

$$I_T = \sqrt{I_R^2 + I_L^2} = \sqrt{83^2 + 67^2} = \sqrt{6889 + 4489}$$
$$= 106.67 \text{ mA}$$

Therefore $\quad Z = \dfrac{E}{I_T} = \dfrac{100 \text{ V}}{106.67 \text{ mA}} = 0.937 \text{ k}\Omega$

17-12
Parallel *RC* Circuits

Parallel *RC* circuits are treated in precisely the same manner as the parallel *RL* circuit outlined in Section 17-11. The current phasor for the capacitive branch will be in the first quadrant, because its current leads

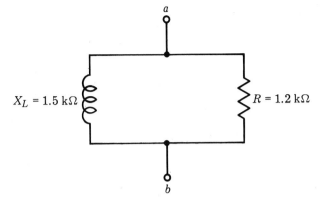

Figure 17–23

Parallel circuit for Example 17–24

the current in the resistive branch. The next example illustrates the method to follow in solving this kind of circuit.

Example 17–25

Determine the impedance of the RC circuit in Figure 17–24.

Solution

Since no supply potential is indicated, we will assume 10 V. Then:

$$I_C = \frac{E}{X_C} = \frac{10 \text{ V}}{200 \text{ }\Omega} = 50 \text{ mA}$$

$$I_R = \frac{E}{R} = \frac{10 \text{ V}}{150 \text{ }\Omega} = 67 \text{ mA}$$

$$I_T = \sqrt{I_R^2 + I_C^2} = \sqrt{67^2 + 50^2} = 83.6 \text{ mA} \qquad \textbf{17–12}$$

Therefore $Z = \dfrac{E}{I_T} = \dfrac{10 \text{ V}}{83.6 \text{ mA}} = 120 \text{ }\Omega$

The phasor diagram for the circuit of Figure 17–24 appears in Figure 17–25. The angle by which the capacitive current leads the assumed supply emf is as follows:

$$\tan \Theta = \frac{I_C}{I_R} = \frac{50}{67} = 0.746$$

$$\tan^{-1} 0.746 = 36.73°$$

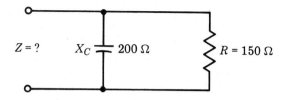

Figure 17–24

Parallel *RC* circuit for Example 17–25

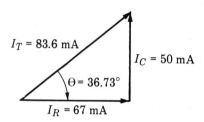

Figure 17–25

Phasor diagram for circuit in Figure 17–24

17–13
Parallel *RCL* Circuits

Circuits containing branches of pure R, C, or L can be readily solved by following the basic procedures outlined in Sections 17–11 and 17–12. Use the following formula:

$$I_T = \sqrt{I_R^2 + (I_C - I_L)^2}$$

17–13

The reactive currents, included within the parentheses of Equation 17–13, are 180° out of phase, as indicated by the fact that one current is subtracted from the other. This equation has the basic form of Equation 17–10. Its application to a parallel *RCL* circuit is given in the following examples.

Example 17–26

Determine the line current in the parallel *RCL* circuit of Figure 17–26.

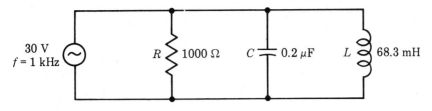

Figure 17–26

Parallel *RCL* circuit for Example 17–26

Solution

The first step is to calculate X_C and X_L:

$$X_C = \frac{1}{6.28 \times 10^3 \times 0.2 \times 10^{-6}} = 796 \ \Omega$$

$$X_L = 6.28 \times 10^3 \times 68.3 \times 10^{-3} = 429 \ \Omega$$

Next, determine the branch currents:

$$I_R = \frac{30 \text{ V}}{1000 \ \Omega} = 0.03 \text{ A}$$

$$I_C = \frac{30 \text{ V}}{796 \ \Omega} = 0.038 \text{ A}$$

$$I_L = \frac{30 \text{ V}}{429 \ \Omega} = 0.070 \text{ A}$$

Finally, the total line current is as follows:

$$I_T = \sqrt{I_R^2 + (I_L - I_C)^2} = \sqrt{0.03^2 + (0.07 - 0.038)^2}$$
$$= \sqrt{0.00192} = 0.0439 \text{ A}$$

The phasor diagram of current for this circuit is shown in Figure 17–27. Since the net reactive current I_X is in the fourth quadrant, the line current is lagging behind the supply voltage by angle Θ.

Example 17–27

In the circuit of Figure 17–28, assume that $X_{C_1} = 200 \text{ k}\Omega$, $R_1 = 15 \text{ k}\Omega$, $X_{L_1} = 50 \text{ k}\Omega$, $X_{C_2} = 150 \text{ k}\Omega$, $X_{L_2} = 33 \text{ k}\Omega$, and $R_2 = 27 \text{ k}\Omega$. Determine Z and Θ for the circuit.

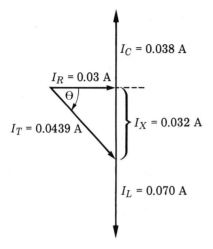

Figure 17–27

Phasor diagram of currents for circuit in Figure 17–26

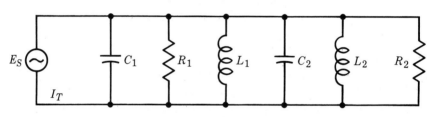

Figure 17–28

Multielement parallel *RCL* circuit for Example 17–27

Solution

First, assume a line voltage of 250 V and calculate all branch currents. The resistive currents are as follows:

$$I_{R_1} = \frac{250 \text{ V}}{15 \text{ k}\Omega} = 16.67 \text{ mA}$$

$$I_{R_2} = \frac{250 \text{ V}}{27 \text{ k}\Omega} = 9.26 \text{ mA}$$

$$I_{RT} = 25.93 \text{ mA}$$

The capacitive currents are as follows:

$$I_{C_1} = \frac{250 \text{ V}}{200 \text{ k}\Omega} = 1.25 \text{ mA}$$

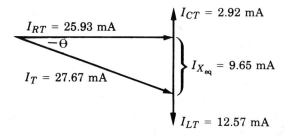

Figure 17–29

Phasor diagram of currents for circuit in Figure 17–28

$$I_{C_2} = \frac{250 \text{ V}}{150 \text{ k}\Omega} = 1.67 \text{ mA}$$

$$I_{CT} = 2.92 \text{ mA}$$

The inductive currents are as follows:

$$I_{L_1} = \frac{250 \text{ V}}{50 \text{ k}\Omega} = 5 \text{ mA}$$

$$I_{L_2} = \frac{250 \text{ V}}{33 \text{ k}\Omega} = 7.57 \text{ mA}$$

$$I_{LT} = 12.57 \text{ mA}$$

Second, calculate the net reactive current:

$$I_{X_{eq}} = I_{LT} - I_{CT} = 12.57 + (-2.92) = 9.65 \text{ mA} \quad \text{(inductive)}$$

Third, draw the phasor diagram of the currents and determine I_T. The phasor diagram is shown in Figure 17–29.

$$I_T = \sqrt{(I_{RT})^2 + (I_{X_{eq}})^2} = \sqrt{25.93^2 + 9.65^2} = 27.67 \text{ mA}$$

Fourth, calculate Z:

$$Z = \frac{250 \text{ V}}{27.67 \text{ mA}} = 9.035 \text{ k}\Omega$$

Finally, determine Θ:

$$\tan \Theta = \frac{-9.65}{25.93} = -0.372$$

$$\tan^{-1} -0.372 = -20.41°$$

17–14
Power in ac Circuits

In dc circuit analysis, the amount of power absorbed by a resistive network is determined by $P = I^2R$. For ac circuits, the calculation of power is a little more involved. Since both current and voltage vary with time, the product of e and i at any moment results in *instantaneous power*. Because e and i are generally out of phase by some angle, their product results in negative values during parts of each cycle. This negative factor is represented in Figure 17–30, for an RL circuit having a 45° lagging phase angle, by the shaded areas below the zero reference axis. This power is reactive and is returned to the source during the positive portions of the power curve. A series RC circuit having a 45° leading phase angle would produce the same overall waveform, except that the negative areas would be advanced 180°.

As the phase angle between e and i increases toward 90°, the negative power areas increase until they resemble the reactive power curves for purely inductive or capacitive circuits, which were shown in Figures 16–7 and 16–14. At this point, there is no resistance left in the circuit. From this discussion, we see that power is consumed only when resistance is present in a circuit.

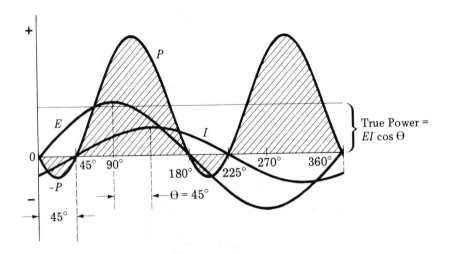

Figure 17–30

Power curve for series *RL* circuit

> **Apparent power**   The product of voltage and current in a single-phase circuit when they are not in phase.

The large positive and negative power curves associated with highly reactive circuits erroneously suggest that large quantities of power are consumed. For example, if 5 A flows through a capacitor connected across a 120 V high-frequency source, it may appear that 600 W is being dissipated. But the power delivered to the capacitor is returned to the source on the next half cycle. This power, which the source must supply (regardless of how much is returned), is called the **apparent power** and is equal to the product of the effective voltage and current. Hence:

$$P_A = EI$$

For the circuit of Figure 17–31, the apparent power is the product of 50 V and 0.707 A, or 35.35 W.

> **True power**   The product of voltage and current in a single-phase circuit times the power factor; the actual power consumed in a purely resistive ac circuit.

Now, apparent power must be distinguished from the actual power, called **true power**, consumed by the resistance. True power may be defined as *the average power dissipated by a circuit over a period of one cycle of input voltage.*

The difference between the areas of positive and negative power loops is the difference between the power delivered to the load and the power returned to the source. It is equal to the power consumed by the resistance of the circuit.

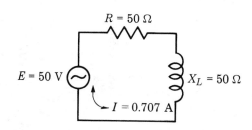

Figure 17–31

Series *RL* circuit

When reactance is present, the apparent power must be multiplied by the ratio R/Z, which is the cosine of the phase angle Θ. Therefore, true power is calculated by the following equation:

$$P_T = EI \cos \Theta \qquad \qquad \text{17-14}$$

If the phase angle of a circuit is 90°, then the true power is zero, since cos 90° equals zero. These conditions apply for the purely reactive circuit, which returns as much power as it receives. If the phase angle is 0° (cos 0° = 1), then the circuit is purely resistive.

Example 17-28

What is the true power consumed in the circuit of Figure 17-31?

Solution

The phase angle of the circuit must be determined. Since X_L and R are given, we can calculate Θ by the tangent function:

$$\tan \Theta = \frac{X_L}{R} = \frac{50 \ \Omega}{50 \ \Omega} = 1$$

$$\tan^{-1} 1 = 45°$$

Then $\quad P_T = EI \cos \Theta = 50 \times 0.707 \times 0.707 = 25 \text{ W}$

Note: An alternate method is to calculate Z, then find the cosine of the relationship R/Z, and finally, multiply the cosine by the product EI.

Example 17-29

Calculate the apparent and true power for the circuit in Figure 17-32.

Solution

First, calculate Z:

$$Z = \sqrt{R^2 + (X_L - X_C)^2} = \sqrt{100^2 + 200^2} \cong 224 \ \Omega$$

Second, determine line current:

$$I = \frac{E}{Z} = \frac{30}{224} = 0.134 \text{ A}$$

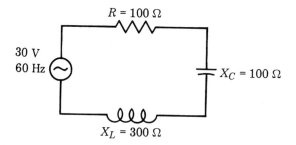

Figure 17–32

Series *RCL* circuit for Example 17–29

Third, calculate the phase angle:

$$\cos \Theta = \frac{R}{Z} = \frac{100}{224} = 0.446$$

$$\cos^{-1} 0.446 \cong 63.5°$$

Fourth, calculate apparent power:

$$P_A = EI = 30 \times 0.134 = 4.02 \text{ W}$$

Finally, calculate true power:

$$P_T = EI \cos \Theta = 4.02 \times 0.446 = 1.794 \text{ W}$$

Thus, 1.794 W is consumed in this circuit, but the generator must supply 4.02 W, 2.22 W of which is returned to the line by the net reactive element.

17–15
Power Factor

Calculation of Power Factor

Power factor   The ratio of the true power of an ac circuit to the apparent power; the ratio of *R/Z*; equal to cosine of the phase angle between applied voltage and circuit current.

The ratio of true power to apparent power in an ac circuit is called the **power factor** and can be expressed as follows:

$$PF = \frac{P_T}{P_A} \qquad 17\text{-}15$$

It is also defined as the ratio of resistance to impedance (series circuit):

$$PF = \frac{R}{Z} \qquad 17\text{-}16$$

Since R/Z is the cosine of the impedance triangle, we can say that the power factor is equal to the cosine Θ:

$$PF = \cos \Theta \qquad 17\text{-}17$$

Because the impedance of a circuit is equal to or greater than the resistance, the power factor is a *decimal fraction* that has a value between 0 and 1. When $\Theta = \pm 90°$, for a purely reactive circuit, PF is 0 because $\cos 90° = 0$. At the other extreme, when $\Theta = 0°$, for a purely resistive circuit, PF is 1 because $\cos 0° = 1$. Sometimes, the power factor is expressed as a percentage (multiplied by 100). The power factor is a dimensionless quantity because it is the ratio of two numbers with the same dimensions.

Example 17-30

What is the power factor for the circuit in Figure 17-32?

Solution

In Example 17-29, we determined the values of P_A and P_T. Therefore:

$$PF = \frac{P_T}{P_A} = \frac{1.794 \text{ W}}{4.02 \text{ W}} = 0.446 \text{ or } 44.6\%$$

In this circuit, X_L predominates. Consequently, this factor is a lagging power factor. If a circuit's impedance is equal to its resistance (circuit has no reactance and is all resistive), then PF would be unity, because the ratio R/Z is 1.

Power Factor Correction

A power factor of unity is desirable because the current and voltage are in phase. However, *in most industrial situations, circuits tend to be somewhat inductive because of the presence of motors, transformers,*

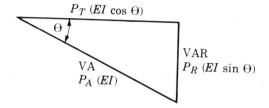

Figure 17–33

Power triangle for parallel RL circuit

and other forms of inductors. The lagging power factor thus produced can be corrected by adding an appropriate amount of capacitance to the circuit. When X_L approximately equals X_C, the circuit's reactance is reduced to a low value and the circuit presents a resistive load to the power line. Utility companies and large industries try to maintain the power factor as near unity as possible in order to keep the size of power transformers and associated wiring to a minimum.

Volt-ampere reactive (VAR) power   The product of E and I in a reactive ac circuit; sometimes called *wattless power.*

The relationship between P_A and P_T is shown by the power triangle in Figure 17–33. The horizontal vector represents true circuit power. The hypotenuse is the product of EI and is called volt-amperes (VA) power. The quadrature (separated by 90°) current in the inductor produces a reactive power component known as the **volt-ampere reactive (VAR) power.** This power is received from the line on one half cycle and returned on the other. It is sometimes called *wattless power* and is equal to $EI \sin \Theta$. The term VAR is used mostly in high-power applications and the electric utility fields.

Loads that have appreciable reactive currents require larger windings in the alternators, transformers, and connecting conductors than loads having power factors at or near unity, because the copper losses in the system (losses in the copper wire due to its resistance) depend on the square of the current (I^2R). Therefore, electric utility companies and large industries find it more economical to maintain their overall power factors as close to unity as possible.

17–16
ac Meters

Several types of basic meter movements are used in measuring ac voltages and currents. Each has specific characteristics; some are suitable for low frequencies only, while others can be used in the radio frequency spectrum. A brief explanation of the principles of operation of several of the more common types is given here.

ac Rectifiers

Any device, such as a silicon diode, that presents a high resistance to current in one direction and low resistance in the other can be used to *rectify* an alternating current. That is, it allows the current to pass during one half cycle but not the other. Such a device, called a *rectifier*, converts ac to dc. It is possible to connect a D'Arsonval meter movement to a rectifier and read ac. See Figure 17–34, where four diodes (rectifiers) are used in a *bridge configuration.*

During one half cycle, when the lower input terminal is negative, current is in the direction of the dotted arrows. (Electron flow is always against the direction of the arrow in the symbols representing the diodes.) Diodes 3 and 2 are in series with the meter during this half cycle, as we can see by tracing the direction of the dotted arrows. In the other half cycle (top input terminal negative), current is in the direction of the solid arrows. Diodes 1 and 4 are in series with the meter during this time interval. The current through the meter is always in the same direction (negative to positive) for each half cycle.

A rectifier meter such as is shown in Figure 17–34 operates on the *average* value of the ac voltage. Because rms values are more useful,

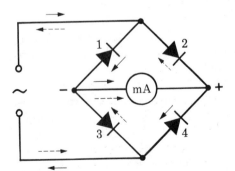

Figure 17–34

Simple bridge rectifier of ac voltmeter

such as in power calculations, the scales are usually calibrated in effective, or rms, values.

RF Ammeters

Conventional ammeters cannot be used to measure RF currents, because a large reactive voltage would develop across the meter coil, causing it to burn out. *Thermocouple ammeters* are used to measure these currents, and they operate in the following manner. If the ends of two dissimilar metallic wires are welded together and this junction is heated, a dc voltage is developed across the two open ends. The voltage developed depends on the kinds of metals used for the wires and on the difference in temperature between the heated junction and the open ends. The junction is heated electrically by the RF current passing through it. The heat produced by this current is proportional to the square of the current ($P = I^2R$). Because the voltage appearing across the two open ends of the wires is proportional to temperature, the movement of the meter element connected across these terminals is proportional to the square of the current through the thermocouple.

The scale of the meter is crowded near the zero end and is progressively less crowded near the maximum end of the scale. For the sake of accuracy, a meter should be selected that will give readings on the more expanded, open part of the scale for the circuit in which it is to be used.

17–17
Wattmeters

Electric power is measured by an instrument called a *wattmeter*. Figure 17–35a shows a typical wattmeter. It consists of a pair of fixed coils, called *current coils* (L_1 and L_2 in the schematic of Figure 17–35b), and one movable coil known as a *potential coil*. The current coils consist of a relatively few turns of large wire, while the voltage (potential) coil has many turns of fine wire. The potential coil is mounted on a shaft and carried in jeweled bearings so that it may turn freely inside the current coils when acted upon by the torque established by currents in both sets of coils. This movable coil carries a needle that moves over a suitably graduated scale. When the meter is deenergized, flat coil springs hold the needle in zero position.

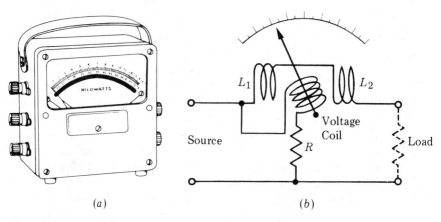

(a) (b)

Figure 17–35

Wattmeter: (a) Commonly used instrument. (b) Schematic.

From Figure 17–35b, we see that the current coils are in series with the load, while the potential coil is across the line. The field strength of the current coil is proportional to the line current and in phase with it. The potential coil has a high resistance (R) in series with it to make it appear as purely resistive as possible, so current in this coil is essentially in phase with the line voltage.

The force acting on the movable coil at any instant is proportional to the product of the instantaneous values of line current and voltage. Since the fixed coils produce a flux dependent on current, and the movable coil produces a flux dependent on voltage, the needle deflection is proportional to power – either dc or ac. For instance, if I is low when E is high (phase angle near 90°), little needle deflection occurs. Meter deflection is proportional to real power and is not affected by a power factor.

A *watt-hour meter* is designed to record the amount of electric power consumed over a particular period of time. The principle of operation is essentially the same as that of the wattmeter. However, it does not have an indicating needle. Instead, the current and voltage coils produce a torque that rotates the armature of a small electric motor, whose instantaneous speed is proportional to the power passing through it. As the armature of the motor rotates, it drives a set of small speed reduction gears to which are attached four (sometimes five) small pointers. Directly behind each pointer is a small circular dial marked off in ten equal divisions. The dials are read from right to left, with the first one reading 1 kW/h for each division of the dial. A complete revolution of the pointer on this dial will move the pointer of the second dial one division and register 10 kW/h. A complete revolution of the pointer of the second dial will move the third pointer one division and register 100 kW/h; and so on.

┌──Summary

The j operator is used to indicate electrical quantities on the y axis (rotated $90°$).

Placing j in front of a quantity implies that it is operating on the quantity to rotate it $90°$ in a ccw direction.

A quantity operated on by $-j$ will be rotated $90°$ in a cw direction.

In a series RL circuit, the current lags the applied voltage by some angle less than $90°$. The angle diminishes as R increases, assuming X_L remains constant.

In a series RL circuit, the voltage across R is in phase with the current, and the voltage across the inductance leads the current by $90°$.

The phasor sum of the inductive and resistive voltages in a series circuit equals the supply voltage:

$$E_S = \sqrt{V_R^2 + V_L^2}$$

Current in a series RC circuit leads the applied voltage. The angle of lead depends upon the ratio of X_C to R and can vary between $90°$ and $0°$.

The phasor sum of the capacitive and resistive voltages in a series circuit equals the supply voltage:

$$E_S = \sqrt{V_R^2 + V_C^2}$$

Impedance is the total opposition that a circuit offers to alternating current and is measured in ohms.

The impedance of a series circuit can be calculated by the following equation:

$$Z = \sqrt{R^2 + (X_L - X_C)^2}$$

Impedance can be expressed in rectangular form as $Z = R \pm jX$.

A purely capacitive circuit would have an impedance of $Z = 0 - jX$, expressed in rectangular notation.

A purely resistive circuit would have an impedance of $Z = R \pm j\,0$, expressed in rectangular form.

Resistive and reactive voltages can also be expressed in rectangular form, as $E_T = V_R \pm jV$.

Impedances can also be expressed in polar form as a phasor of a certain magnitude at an angle Θ; for example, $Z = 10 \angle 53° \; \Omega$.

The phase angle of an impedance in polar form has the same sign as that of j in rectangular form.

Ohm's law for the ac circuit is $I = E/Z$.

In a series RCL circuit, the net reactance is equal to the difference between X_L and X_C.

In series circuits, current is the reference vector.

In parallel circuits, the source voltage is the reference vector.

Total current in a parallel circuit is the algebraic sum of the branch phasors.

The total current in a parallel circuit can be calculated by the following formula:

$$I_T = \sqrt{I_R^2 + (I_L - I_C)^2}$$

The impedance of a parallel RCL circuit is determined by dividing the line current into the applied voltage (assumed voltage if no line voltage is given).

Power is consumed only in resistive elements of ac circuits.

Apparent power P_A equals EI.

True power P_T equals $EI \cos \Theta$.

The power factor is the ratio of true power to apparent power. It is calculated as $PF = \cos \Theta$.

Ideally, the power factor should be as near 1 (unity) as possible.

The quadrature currents in a parallel ac circuit are the result of reactive components. The reactive power component is known as VAR (volt-ampere reactive).

Rectifier ac voltmeters use diodes (usually in a bridge configuration) to rectify the ac voltage so that it can operate a D'Arsonval meter movement.

Thermocouple ammeters are used to measure radio frequency currents.

Wattmeters read true power and are not affected by the power factor.

Watt-hour meters are designed to read power consumed over a period of time.

⌐Progress Test

The bracketed number after each question indicates the section of this chapter where the answer can be found.

1. The j operator is used to indicate: (a) negative values on the x axis. (b) positive values on the x axis. (c) quantities that are 90° from the x axis. (d) quantities that are rotated between 0° and 90° from the real axis. [17–1]

2. An inductive reactance is written as: (a) jX. (b) Xj. (c) $-jX$. (d) $-Xj$. [17–1 and 17–6]

3. Adding series resistance to a pure inductance: (a) reduces the power factor. (b) increases the impedance. (c) increases the phase angle. (d) changes the j value. [17–2]

4. If $V_R = 8.6$ V and $V_L = 13.2$ V in a series circuit, the supply potential is: (a) 15.75 V. (b) 15.15 V. (c) 14.72 V. (d) 14.05 V. [17–2]

5. The phase angle between line current and supply voltage in Question 4 is: (a) 49.6°. (b) 51.7°. (c) 54.3°. (d) 56.9°. [17–3]

6. A series RC circuit is connected across a 45 V ac supply. If $V_R = 22$ V, then V_C is: (a) 41.52 V. (b) 39.26 V. (c) 36.09 V. (d) 33.84 V. [17–4]

7. The combined impedance of 1 kΩ of resistance in parallel with 1 kΩ of capacitive reactance is: (a) 2 kΩ. (b) 1 kΩ. (c) 707 Ω. (d) 500 Ω. [17–12]

8. The total impedance of a series circuit of $R = 1000$ Ω and $jX = 1000$ Ω is: (a) 2000 Ω. (b) 1414 Ω. (c) 1000 Ω. (d) 707 Ω. [17–6]

9. A series RCL circuit has the following values: $R = 20$ Ω, $X_C = 40$ Ω, and $X_L = 30$ Ω. The impedance in rectangular form is: (a) 20

$+ j$ 70 Ω. (b) $-j$ 20 $+$ 10 Ω. (c) 20 $+ j$ 10 Ω. (d) 20 $- j$ 10 Ω.   [17–6 and 17–10]

10.   For Question 9, the impedance in polar form is: (a) 22.36 $\angle -26.57°$ Ω. (b) 20.36 $\angle -22.57°$ Ω. (c) 24.52 $\angle -26.37°$ Ω. (d) 26.32 $\angle -26.36°$ Ω.   [17–8 and 17–10]

11.   A series circuit has an impedance of 30 $\angle 25°$ Ω. If 100 V ac is applied, the line current is: (a) 4.21 A. (b) 3.66 A. (c) 3.33 A. (d) 3.03 A.   [17–9]

12.   In a parallel circuit having a lagging power factor, the power factor can be improved by: (a) adding capacitance. (b) adding inductance. (c) increasing the resistance. (d) reducing the inductance.   [17–15]

Problems

Answers to odd-numbered problems are at the back of the book.

1.   Voltmeter readings across a series RL circuit indicate the following effective values: V_R = 27 V and V_L = 36 V. What is the value of the supply potential?

2.   What is the phase angle between the line current and voltage for the circuit described in Problem 1?

3.   Voltage readings in a series RCL circuit are V_R = 13 V, V_L = 29 V, and V_C = 12 V. Calculate the supply potential.

4.   A series RC circuit has a supply voltage of 120 V ac. If V_R = 69 V, what is the value of V_C?

5.   Draw the phasor diagram of the circuit described in Problem 4. What is the phase angle between line voltage and V_R?

6.   Calculate the impedance of a series circuit with parameters of X_L = 239 Ω and R = 192 Ω.

7.   What is the impedance of a series circuit having R = 83 Ω, X_L = 161 Ω, and X_C = 45 Ω?

8.   Express the impedance of the circuit described in Problem 7 in rectangular form.

9.   Draw the phasor diagram of the circuit described in Problem 7. What is the impedance in polar notation?

10.   Refer to the circuit of Figure 17–36. (a) Calculate the circuit impedance. (b) Express the impedance in rectangular form. (c) Express the impedance in polar notation.

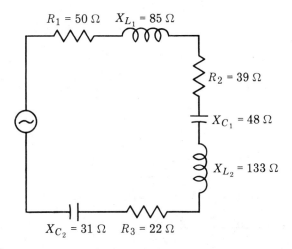

Figure 17-36

Circuit for Problems 10 and 11

11.   Refer to the circuit of Figure 17–36. (*a*) Calculate the line current if 50 V ac is applied to the circuit. (*b*) What is the value of the reactive voltage across X_{L_2}?

12.   Calculate the impedance of a parallel combination of 200 Ω resistance and 150 Ω inductive reactance.

13.   A capacitive reactance of 2.2 kΩ is shunted by a 3.3 kΩ resistor. (*a*) Calculate the impedance of the combination. (*b*) What is the phase angle between supply voltage and line current?

14.   What is the impedance of a parallel circuit consisting of R = 100 Ω, X_L = 180 Ω, and X_C = 120 Ω? What is the phase angle between supply voltage and line current?

15.   Find the circuit impedance and phase angle of a parallel circuit such as is shown in Figure 17–28, where X_{C_1} = 41 kΩ, R_1 = 18 kΩ, X_{L_1} = 68 kΩ, X_{L_2} = 27 kΩ, R_2 = 27 kΩ, X_{C_2} = 22 kΩ, and E_S = 230 V.

16.   A 2 H inductor is connected in series with a resistor. If 220 V, 60 Hz is connected across the combination and the current is 0.2 A, what is the ohmic value of the resistor?

17.   How much capacitance must be inserted in series with a 36 Ω load resistor for a current of 1.5 A from a 60 V, 2.5 MHz radio frequency transmitter?

18.   Calculate the apparent and true power for the circuit described in Problem 17.

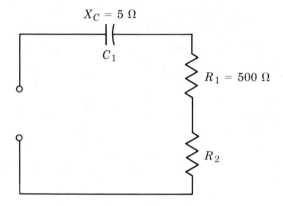

Figure 17–37

Circuit for Problem 23

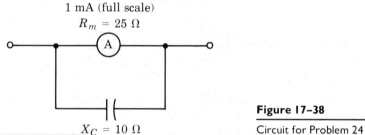

Figure 17–38

Circuit for Problem 24

19. A 220 pF capacitor of negligible resistance is in series with a 1 H inductor having an effective resistance of 250 Ω. How much apparent power is drawn from a source of 100 V at 10 kHz?

20. Calculate the power factor of the circuit described in Problem 19.

21. A series RCL circuit has the following reactive elements: $25 + j\,30$ Ω, $32 - j\,6$ Ω, $12 - j\,10$ Ω, and $+j\,2$ Ω. What is the impedance of the circuit in rectangular notation?

22. What is the phase angle of the circuit described in Problem 21?

23. In the circuit of Figure 17–37, if C_1 is a pure capacitance and the impedance looking into the circuit is $1000 + j\,5$ Ω, what is the value of R_2? If the resistors are wire-wound, what amount of X_L would they contribute to the circuit?

24.   In the circuit of Figure 17-38, if a pure capacitance with $X_C = 10 \ \Omega$ is connected across a resistive meter movement as shown, what impedance will the circuit have?

25.   What is the power factor of a series RCL circuit with $X_L = 60 \ \Omega$, $X_C = 40 \ \Omega$, and $R = 20 \ \Omega$?

26.   The power factor of a series circuit with a reactive phasor of $j \ 8 \ \Omega$ is 0.8. What resistance is in the circuit?

27.   A circuit with a PF of 0.83 draws 5 A when 120 V ac is applied. What are P_A, P_T, and P_R (VAR)?

28.   For the circuit in Figure 17-39, what is the impedance of the parallel network consisting of L_1 and C_2?

29.   What is the impedance of the entire circuit in Figure 17-39 in rectangular notation? In polar form?

30.   What are PF and P_T of the circuit in Figure 17-39?

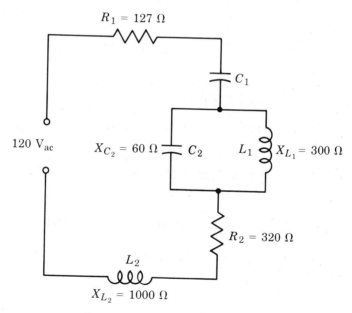

Figure 17-39

Circuit for Problems 28-30

18

Network Analysis of ac Circuits

Objectives

After studying this chapter, you will be able to:

Calculate the impedance of alternating current networks in electronic circuits.

Discuss the effects of series and parallel reactances and impedances in the output circuit of active devices (such as transistors).

Introduction

In the real world of applied electronics, circuit elements such as inductors have series resistance in their windings and distributed capacity across them. Capacitors include small series and large shunt resistances among their major characteristics, and carbon composition resistors have small shunt capacities across them. At low frequencies, these series and shunt elements may not be significant; but at high frequencies, their presence can be pronounced and troublesome. To determine the impedance of a circuit accurately, we must consider the effect of these series and parallel elements.

Applications

When parallel circuits contain branches made up of series arrangements of R, C, and L, we must first consider the phase relationships of
these branches in order to calculate the impedance of the entire circuit.
The phase angles necessitate the use of polar and rectangular notations.
While the techniques described in this chapter can be used in analyzing
any kind of ac circuit, they are particularly helpful in determining the
equivalent series circuit of a given parallel combination.

Technicians and engineers alike use hand-held scientific calculators to perform conversions from rectangular to polar coordinates,
thus simplifying the analysis of complex circuits. It may not be immediately appreciated, however, that the basic element of these calculators is itself a microminiature complex circuit, or *chip*, which contains
the equivalent of hundreds or thousands of resistors and transistors.

18–1
Vector Algebra

Phasor quantities are easily handled by trigonometric functions when
the values involved are in phase or in quadrature (at a 90° angle). However, when the circuit's parameters are at some angle other than 90°,
the solution becomes more difficult. Graphical solutions, for example,
are not sufficiently accurate. For efficient solution, the phasors must be
broken down into their real and reactive components and the like terms
added. The square root of the sum of their squares will then yield the
resultant value.

The use of vector algebra simplifies the solution of problems
involving phasors that are at angles other than 90°. There are two
commonly used systems of vector algebra. One uses rectangular coordinates and the other uses polar notation, both of which were studied
and used in Chapter 17. There are advantages and disadvantages –
perhaps limitations is a better word – to each system. Addition and
subtraction of vector quantities is easily achieved by using the rectangular coordinates system. However, multiplication, division, and
squaring are difficult to accomplish. The reverse is true of polar coordinates. So the choice of system clearly depends on what arithmetic
process is needed in a particular problem.

With the popularity of the hand-held scientific calculator, the problem of converting rectangular coordinates to polar or polar coordinates to rectangular is a matter of only a few manipulations. However, for those who are interested in the actual mathematical processes or who do not have access to these calculators, the step-by-step procedure to the solutions of these problems is detailed in this chapter.

Polar quantities that need to be added must first be converted to rectangular coordinates. They may then be added in their quadrature forms and finally converted to polar notation. If rectangular quantities are to be multiplied or divided, they must first be converted to polar form so that these operations can be performed. Then, the resultant can be converted to rectangular coordinates.

18–2
Addition and Subtraction by Vector Algebra

Addition of Vector Quantities

Voltages in a series ac circuit can be vectorially added or subtracted to determine the supply potential. Impedances can be added vectorially to find the total impedance, and in parallel circuits, the branch currents must be vectorially added to find the line current. In each of these cases, *the values must be expressed in rectangular form.* First, the real terms are added; then, the quadrature components are added. In summary, *for addition or subtraction of numbers in vector algebra, the numbers must be in rectangular form.*

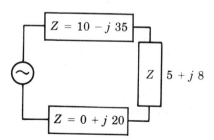

Figure 18–1

Circuit values for Example 18–1

Example 18–1 ————————————————

Calculate the total impedance, in rectangular form, of the circuit indicated in Figure 18–1.

Solution

Add the real and reactive terms separately:

$$Z_T = (10 - j\,35) + (5 + j\,8) + (0 + j\,20) = 15 - j\,7\;\Omega$$

Subtraction of Vector Quantities

Subtraction of two vector quantities expressed in rectangular coordinates requires only simple subtraction of the real and quadrature components. The following example illustrates the procedure.

Example 18–2 ————————————————

Subtract $6 + j\,3$ from $8 - j\,4$.

Solution

Subtract the real and quadrature components:

$$(8 - j\,4) - (6 + j\,3) = 2 - j\,7$$

The vector diagram in Figure 18–2 illustrates this operation, with the resultant equal to the difference between the two rectangular quantities.

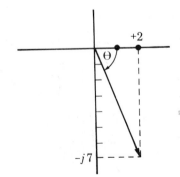

Figure 18–2

Vector diagram for Example 18–2

18–3
Multiplication and Division by Vector Algebra

Multiplication of Vector Quantities

Vector quantities expressed in rectangular form can be multiplied, but some skill is required in handling the algebraic terms. The simplest procedure is to carry out the multiplication with *the quantity expressed in polar form*, converting from rectangular if necessary. The procedure for accomplishing this operation will be detailed in Section 18–4. The following rule applies to multiplication: *Polar quantities can be multiplied by multiplying their magnitudes and algebraically adding their phase angles.* Expressed mathematically, this rule is as follows:

$$Z_1 \ \angle \theta_1 \times Z_2 \ \angle \theta_2 = Z_1 Z_2 \ \angle \theta_1 + \theta_2 \qquad \textbf{18–1}$$

Example 18–3 ───────────────────────────────────

Find the product of two impedances with polar values of $Z_1 = 50 \ \angle 23° \ \Omega$ and $Z_2 = 70 \ \angle 15° \ \Omega$.

Solution

Substitute these values into Equation 18–1:

$$Z_{eq} = Z_1 Z_2 \ \angle \theta_1 + \theta_2 = 50 \times 70 \ \angle 23° + 15° \ \Omega$$
$$= 3500 \ \angle 38° \ \Omega$$

Example 18–4 ───────────────────────────────────

Find the product of the following phasor voltages in a series circuit: $E_1 = 8.5 \ \angle 68° \ V$ and $E_2 = 5.7 \ \angle -22.5° \ V$.

Solution

This problem is solved by using Equation 18–1 and substituting E for Z:

$$E_{eq} = E_1 E_2 \ \angle \theta_1 + \theta_2 = 8.5 \times 5.7 \ \angle 68° - 22.5° \ V$$
$$= 48.45 \ \angle 45.5° \ V$$

If a phasor is multiplied by a real number, the result is *the product of their magnitudes*. Since the real number has an angle of $0°$, the phase angle is that of the phasor. For example:

$$5 \times 8 \angle 45° = 40 \angle 45°$$
$$7 \times 3.5 \angle -17° = 24.5 \angle -17°$$

Division of Vector Quantities

While vector quantities can be divided in rectangular form, the operation is simplified if they are in polar form. The following rule applies for division: *Polar quantities can be divided by dividing their magnitudes and algebraically subtracting their phase angles.* Expressed mathematically, this rule is as follows:

$$Z_1 \angle \Theta_1 \div Z_2 \angle \Theta_2 = \frac{Z_1}{Z_2} \angle \Theta_1 - \Theta_2 \qquad \text{18-2}$$

By substituting E for Z, we can use the same formula with voltages. The next example shows the procedure.

Example 18–5 ——————————————————————

Divide the following voltage phasors:

$$36 \angle 50° \text{ V} \div 6 \angle 20° \text{ V}$$
$$100 \angle 75° \text{ V} \div 8 \angle 35° \text{ V}$$
$$60 \angle -40° \text{ V} \div 20 \angle 30° \text{ V}$$

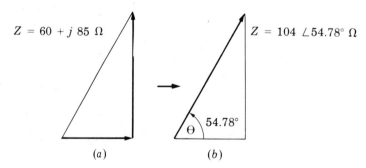

$Z = 60 + j\ 85\ \Omega$

$Z = 104 \angle 54.78°\ \Omega$

$54.78°$

Θ

(a) (b)

Figure 18–3

Equal impedances: (a) Rectangular form. (b) Polar form.

Solution

$$\frac{36}{6} \angle 50° - 20° = 6 \angle 30°$$

$$\frac{100}{8} \angle 75° - 35° = 12.5 \angle 40°$$

$$\frac{60}{20} \angle -40° - 30° = 3 \angle -70°$$

18–4
Rectangular-to-Polar Conversion

In the solution of complex circuits, it is frequently necessary or convenient to convert from one system of notation to another. If a quantity is in polar form, it can be multiplied or divided more easily than if it is in rectangular form. Therefore, it is expedient to convert to polar form for these operations. This conversion was outlined in Chapter 17, but it will be reviewed here with the aid of two examples.

Example 18–6

Convert the impedance $Z = 60 + j\,85\ \Omega$ to polar form.

Solution

The rectangular form of this impedance is shown in Figure 18–3a. Convert to polar form:

$$Z = \sqrt{60^2 + 85^2} = 104\ \Omega$$

$$\theta = \tan^{-1}\frac{85}{60} = 1.42 = 54.78°$$

$$Z = 104\ \angle 54.78°\ \Omega$$

This polar form is shown in Figure 18–3b.

An alternative procedure to the one shown in Example 18–6 would be to calculate Z by trigonometric functions.

Example 18-7

Convert the current $I = 1.4 - j\,0.8$ A to polar notation.

Solution

$$I_T = \sqrt{1.4^2 + 0.8^2} = 1.61 \text{ A}$$

$$\Theta = \tan^{-1}\frac{0.8}{1.4} = 29.74°$$

$$I_T = 1.61 \; \angle -29.74° \text{ A}$$

18-5
Polar-to-Rectangular Conversion

Solving complex circuits frequently requires the addition or subtraction of vector quantities. These operations can be easily accomplished when the quantities are in rectangular form. But when the quantities are in polar form, they cannot be added or subtracted directly. To convert from polar to rectangular form, we must *solve for the in-phase and quadrature components* of the phasor quantity involved. Then, we can add all real components by simple algebraic addition. Similarly, all vertical or quadrature components (j and $-j$) can be added algebraically. The net result of these additions is that one plus or minus real coordinate and one $+j$ or $-j$ coordinate will remain. These elements are the *rectangular coordinates of the algebraic sum of the several phasors.*

The in-phase component is equal to the *product* of the phasor and the cosine of its angle (magnitude $\times$ cos Θ). The quadrature component is the *product* of the phasor and the sine of its angle (magni-

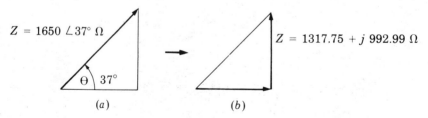

$Z = 1650 \angle 37° \; \Omega$ $Z = 1317.75 + j\,992.99 \; \Omega$

Θ 37°

(*a*)   (*b*)

Figure 18-4

Equal impedances: (*a*) Polar form. (*b*) Rectangular form.

tude $\times \sin \Theta$). For positive angles, the quadrature component is positive $(+j)$, and for negative angles, it is negative $(-j)$. The following two examples illustrate this procedure.

Example 18–8

Convert the impedance $Z = 1650 \angle 37° \, \Omega$ to rectangular form.

Solution

The polar form of this impedance is shown in Figure 18–4a. Convert to rectangular form:

$$R = Z \cos \Theta = 1650 \times 0.799 = 1317.75 \, \Omega$$
$$X_L = Z \sin \Theta = 1650 \times 0.602 = 992.99 \, \Omega$$
$$Z = R + jX = 1317.75 + j \, 992.99 \, \Omega$$

This rectangular form is shown in Figure 18–4b.

Example 18–9

Find the rectangular components of a voltage with a polar value of $E_T = 120 \angle -51° \, V$.

Solution

$$E_R = E_T \cos \Theta = 120 \times 0.629 = 75.52 \, V$$
$$E_C = E_T \sin \Theta = 120 \times 0.777 = 93.26 \, V$$
$$E_T = 75.52 - j \, 93.26 \, V$$

18–6
Impedance of Series Circuits

> **Complex impedance**   An impedance made up of combinations of $R, C,$ and L.

Complex impedances (any combination of $R, C,$ and L) can be represented in rectangular or polar form. In Figure 18–5a, two complex impedances are represented in series. By adding the impedance vectors

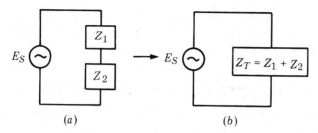

Figure 18–5

Complex impedances: (a) Impedances in series. (b) Vector addition of series impedances.

of Z_1 and Z_2, we can find the total circuit impedance seen by the source E_S. Expressed mathematically, the total impedance is as follows:

$$Z_T = Z_1 + Z_2 \qquad\qquad \textbf{18–3}$$

Figure 18–5b represents the vector addition of Z_1 and Z_2.

If the impedances are in rectangular form, they can be algebraically added. If in polar notation, they must be converted to rectangular form before being added.

Example 18–10

Calculate the vector impedance Z_T for the impedances shown in Figure 18–6.

Solution

The impedances must be in rectangular form before they can be added.

$$Z_1 = 20 + j\,15 \ \Omega$$
$$Z_2 = 12 \ \angle\,30° = 12 \cos 30° + j\,12 \sin 30°$$
$$= 12 \times 0.866 + j\,12 \times 0.5 = 10.39 + j\,6 \ \Omega$$
$$Z_T = Z_1 + Z_2 = (20 + j\,15 \ \Omega) + (10.39 + j\,6 \ \Omega)$$
$$= 30.39 + j\,21 \ \Omega$$

Example 18–11

Determine the magnitude and phase angle of the current in the circuit of Figure 18–6.

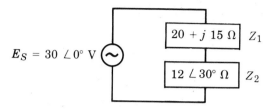

Figure 18–6

Series complex impedances for Example 18–10

Solution

We must first convert the total impedance to polar form so that it may be divided into the supply voltage E_S:

$$Z_T = \sqrt{30.39^2 + 21^2} = 36.94 \; \Omega$$

$$\theta = \tan^{-1} \frac{21}{30.39} = 34.65°$$

Hence
$$Z_T = 36.94 \; \angle 34.65° \; \Omega$$

The circuit is inductive because of the positive phase angle. Therefore:

$$I_T = \frac{E_S}{Z_T} = \frac{30 \; \angle 0° \; V}{36.94 \; \angle 34.65° \; \Omega} = 0.81 \; \angle -34.65° \; A$$

Example 18–12

From the data obtained in Examples 18–10 and 18–11, calculate the voltage across each impedance in Figure 18–6.

Solution

The voltage across Z_1 is as follows:

$$V_1 = I_T Z_1$$

Since Z_1 is expressed in rectangular notation, it must be converted to polar form so that it may be multiplied by I_T:

$$Z_1 = 20 + j\,15 \; \Omega = 25 \; \angle 36.87° \; \Omega$$
$$V_1 = 0.81 \; \angle -34.65° \; A \times 25 \; \angle 36.87° \; \Omega$$
$$= 20.25 \; \angle 2.22° \; V$$

The voltage across Z_2 is as follows:

$$V_2 = I_T Z_2 = 0.81 \angle -34.65° \text{ A} \times 12 \angle 30° \text{ } \Omega = 9.72 \angle -4.65° \text{ V}$$

Because the phase angles of V_1 and V_2 are so small, the voltages are nearly in phase. Consequently:

$$V_1 + V_2 \cong E_S$$

18–7
Impedance of Parallel Circuits

When a purely capacitive reactance and inductive reactance are parallel-connected, as shown in Figure 18–7, the reactive currents will be 180° out of phase. The capacitive current is as follows:

$$I_C = \frac{120 \text{ V}}{50 \text{ } \Omega} = 2.4 \text{ A}$$

The inductive current is as follows:

$$I_L = \frac{120 \text{ V}}{100 \text{ } \Omega} = 1.2 \text{ A}$$

In series circuits, the current is used as the reference phasor because the current is the same in all parts of the circuit. In parallel cir-

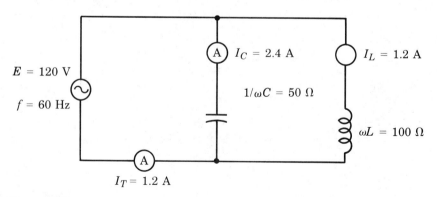

Figure 18–7

Parallel *CL* circuit

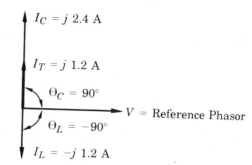

Figure 18–8

Phasor diagram for circuit of Figure
18–7

cuits, there are different values of currents in various parts of a circuit. Therefore, the current cannot be used as the reference phasor.

Since the same voltage exists across two or more parallel branches of a circuit, the applied voltage can be used as the reference phasor, as illustrated in Figure 18–8. Observe that the current I_C through the capacitor is leading the applied voltage by 90°, and the current through the inductor is lagging the voltage by 90°. The total line current I_T, which is the phasor sum of the branch currents, is leading the source voltage by 90°. This result can be shown in rectangular phasor notation:

$$I_C = 0 + j\,2.4\text{ A}$$
$$\underline{I_L = 0 - j\,1.2\text{ A}}$$
$$I_T = 0 + j\,1.2\text{ A} = 1.2\ \angle\,90°\text{ A}$$

Since the line current leads the supply voltage by 90°, the equivalent series circuit consists of the following capacitive reactance:

$$X_C = \frac{E}{I_T} = \frac{120\text{ V}}{1.2\text{ A}} = 100\ \Omega$$

From this result, we determine that the parallel circuit could be replaced with a 26.54 μF capacitor, which would result in 1.2 A leading the voltage by 90°. The supply source could not distinguish between the original circuit and the equivalent series circuit.

Rules for Calculating Impedance

From a practical standpoint, most parallel ac circuits are made up of branches of complex impedances. For calculation of the total impedance of the parallel network, it is convenient to use vector algebra (polar and rectangular notations).

Before attempting to find the equivalent impedance of a parallel network, let us review the procedures that must be followed in pursuing an orderly solution (refer to the block diagram in Figure 18–9):

1. Calculate the current through *each branch* by the basic equation $I = E/Z$. The impedance *must* be in polar form so that it can be divided into the supply voltage. (If no voltage is given, assume any convenient value.)
2. Find I_T, the vector sum of the branch currents. (Currents must be in rectangular form so that they can be added.)
3. Convert I_T into polar form so that it can be divided into the supply voltage.
4. Calculate the circuit impedance: $Z_T = E/I_T$.

The following example demonstrates how we calculate the total impedance of two parallel impedances.

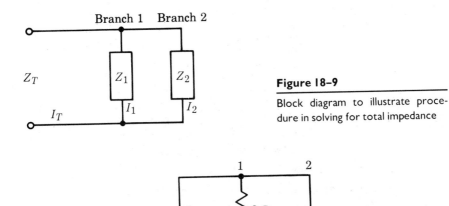

Figure 18–9

Block diagram to illustrate procedure in solving for total impedance

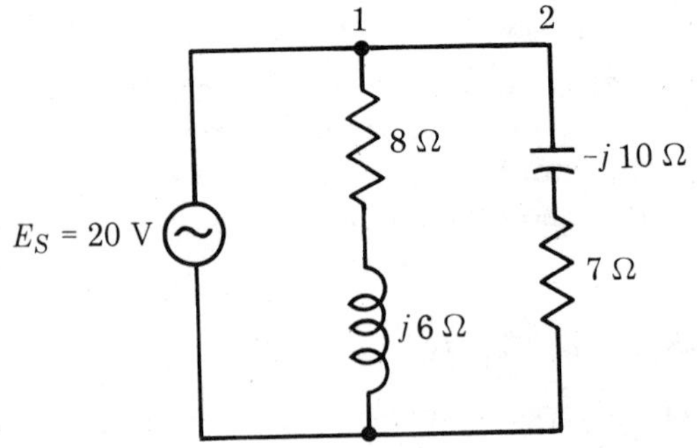

Figure 18–10

Parallel circuit for Example 18–13

Example 18-13 ─────────────────────────────

Calculate the circuit impedance of the circuit in Figure 18-10.

Solution

1. Convert each branch impedance to polar form.

$$Z_1 = 8 + j\,6\;\Omega = 10\;\angle\,36.87°\;\Omega$$
$$Z_2 = 7 - j\,10\;\Omega = 12.21\;\angle -55°\;\Omega$$

Each branch current is as follows:

$$I_1 = \frac{E}{Z_1} = \frac{20\;\angle\,0°}{10\;\angle\,36.87°} = 2\;\angle -36.87°\;A$$

$$I_2 = \frac{E}{Z_2} = \frac{20\;\angle\,0°}{12.21\;\angle -55°} = 1.64\;\angle\,55°\;A$$

2. Convert the branch currents to rectangular form so that they can be added.

$$I_1 = 2\;\angle -36.87°\;A = 1.60 - j\,1.20\;A$$
$$I_2 = 1.64\;\angle\,55°\;A = 0.94 + j\,1.34\;A$$
$$\overline{\qquad I_T = 2.54 + j\,0.14\;A}$$

The angle is *leading* if the j term is positive; it is *lagging* if the j term is negative.

3. Convert I_T into polar form so that it can be divided into the supply voltage.

$$I_T = 2.54 + j\,0.14\;A = 2.54\;\angle\,3.15°\;A$$

From the small phase angle, we see that the circuit is very nearly all resistive.

4. Calculate the circuit impedance:

$$Z_T = \frac{E_S}{I_T} = \frac{20\;\angle\,0°}{2.54\;\angle\,3.15°} = 7.87\;\angle -3.15°\;\Omega$$

Product-over-Sum method of Calculating Z_T

In cases where only two parallel impedances are involved, the product-over-sum formula may be used:

$$Z_T = \frac{Z_1 Z_2}{Z_1 + Z_2} \qquad\qquad \text{18-4}$$

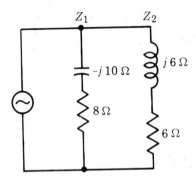

Figure 18-11

Parallel circuit for Example 18-14

Because many electronic circuits contain only two parallel branches, or can be easily reduced to two, this formula offers a convenient method of determining the total impedance. The next example illustrates the procedure.

Example 18-14

Calculate the impedance of the circuit in Figure 18-11, using the product-over-sum formula.

Solution

$$Z_1 = 8 - j\,10 = 12.81 \;\angle -51.34° \; \Omega$$
$$Z_2 = 6 + j\,6 = 8.49 \;\angle\,45° \; \Omega$$
$$Z_1 + Z_2 = 14 - j\,4 = 14.56 \;\angle -15.95° \; \Omega$$
$$Z_T = \frac{Z_1 Z_2}{Z_1 + Z_2} = \frac{108.76 \;\angle -6.34°}{14.56 \;\angle -15.95°} = 7.47 \;\angle\,9.61° \; \Omega$$

18-8
Equivalent Series Circuits

It is frequently convenient, when we are analyzing parallel ac circuits, to replace the parallel network with its equivalent series circuit. If the magnitude of the current and its phase angle is unchanged with respect to the supply, then it doesn't matter whether the load is a parallel or a series circuit.

Because a parallel impedance can be represented in polar form as $Z_T \angle \Theta°$, the equivalent series resistance is as follows:

$$R = Z_T \cos \Theta$$

The equivalent series reactance is as follows:

$$X = Z_T \sin \Theta$$

If Θ is *positive*, use the following equivalent series inductance:

$$L = \frac{X}{2\pi f}$$

If Θ is *negative*, use the following equivalent series capacitance:

$$C = \frac{1}{2\pi f X}$$

The following example shows this procedure.

Example 18–15

A parallel circuit has an impedance and phase angle of $Z = 100 \angle 40°$ Ω. Find the equivalent series circuit parameters if the frequency of the supply is 400 Hz.

Solution

Refer to the phasor diagram in Figure 18–12a. The effective R_{eq} and X_{eq} can be calculated as follows:

$$R_{eq} = Z_T \cos \Theta = 100 \times 0.766 = 76.6 \ \Omega$$
$$X_{eq} = Z_T \sin \Theta = 100 \times 0.643 = 64.3 \ \Omega$$

Since the phase angle is positive, the reactance must be inductive. The circuit would be as shown in Figure 18–12b. Calculate the value of L:

$$L = \frac{X_L}{2\pi f} = \frac{64.3}{6.28 \times 400} = 0.0256 \text{ H}$$

Hence, the equivalent series circuit is as shown in Figure 18–12c.

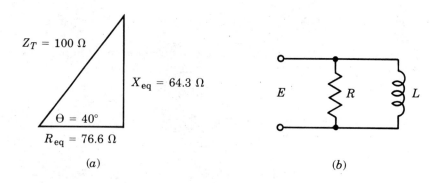

(a)   (b)

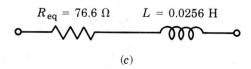

(c)

Figure 18–12

Parallel circuit for Example 18–15: (a) Phasor diagram. (b) Basic circuit of RL. (c) Equivalent series circuit.

Example 18–16 ───────────────────────────────

Refer to the parallel circuit in Figure 18–13. Calculate the equivalent series circuit at a frequency of 5 MHz.

Solution

$$X_L = \omega L = 6.28 \times 5 \times 10^6 \times 9 \times 10^{-6} = 283 \ \Omega$$
$$X_C = \frac{1}{\omega C} = \frac{1}{6.28 \times 5 \times 10^6 \times 100 \times 10^{-12}} = 318 \ \Omega$$

Then
$$I_R = \frac{E}{R} = \frac{100 \ V}{2 \ k\Omega} = 50 \ mA$$

$$I_L = \frac{E}{X_L} = \frac{100 \ V}{283 \ \Omega} = 353 \ mA$$

$$I_C = \frac{E}{X_C} = \frac{100 \ V}{318 \ \Omega} = 314 \ mA$$

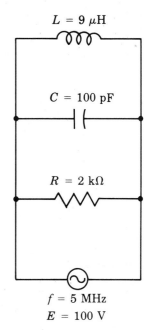

$L = 9\ \mu\text{H}$

$C = 100\ \text{pF}$

$R = 2\ \text{k}\Omega$

$f = 5\ \text{MHz}$

$E = 100\ \text{V}$

Figure 18-13

Parallel *RCL* circuit for Example 18-16

The total current I_T is the phasor sum of the three branch currents and is shown in the phasor diagram of Figure 18-14. Add currents in their rectangular form:

$$
\begin{aligned}
I_R &= 50 \ + j \quad 0 \ \text{mA} \\
I_L &= \ \ 0 \ - j\ 353 \ \text{mA} \\
I_C &= \ \ 0 \ + j\ 314 \ \text{mA} \\
\hline
I_T &= 50 \ - j \quad 39 \ \text{mA} = 63.4 \ \angle -37.95° \ \text{mA}
\end{aligned}
$$

Now, calculate the power factor:

$$PF = \cos \angle -37.95° = 0.788 \quad \text{(lagging)}$$

The circuit Z_T is as follows:

$$Z_T = \frac{E}{I_T} = \frac{100\ \text{V}}{0.0634\ \text{A}} = 1577\ e$$

Since the current is lagging the supply voltage, the equivalent series circuit consists of a resistance and inductive

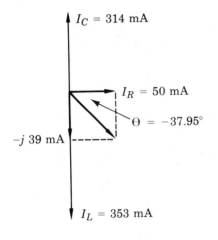

Figure 18–14

Phasor diagram for circuit in Figure 18–13

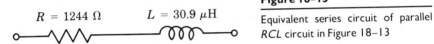

Figure 18–15

Equivalent series circuit of parallel RCL circuit in Figure 18–13

reactance. The phasor sum of this combination is 1577 Ω at a phase angle of 37.95°. Therefore:

$$Z_T = 1577 \angle 37.95° = 1244 + j\,970\ \Omega$$

Hence, the equivalent series circuit consists of $R = 1244\ \Omega$ and $X_L = 970\ \Omega$. Solve for L:

$$L = \frac{970}{6.28 \times 5 \times 10^6} = 30.9\ \mu\text{H}$$

Therefore, the equivalent series circuit is as shown in Figure 18-15.

18–9
Conductance, Susceptance, and Admittance

Our calculations of circuit impedance thus far have dealt with terms that refer to the opposition that various circuit components offer to ac. An alternative, and frequently easier, approach involves terms that indicate how easily current is conducted through the several circuit parameters. For example, in the case of the purely resistive circuit, you learned that *conductance* is the reciprocal of resistance (see Section 2–6). The SI unit for conductance is the *siemens*, abbreviated S.

> **Susceptance (B)**   The reciprocal of reactance, measured in siemens; B_L designates inductive susceptance and B_C capacitive susceptance.

In the simple parallel circuit of Figure 18–16, we can show that conductance also applies to the ac circuit. Thus, branch 1 has a conductance of $G = 1/R$. In the case of branches 2 and 3, their abilities to allow current flow are expressed in terms of the reciprocals of their oppositions. We cannot use the term *conductance* because of *the phase relationship* between X_C and X_L. The condition described is called **susceptance** (symbol B) and is measured in siemens. To differentiate between inductive and capacitive susceptance, subscripts are used. Thus, for pure C and L, we have the following equations:

$$B_C = \frac{1}{X_C} \qquad\qquad \text{18–5}$$

$$B_L = \frac{1}{X_L} \qquad\qquad \text{18–6}$$

Just as X_C and X_L are opposing quantities, so are B_C and B_L. Consequently, the net susceptance of parallel LC branches is $B_C - B_L$. These susceptances are *vector quantities* and are drawn 90° from the real axis. Because these quantities are the opposite of their reactances, they are drawn in opposite directions to X_C and X_L. Capacitive susceptance B_C is drawn upward, while inductive susceptance B_L is drawn downward.

> **Admittance (Y)**   The reciprocal of a circuit's impedance, measured in siemens.

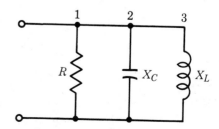

Figure 18–16

Simple parallel circuit to explain *G, B,* and *Y*

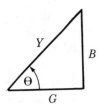

Figure 18–17

Phasor diagram showing relationship of Y, G, and B

When conductance and susceptance are present in a parallel circuit, their combined effect is called **admittance** (symbol Y). Since G and B are reciprocals of R and X, so is Y the reciprocal of impedance. Therefore, $Y = 1/Z$ and is measured in siemens. These three terms are related as shown in the admittance phasor diagram in Figure 18–17. Expressed mathematically, this relationship is as follows:

$$Y = G \pm jB \qquad\qquad \text{18-7}$$

This equation is in the familiar rectangular form. Conversion can be made to polar form by using the procedures outlined earlier in this chapter. The phase angles of B and Y are opposite those for X and Z (because of their reciprocal natures). Therefore, capacitive branches have susceptances of $+jB$, and inductive branches have susceptances of $-jB$.

Example 18–17 ──────────────────────────────

For the circuit of Figure 18–16, calculate G, B, and Y if the circuit values are $R = 500\ \Omega$, $X_C = 200\ \Omega$, and $X_L = 300\ \Omega$.

Solution

First, find the conductance and susceptance of each branch.

$$G = \frac{1}{R} = \frac{1}{500} = 0.002\ \text{S}$$

$$B_C = \frac{1}{X_C} = \frac{1}{200} = 0.005\ \text{S}$$

$$B_L = \frac{1}{X_L} = \frac{1}{300} = 0.0033\ \text{S}$$

Next, find the equivalent susceptance.

$$B_{eq} = B_C - B_L = 0.005 - 0.0033 = +j\,0.0017\text{ S}$$

Finally, calculate the admittance.

$$Y = G + jB = 0.002 + j\,0.0017$$
$$= \sqrt{0.002^2 + 0.0017^2} = 0.0026\text{ S}$$

Example 18–18

Refer to Example 18–17. Calculate the admittance in polar form and determine the impedance in polar form.

Solution

First, calculate the phase angle Θ_Y:

$$\Theta_Y = \tan^{-1}\frac{B}{G} = \tan^{-1}\frac{0.0017}{0.002} = 40.36°$$

Therefore $\quad Y = 0.0026\ \angle\,40.36°\text{ S}$

The phasor diagram of this admittance is shown in Figure 18–17. The impedance in polar form is as follows:

$$Z = \frac{1}{Y} = \frac{1}{0.0026\ \angle\,40.36°} = 384.62\ \angle-40.36°\ \Omega$$

Because admittance is the reciprocal of impedance, the current in a circuit can be determined by multiplying voltage by admittance:

$$I = EY = E\,(G \pm jB) \qquad\qquad\qquad \textbf{18–8}$$

Example 18–19

Suppose the circuit in Figure 18–16 has 100 V applied. Calculate the line current.

Solution

$$I = EY = (100\ \angle\,0°)\,(0.0026\ \angle\,40.36°) = 0.26\ \angle\,40.36°\text{ A}$$

18–10
Series-Parallel Impedances

Calculating the total impedance of a series-parallel network is somewhat more involved than finding the impedance of a parallel or series circuit alone. The following steps outline the procedure to follow:

1. Calculate the equivalent impedance of the parallel combination in rectangular form.
2. Add the series impedance to that found in step 1 to obtain Z_T.
3. Convert Z_T to polar notation to determine I_T.

This procedure is demonstrated in the following example.

Example 18–20

Determine the total circuit impedance and line current in the series-parallel circuit of Figure 18–18.

Solution

1. Calculate Z_{eq} of the parallel combination, converting to rectangular form so that they can be added.

$$Z_2 = 85 \angle 50° \ \Omega = 54.64 + j\,65.11 \ \Omega$$
$$Z_3 = 60 \angle -27° \ \Omega = 53.46 - j\,27.24 \ \Omega$$

$$Z_{eq} = \frac{85 \angle 50° \times 60 \angle -27°}{108.1 + j\,37.87} \cong 44.54 \angle 3.69° \ \Omega$$

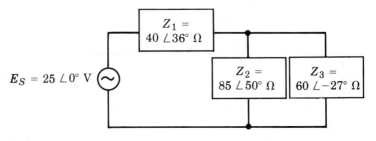

Figure 18–18

Series-parallel circuit for Example 18–20

2. Add the series impedance to Z_{eq} to find Z_T.

$$Z_1 = 40 \angle 36° \ \Omega = 32.36 + j\,23.51 \ \Omega$$

Therefore

$$\begin{aligned}
Z_T &= Z_{eq} + Z_1 = (44.47 + j\,2.86) + (32.36 + j\,23.51) \\
&= 76.83 + j\,26.37 \ \Omega
\end{aligned}$$

3. Convert Z_T to polar form.

$$Z_T = 81.23 \angle 18.9° \ \Omega$$

So $\qquad I_T = \dfrac{E_S}{Z_T} = \dfrac{25 \angle 0°}{81.23 \angle 18.9°} = 0.308 \angle -18.9° \ \text{A}$

The voltage drop across each impedance is simply the product of the current and the respective impedance. The following example illustrates the calculation.

Example 18–21

Find the voltage drop across Z_1 and Z_3 in the circuit of Figure 18–18.

Solution

$$V_1 = I_T \times Z_1 = 0.308 \angle -18.9° \times 40 \angle 36° = 12.32 \angle 17.1° \ \text{V}$$

The voltage drop across Z_3 will be the same as $I_1 \times Z_{eq}$.

$$Z_{eq} = 44.54 \angle 3.69° \ \Omega$$

Therefore $\qquad \begin{aligned} V_3 &= I_T \times Z_{eq} = 0.308 \angle -18.9° \times 44.54 \angle 3.69° \\ &= 13.72 \angle -15.21° \ \text{V} \end{aligned}$

18–11
Thévenin's Theorem for ac Circuits

In Chapter 8, you learned that any linear resistive network could be replaced by its equivalent Thévenin circuit. This theorem can be extended to apply to ac networks when we wish to analyze a complex circuit under different load conditions. A typical example is determining the optimum load for an active device such as a transistor. Recall that calculating the Thévenin equivalent circuit consists of determining the open-circuit voltage E_{TH} across the load terminals but with the load disconnected. The Thévenin impedance Z_{TH} is determined as the total equivalent impedance seen looking into the network from the load terminals. All voltage sources must be replaced by their internal impedances. If the source impedances are small compared with the network values, they may be replaced by short circuits. The next example illustrates this procedure.

Example 18–22

Compute the current through and the voltage across R_L in the circuit of Figure 18–19, using Thévenin's theorem.

Solution

It is convenient to group the network impedances as shown in Figure 18–20 and proceed to calculate E_{TH}. Because some of the calculations will require Z_1 and Z_2 to be in polar form, we will perform these now.

$$Z_1 = 70 - j\,30 = 76.16 \angle -23.2° \; \Omega$$
$$Z_2 = 45 + j\,100 = 109.66 \angle 65.8° \; \Omega$$

Now, calculate E_{TH}. This voltage will be the same as V_{Z_2}

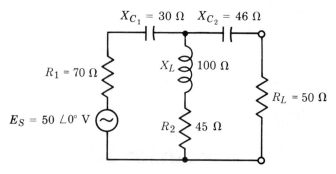

Figure 18–19

Series-parallel circuit for Example 18–22

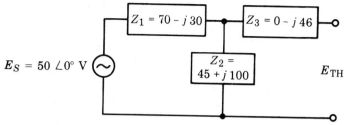

Figure 18–20

Calculating E_{TH} for Example 18–22

since there is no voltage drop across V_{Z_3} with R_L removed. Use the voltage divider formula.

$$E_{TH} = E_S \left(\frac{Z_2}{Z_1 + Z_2}\right) = 50 \angle 0° \left[\frac{109.66 \angle 65.8°}{(70 - j\,30) + (45 + j\,100)}\right]$$

$$= 50 \angle 0° \left(\frac{109.66 \angle 65.8°}{115 + j\,70}\right)$$

Convert the denominator to polar form so that division can be performed.

$$E_{TH} = 50 \angle 0° \left(\frac{109.66 \angle 65.8°}{134.63 \angle 31.3°}\right)$$

$$= 50 \angle 0° \times 0.815 \angle 34.5° = 40.75 \angle 34.5° \text{ V}$$

Next, calculate Z_{TH}, which is the internal impedance of the network as seen from the load terminals, with the source replaced by a short. Since Z_1 is in parallel with Z_2, the Thévenin impedance is as follows:

$$Z_{TH} = Z_3 + \frac{Z_1 Z_2}{Z_1 + Z_2}$$

Substitute values into this equation.

$$Z_{TH} = 0 - j\,46 + \frac{(76.16 \angle -23.2°)(109.66 \angle 65.8°)}{(70 - j\,30) + (45 + j\,100)}$$

$$= 0 - j\,46 + \frac{8351.7 \angle 42.6°}{115 + j\,70}$$

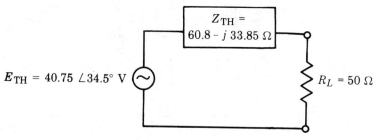

Figure 18–21

Thévenin equivalent circuit for Example 18–22

Express the denominator in polar form for division (this conversion was previously done for E_{TH}).

$$Z_{TH} = 0 - j\,46 + \frac{8351.7 \ \angle\,42.6°}{134.63 \ \angle\,31.3°} = 0 - j\,46 + 62 \ \angle\,11.3° \ \Omega$$

Convert the last term to rectangular form for addition.

$$62 \ \angle\,11.3° = 60.8 + j\,12.15 \ \Omega$$

So

$$Z_{TH} = 0 - j\,46 + 60.8 + j\,12.15 = 60.8 - j\,33.85 \ \Omega$$

Now that we have calculated E_{TH} and Z_{TH}, the equivalent Thévenin circuit is as shown in Figure 18–21. The voltage across R_L is calculated by using the voltage divider formula.

$$V_L = E_{TH} \left(\frac{R_L}{Z_{TH} + R_L}\right) = \frac{(40.75 \ \angle\,34.5°) \ (50 \ \angle\,0°)}{(60.8 - j\,33.85) \ (50 + j\,0)}$$

$$= \frac{2037.5 \ \angle\,34.5°}{110.8 - j\,33.85}$$

Convert the demoninator to polar form for division.

$$110.8 - j\,33.85 = 115.86 \ \angle -17°$$

So

$$V_L = \frac{2037.5 \ \angle\,34.5°}{115.86 \ \angle -17°} = 17.6 \ \angle\,51.5° \ V$$

The load current is determined by Ohm's law.

$$I_L = \frac{V_L}{R_L} = \frac{17.6 \ \angle\,51.5°}{50 \ \angle\,0°} = 0.35 \ \angle\,51.5° \ A$$

18–12
Norton's Theorem for ac Circuits

Norton's equivalent circuit is frequently used in conjunction with Thévenin's equivalent circuit to find the current through a load. The equivalent Norton circuit consists of a constant-current source in parallel with an ac impedance, as shown in Figure 18–22. The constant-current generator represents a Norton current source I_N that supplies current to any load connected to the Norton equivalent circuit. Since Z_N is in parallel with the load, I_N will divide between them.

The value of I_N is determined from the following equation. The shunt impedance Z_N is equal to the Thévenin impedance Z_{TH}. Expressed mathematically,

$$I_N = \frac{E_{TH}}{Z_{TH}} \qquad \textbf{18–9}$$

where E_{TH} = Thévenin voltage of the circuit
                   Z_{TH} = Thévenin impedance of the circuit

Example 18–23

Use Norton's theorem and the circuit in Figure 18–19 to calculate the current through the 50 Ω resistive load.

Solution

Use the values of E_{TH} and Z_{TH} calculated in Example 18–22 and solve for I_N.

$$I_N = \frac{E_{TH}}{Z_{TH}} = \frac{40.75 \angle 34.5°}{60.8 - j\,33.85}$$

Convert the denominator to polar form for division.

$$60.8 - j\,33.85 = 69.59 \angle -29.11°$$

So $I_N = \dfrac{40.75 \angle 34.5°}{69.59 \angle -29.11°} = 0.59 \angle 63.61° \text{ A}$

Use the current divider formula to find I_L.

$$I_L = I_N \frac{Z_{TH}}{Z_{TH} + Z_L} = 0.59 \angle 63.61° \left(\frac{69.59 \angle -29.11°}{60.8 - j\,33.85 + 50 + j\,0} \right)$$

$$= 0.59 \angle 63.11° \left(\frac{69.59 \angle -29.11°}{110.8 - j\,33.85} \right)$$

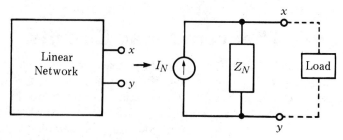

Figure 18–22

Analysis of linear network by Norton's theorem

Convert the denominator to polar form for division.

$$110.8 - j\,33.85 = 115.86 \angle -16.99°$$

Then
$$I_L = 0.59 \angle 63.11° \left(\frac{69.59 \angle -29.11°}{115.86 \angle -16.99°} \right)$$

$$= 0.59 \angle 63.11° \times 0.60 \angle -12.12°$$

$$= 0.35 \angle 51.49° \text{ A}$$

This result agrees with the solution obtained in Example 18–22 using Thévenin's theorem.

Summary

Vector algebra makes the solution of problems involving phasors, at angles other than 90°, relatively simple.

The two commonly used systems of vector algebra are rectangular and polar coordinates.

The addition and subtraction of vector quantities are most easily handled when they are in rectangular form.

The multiplication and division of vector quantities are most easily handled when they are in polar form.

Polar quantities that need to be added should first be converted to rectangular form.

Rectangular quantities that are to be multiplied or divided should first be converted to polar form.

Polar quantities can be multiplied by multiplying their magnitudes and algebraically adding their phase angles.

Polar quantities can be divided by dividing their magnitudes and algebraically subtracting their phase angles.

In the conversion of a polar impedance to rectangular form, the real term is $R = Z \cos \Theta$ and the quadrature form is $X = Z \sin \Theta$.

Parallel impedances can be calculated by the following formula:

$$Z_T = \frac{Z_1 Z_2}{Z_1 + Z_2}$$

The equivalent series circuit of a parallel impedance in polar form can be represented by its in-phase and quadrature components.

Parallel circuits can be solved by the admittance method, using Y, G, and B, where $Y = 1/Z$, $G = 1/R$, and $B = 1/X$, and the resulting values are expressed in siemens (S).

The admittance of a circuit can be calculated by the following formula:

$$Y = G \pm jB$$

Thévenin's theorem can be applied to linear ac circuits.

The load current in an ac circuit can be calculated by using Norton's theorem.

Progress Test

The bracketed number after each question indicates the section of this chapter where the answer can be found.

    1.   Which of the following statements is true? (a) Addition and subtraction of vector quantities is difficult with rectangular coordinates. (b) Polar coordinates are easily added and subtracted. (c) The practical way to multiply and divide vector quantities is by polar coordinates. (d) Multiplication and division of vector quantities is easily done with rectangular coordinates.   [18-1]

2. To add polar quantities: (a) first convert to rectangular form, then add real and reactive components. (b) add their real components and subtract their quadrature components. (c) add their phase angles and reactive terms. (d) add their real terms and phase angles. [18-1]

3. The total impedance of two series-connected impedances of $13 - j\,21$ Ω and $7 + j\,12$ Ω is: (a) $6 - j\,9$ Ω. (b) $20 - j\,9$ Ω. (c) $20 - j\,33$ Ω. (d) $j\,9 - 20$ Ω. [18-6]

4. The total impedance of two parallel-connected impedances of $14 \angle 21.5°$ Ω and $9 \angle 37.6°$ Ω is: (a) $23 \angle 59.1°$ Ω. (b) $126 \angle 59.1°$ Ω. (c) $126 \angle 16.1°$ Ω. (d) $23 \angle 16.1°$ Ω. [18-7]

5. The result of dividing a voltage phasor of $169 \angle 70°$ V by a phasor of $27 \angle 48°$ V is: (a) $6.26 \angle 22°$ V. (b) $6.62 \angle -22°$ V. (c) $142 \angle 22°$ V. (d) $142 \angle 1.46°$ V. [18-3]

6. An impedance of $37 + j\,29$ Ω expressed in polar form is: (a) $65 \angle 44°$ Ω. (b) $59 \angle 40°$ Ω. (c) $51 \angle 44°$ Ω. (d) $47 \angle 38°$ Ω. [18-4]

7. An impedance is designated as $1200 \angle 69°$ Ω. Expressed in rectangular notation, this impedance is: (a) $523 + j\,1072$ Ω. (b) $409 - j\,1120$ Ω. (c) $430 + j\,1120$ Ω. (d) $396 + j\,972$ Ω. [18-5]

8. An impedance of $18 \angle 35°$ Ω is in series with an impedance of $42 + j\,22$ Ω. Their combined impedance is: (a) $41.61 + j\,27.38$ Ω. (b) $45.27 + j\,29.04$ Ω. (c) $51.58 + j\,31.65$ Ω. (d) $56.74 + j\,32.32$ Ω. [18-6]

9. Two impedances of $10 + j\,14$ Ω and $6 - j\,8$ Ω are parallel-connected. Their combined impedance is: (a) $10.1 \angle -19.44°$ Ω. (b) $9.6 \angle 23.4°$ Ω. (c) $8.8 \angle -16.51°$ Ω. (d) $7.9 \angle 25.7°$ Ω. [18-7]

10. A parallel circuit consists of a conductance of 0.045 S and a capacitive susceptance of 0.080 S. The circuit admittance is: (a) 0.0061 S. (b) 0.0916 S. (c) 0.0847 S. (d) 0.0519 S. [18-9]

Problems

Answers to odd-numbered problems are at the back of the book.

1. Calculate the net impedance in polar form of two parallel-connected polar quantities of $88 \angle 14.2°$ and $52 \angle 31.8°$ Ω.

2. What is the total voltage, in rectangular form, of a series circuit with individual voltages of $10 + j\,0$ and $0 + j\,17$ V?

3.   Calculate the net rectangular impedance of a series circuit with individual components of the following values: $Z_1 = 274 - j\,126$ Ω, $Z_2 = 168 + j\,319$ Ω, and $Z_3 = 89 - j\,71$ Ω.

4.   The currents in the branches of a parallel circuit are $2.18 - j\,0.77$, $1.41 + j\,2.55$, and $0.83 - j\,1.07$ A. Express the line current in rectangular notation.

5.   Branch currents in a parallel circuit are $28 + j\,37$ and $19 - j\,23$ mA. Calculate the total current in polar notation.

6.   Calculate the quotient of the following voltage phasors: (a) $39 \angle 21°$ V $\div$ $13 \angle 7°$ V (b) $20 \angle 54°$ V $\div$ $3.5 \angle 17.5°$ V (c) $126 \angle 62°$ V $\div$ $33 \angle 26.7°$ V.

7.   Find the net impedance, in polar notation, of the following impedances: (a) $1200 \angle 45°$ Ω $\div$ $422 \angle -36°$ Ω (b) $75 \angle -18°$ Ω $\div$ $16.5 \angle 29°$ Ω.

8.   Convert the following impedances to polar form: (a) $350 + j\,250$ Ω (b) $15 - j\,20$ Ω (c) $220 + j\,1750$ Ω.

9.   Express the following voltages in rectangular notation: (a) $16 \angle -57.3°$ V (b) $590 \angle 37.2°$ V (c) $85 \angle 18.9°$ V.

10.   An impedance of $33 + j\,26$ Ω is in series with an impedance of $29 \angle 50°$ Ω. What is the vector impedance of the combination?

11.   Calculate the total impedance of the circuit in Figure 18–23 in polar form.

12.   Determine the magnitude and phase angle of the current for the circuit of Problem 11.

13.   An impedance of $37 + j\,26$ Ω is in series with an impedance of $42 \angle 32°$ Ω. Calculate the total circuit impedance: (a) in rectangular form (b) in polar form.

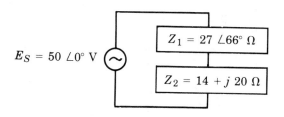

$E_S = 50 \angle 0°$ V

$Z_1 = 27 \angle 66°$ Ω

$Z_2 = 14 + j\,20$ Ω

Figure 18–23

Circuit for Problem 11

14. Suppose 0.45 ∠−33.6° A is passing through the circuit described in Problem 13. (a) What is the voltage across each impedance? (b) What is the supply potential?

15. What is the equivalent impedance of two impedances that are parallel-connected, where $Z_1 = 131 ∠8.5°$ Ω and $Z_2 = 76 ∠48.2°$ Ω?

16. Calculate the equivalent impedance of two parallel-connected impedances, where $Z_1 = 63.8 − j\,37.7$ Ω and $Z_2 = 40 + j\,50$ Ω.

17. What impedance must be connected in parallel with $60.5 + j\,40.7$ Ω to produce $41.3 + j\,150$ Ω?

18. Refer to Figure 18–24. If $Z_S = 8.6 + j\,10.2$ Ω, $Z_1 = 66.6 − j\,36$ Ω, and $Z_2 = 37.4 + j\,40.2$ Ω, what is the total circuit impedance Z_T?

19. Calculate the equivalent series circuit of a parallel circuit whose impedance is $250 ∠54.5°$ Ω if the frequency is 1 kHz.

20. Refer to Figure 18–16. Calculate the circuit's admittance if $R = 200$ Ω, $X_C = 300$ Ω, and $X_L = 450$ Ω.

21. Refer to Problem 20. Calculate the admittance in polar form.

22. A parallel circuit has a conductance G of 0.045 S and a B_C of 0.032 S. Determine the circuit current if 26.5 V ac is applied.

23. An impedance $Z_1 = 60 ∠35°$ Ω is in parallel with $Z_2 = 42 ∠− 65°$ Ω. The combination is in series with an impedance of $27 ∠21°$ Ω. (a) Calculate the total circuit impedance in polar form. (b) What is the total current if the supply is $50 ∠0°$ V?

24. Determine the current through and the voltage across R_L in the circuit of Figure 18–25, using Thévenin's theorem.

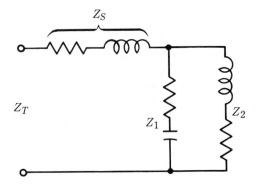

Figure 18–24

Circuit for Problem 18

25. Find the Thévenin voltage across R_L in the circuit of Figure 18–26 if the supply frequency is 30 Hz.

26. Using the values of E_{TH} and Z_{TH} found in Problem 25, calculate the current through the 5 kΩ load resistor in the circuit of Figure 18–26.

27. An impedance of $25 + j\ 32$ Ω is in series with an impedance of $12\ \angle -8°$ Ω. Both of these impedances are in parallel with an impedance of $46 - j\ 53$ Ω. See Figure 18–27. What is the Z_T for this entire network? Express Z_T in both polar and rectangular form.

28. In the circuit of Figure 18–27, if the source voltage is 30 $\angle 0°$ V, what is the voltage drop across Z_2?

29. For the circuit of Figure 18–28, draw and label the Norton equivalent circuit.

30. Check your work for Problem 29 by calculating E_{TH} for the circuit by using $I_N \times Z_N$ and by using the voltage divider method.

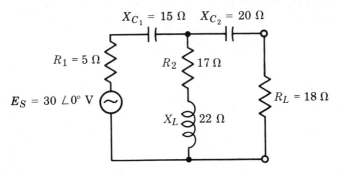

Figure 18–25

Circuit for Problem 24

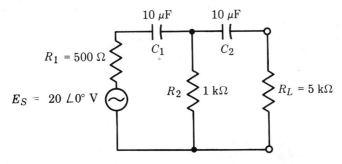

Figure 18–26

Circuit for Problem 25

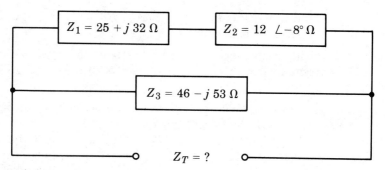

Figure 18–27

Circuit for Problems 27 and 28

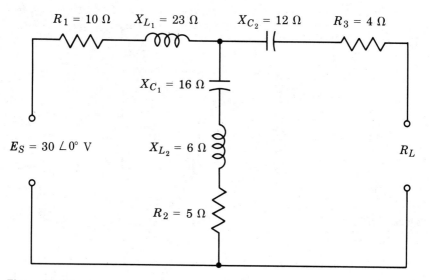

Figure 18–28

Circuit for Problem 29

19

Tuned Circuits and Resonance

Objectives

After studying this chapter, you will be able to:

Explain how tuned circuits are used in communication devices to select a desired frequency.

Calculate the resonant frequency of an LC combination.

Calculate the required value of C to resonate with a given L to produce the required resonant circuit.

Introduction

The basic principles of series and parallel ac circuits were explored in the previous chapter. However, we did not consider the very special case when inductive and capacitive reactances are equal. Circuits exhibit very distinctive qualities when they contain series X_L and X_C components that are equal at a particular frequency. When these same elements are connected in parallel and have equal reactances, the circuit's characteristics are completely changed. This chapter will investigate that special series and parallel circuit state known as *resonance*.

559

Applications

The phenomenon of resonance occurs in many fields other than electronics. A weight attached to a spring, for example, will have a natural frequency determined by the mass of the weight and the stiffness of the spring. A friction-producing shock absorber connected to such a resonant device will have the same dampening and broadening effects as a resistor in a resonant circuit. In fact, there is a direct analog between resistance, capacitance, and inductance, on the one hand, and friction, spring stiffness, and mass, on the other. The mathematical descriptions of all resonant phenomena are almost identical.

The ability of a series or parallel combination of X_L and X_C to produce the condition known as resonance provides a number of unique electronics applications. These resonant circuits make it possible for a radio receiver to tune in, or receive, a particular frequency. When you want to change stations on a radio, the tuning knob operates a variable capacitor. The tuning circuit, consisting of a coil of wire in parallel with a capacitor, impedes current of all frequencies except one particular resonant frequency. Various resonant frequencies can be tuned in by varying the capacity. Thus, tuners enable radios and televisions to select one station's signal from among all the other signals present in the air at different frequencies.

In a television set, the VHF channels are so far apart that separate resonant circuits are switched in for each channel, but a variable condenser is still used for fine tuning of the set to the exact frequency. Tuned circuits, when they are properly connected in a circuit, can also be used to reject unwanted signals. Thus, FM receivers use tuned circuits as discriminators to recover the transmitted intelligence. Tuned circuits are vital to the operation of radio transmitters, amateur radio, and many other forms of communications equipment such as radar and telemetry.

19-1
Definition of Resonance

> **Resonance** The condition that occurs when a circuit's inductive and capacitive reactances are balanced.

In Chapter 16, we saw that inductive reactance increases directly with frequency, while capacitive reactance varies inversely with frequency. If both L and C are present in an ac circuit, there will be some particular frequency where their reactance will be equal but opposite. This condition is called **resonance**. A circuit exhibiting this characteristic is called a *resonant circuit*. Both L and C must be present for this condition to occur.

In any resonant circuit, some resistance is usually present to a greater or lesser degree. Resistance does not influence the resonant frequency, although it does have an effect on other resonant circuit parameters (which will be explained in later sections of this chapter).

The values of L and C determine the specific frequency of resonance. Changing either or both of them will result in a different resonant frequency. Generally speaking, the larger L or C becomes, the lower is the frequency at which it will resonate. Conversely, making these circuit elements smaller increases the resonant frequency. For example, the amount of inductance and capacitance necessary to tune (resonate) a TV receiver to a particular channel is considerably less than that required to tune an AM receiver to a particular broadcast station. In fact, at extremely high frequencies, the *distributed capacity* across the several turns of the inductance is frequently enough to cause the circuit to resonate.

Above and below the resonate frequency of any particular LC combination, the circuit behaves as does any standard ac circuit. Standard ac circuit behavior has been explored in previous chapters.

Resonance is generally desirable only with radio frequencies in tuning receivers and transmitters and certain other industrial and test equipment. It is unwanted in audio amplifiers and power supplies, although it sometimes inadvertently occurs because of improper design or faulty components. Resonant circuits are rarely used in the audio band of frequencies.

19–2
Series-Resonant Circuit

There are two basic kinds of resonant circuits: *series* and *parallel*. Series circuits will be investigated in this section. Parallel circuits will be studied in Sections 19–5 through 19–8.

> ┌─ **Series resonance**   The condition that occurs when the current in
>   a series circuit consisting of inductance and capacitance is in
>   phase with the voltage across the circuit; characterized by low
>   impedance at the resonant frequency and high attenuation of
>   all others; the condition when the inductive and capacitive reac-
>   tances cancel.

When L and C are series-connected and their reactances are equal, **series resonance** occurs. Because X_L is at $90°$ with respect to the real, or resistive, axis, and X_C is at $270°$, they are $180°$ out of phase.

Graph of X_L and X_c

Since the condition of resonance is that $X_L = X_C$, it follows that they completely cancel because they are $180°$ out of phase; see Figure 19–1. At the resonant frequency f_r, X_L is equal to X_C. The opposing nature of their reactances is clearly evident.

If L were changed and not C, the *slope* of the X_L curve would change, becoming *steeper* for larger values of L (as represented by the

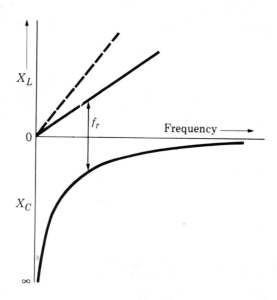

Figure 19–1

Graph of X_L and X_C showing equal values at resonance

dotted lines in the figure). This change *reduces* the resonant frequency, as indicated by the fact that we must move to the left on the frequency axis in order to find a new f_r where $X_L = X_C$. Of course, changing C also changes the slope of the X_C curve and establishes a new resonant frequency. Hence, any change of C or L, or both, results in a new f_r. The resonant frequency *increases* as either, or both, of these circuit parameters becomes *smaller.*

Impedance of Series-Resonant Circuit

A series-resonant circuit is shown in Figure 19–2. The resistance r_s represents the internal series resistance of the inductor. Resonance is indicated by the fact that X_L and X_C are equal. The impedance of the circuit can be calculated by the following familiar formula:

$$Z = \sqrt{r_s^2 + (X_L - X_C)^2}$$

The two reactances cancel at resonance. The impedance, then, is as follows:

$$Z = \sqrt{r_s^2} \quad \text{or} \quad Z = r_s$$

Hence, at resonance, the source experiences a *pure resistance.* The impedance of the series circuit is limited only by the value of r_s. Because the reactances cancel, the line current is in phase with the source voltage.

Line Current Maximum at Series Resonance

The line current in the series LC circuit is *maximum at resonance* because its *impedance is minimum.* The power factor of the circuit is unity. On either side of the resonant frequency, the impedance rises rapidly; consequently, the line current quickly drops off. The line current, at f_r, is determined by the ratio E_S/r_s.

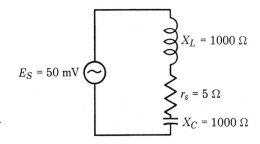

Figure 19–2

Series-resonant circuit

Example 19–1 ──

Determine the resonant line current for the series circuit in Figure 19–2.

Solution

Since $X_L = X_C$, the reactances cancel.

$$Z = r_s$$

Then
$$I_T = \frac{E_S}{r_s} = \frac{50 \times 10^{-3}\ \text{V}}{5\ \Omega} = 10\ \text{mA}$$

This current is the maximum current that can flow for the value of E_S indicated.

Impedance Minimum at Series Resonance

Figure 19–3 shows a graph (solid curve) of line current versus frequency for the series-resonant circuit. Observe the rapid falloff of current on either side of resonance. The dotted curve indicates the variation of circuit impedance with frequency. Below and above resonance, the impedance is high and drops to a very low value at f_r. At resonance, the impedance is determined by the amount of series resistance, usually that of the inductive winding.

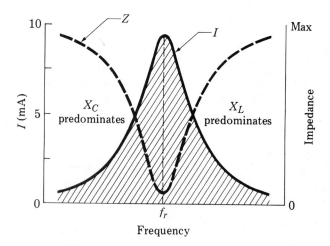

Figure 19–3

Variation of line current (solid line) and impedance (dotted line) with frequency for series-resonant circuit

Below resonance, the circuit has more X_C than X_L. Consequently, the circuit looks *capacitive* and has a leading power factor. The predominance of X_C can be verified by examining the curves in Figure 19–1. Above the resonant frequency, conditions are reversed. There is more X_L than X_C, and therefore, the circuit is inductive and has a lagging power factor.

Maximum Voltage across L and C at Resonance

Because $X_L = X_C$ at resonance, we should not assume that each of these reactances is zero. Quite the opposite is true. The reactances may even be fairly high, as indicated by the values of the two circuit elements in Figure 19–2. Now, because the circuit current is maximum at f_r, and since X_L and X_C may be large, the voltages developed across these reactances can be significant. These reactive voltages can exceed the supply voltage many times. The following example illustrates such a situation.

Example 19–2

For the circuit of Figure 19–2, determine the voltages appearing across each of the reactances.

Solution

In Example 19–1, we determined the line current at resonance to be equal to 10 mA. Therefore:

$$E_L = I_T X_L = 10 \text{ mA} \times 1 \text{ k}\Omega = 10 \text{ V}$$

Voltage E_C will be the same because $X_L = X_C$.

Table 19–1   Calculations for Series-Resonant Circuit in Figure 19–4

Frequency (kHz)	X_L (Ω)	X_C (Ω)	$X_L - X_C$ (Ω)	Z_T (Ω)	I_T (μA)	E_L (mV)	E_C (mV)
600	600	1667	1067	1067	47	28	78
800	800	1250	450	450	111	89	139
1000	1000	1000	0	5*	10,000	10,000	10,000
1200	1200	833	367	367	136	163	113
1400	1400	714	686	686	73	102	52

*Z and I calculations made without r_s, except at f_r.

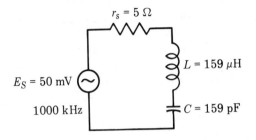

$r_S = 5\ \Omega$

$L = 159\ \mu H$

$E_S = 50\ mV$

1000 kHz

$C = 159\ pF$

Figure 19-4

Series-resonant circuit for the calculations in Table 19-1

The data in Table 19-1 further illustrates and summarizes the information presented in this section. Calculations are based on the circuit shown in Figure 19-4, which is typical of a resonant circuit in a standard broadcast receiver. The resonant frequency is 1000 kHz, which occurs at the approximate center of this band. Observe the increase in X_L and decrease in X_C as the frequency increases from 600 to 1400 kHz. At resonance, the circuit impedance is 5 Ω, which is presumed to be the resistance of the windings in the inductor. It is significant to note the relatively large voltages developed across L and C at resonance and how they quickly drop off on either side of f_r.

19-3
Calculating the Resonant Frequency

The condition of resonance in any circuit is determined by the equal and opposite reactances in the circuit. Since both X_L and X_C depend on the frequency of the applied voltage, the frequency at which resonance occurs can be calculated as follows:

$$2\pi f L = \frac{1}{2\pi f C}$$

Use f_r to indicate the resonant frequency. Then:

$$4\pi^2 f_r^2 LC = 1$$

Solve for f_r

$$f_r^2 = \frac{1}{4\pi^2 LC}$$

$$f_r = \frac{1}{2\pi\sqrt{LC}} \qquad \textbf{19-1}$$

where
$$f_r = \text{resonant frequency in hertz}$$
$$L = \text{inductance in henrys}$$
$$C = \text{capacitance in farads}$$

A convenient form of this equation, using the reciprocal of 2π, is as follows:

$$f_r = \frac{0.159}{\sqrt{LC}} \qquad\qquad \text{19-2}$$

Example 19-3

Calculate the resonant frequency of a 350 pF capacitor in series with a 25 μH inductor.

Solution

$$f_r = \frac{0.159}{\sqrt{LC}} = \frac{0.159}{\sqrt{25 \times 10^{-6} \times 350 \times 10^{-12}}}$$

$$\cong 1.7 \times 10^6 \text{ Hz} = 1.7 \text{ MHz}$$

Product of *LC* Determines *f$_r$*

An examination of Equations 19-1 or 19-2 indicates that, for any value of L and C, there is but one resonant frequency. This frequency depends on the product of L and C alone, as all other factors are constant. Thus, various combinations of L and C may be used in a circuit to achieve resonance at a particular frequency as long as the product of their values is the same. A large value of L and a small value of C may resonate at the same frequency as a large value of C and a small value of L. For example, if the values of L and C in Figure 19-4 were changed to 101 μH and 250 pF, respectively, the resonant frequency would still be about 1 MHz. Another possible combination resonating at approximately 1 MHz would be 80 pF and 315 μH.

f$_r$ Varies Inversely with *L* or *C*

If the product of L and C *increases*, the resonant frequency is *reduced*. Both circuit elements need not be increased to lower the frequency, only one or the other. The opposite also holds true. Any decrease in L or C reduces their product and increases f_r. Hence, the resonant frequency of a series LC circuit varies inversely with their product.

It is a relatively simple procedure to calculate either L or C if the desired resonant frequency and one of these two circuit parameters are known. Using the resonance formula (Equation 19–1), we can solve for either L or C as follows:

$$f_r = \frac{1}{2\pi\sqrt{LC}}$$

Square both sides to remove the radical sign:

$$f_r^2 = \frac{1}{4\pi^2 LC}$$

Transpose f_r^2 and L:
$$L = \frac{1}{4\pi^2 f_r^2 C} \qquad \text{19–3}$$

If we use the reciprocal of $4\pi^2$, the equation can be simplified as follows:

$$L = \frac{0.0253}{f_r^2 C} \qquad \text{19–4}$$

where
$$L = \text{inductance in henrys}$$
$$f_r = \text{resonant frequency in hertz}$$
$$C = \text{capacitance in farads}$$

We can transpose the resonance formula and solve for C as follows:

$$C = \frac{1}{4\pi^2 f_r^2 L} \qquad \text{19–5}$$

This equation can be further simplified:

$$C = \frac{0.0253}{f_r^2 L} \qquad \text{19–6}$$

Example 19–4

Calculate the value of capacitance required to resonate the secondary winding of an FM receiver's intermediate-frequency (IF) transformer if $f_r = 10.7$ MHz and $L = 23$ μH.

Solution

Use Equation 19-6.

$$C = \frac{0.0253}{f_r^2 L} = \frac{0.0253}{(10.7 \times 10^6)^2 \times 23 \times 10^{-6}}$$
$$= 9.6 \times 10^{-2}\,F = 9.6\ \text{pF}$$

Example 19-5 ─────────────────────────────────────

How much inductance is required to produce resonance with a 100 pF capacitor at 456 kHz?

Solution

Use Equation 19-4.

$$L = \frac{0.0253}{f_r^2 C} = \frac{0.0253}{(4.56 \times 10^5)^2 \times 100 \times 10^{-12}} = 1.2\ \text{mH}$$

19-4
Circuit Q

Quality of an inductor (Q)   The ratio of inductive reactance to resistance of a coil; also called *figure of merit*.

The ratio of the energy stored in an inductor during the time the magnetic field is established to its losses during the same time is known as the **quality of the inductor**, or Q. It is also called the *figure of merit.* Quality is *the ratio of the inductive reactance of a coil to its resistance.* Expressed mathematically, quality is as follows:

$$Q = \frac{X_L}{r_s} \qquad\qquad \text{19-7}$$

where
Q = quality
X_L = inductive reactance at f_r
r_s = resistance in series with X_L

Quality Q is a *dimensionless quantity* because X_L and r_s are both measured in ohms and they cancel.

Example 19-6

Calculate Q for the inductor in the series-resonant circuit of Figure 19-4.

Solution

From Table 19-1, we see that $X_L = 1000 \, \Omega$ at $f_r = 1000$ kHz. Substitute these values into Equation 19-7:

$$Q = \frac{1000}{5} = 200$$

The Q value calculated in Example 19-6 is considered a high Q. For receiving circuits, typical values of Q range from 50 to about 250. In contrast, the Q of a parallel-resonant circuit (covered in the next section) in the output stage of a typical transmitter can vary from 6 to 20.

We can determine Q for a capacitor by the same formula since, at resonance, its reactance equals that of the inductor. It is customary, however, to use X_L for the calculations because the series resistance is usually in the inductance. Therefore, Q for the inductor and Q for the series-resonant circuit are equal. If resistance should be added to the circuit, the circuit's Q would decrease, while the coil's Q would be unchanged.

Skin effect   The tendency of high-frequency currents to flow near the surface of a conductor, thus effectively increasing its resistance.

The actual resistance of an inductor can be considerably greater than its dc resistance when it is connected in a high-frequency circuit, because high-frequency currents do not travel uniformly throughout the cross section of the conductor but have a tendency to travel closer to the surface as the frequency increases. This phenomenon is called **skin effect**.

The portion of the conductor that carries current is decreased because of skin effect. This decrease, in turn, increases the ratio of ac-to-dc resistance of the conductor and is known as the *resistance ratio*. Many coils are wound with *Litz wire* to reduce the resistance ratio. This wire consists of a number of very small interwoven strands of separately insulated wire (about #40) connected together only at the ends of the coil.

Example 19-7

A series circuit, resonant at 5 MHz, has a 40μH inductor with a Q of 120. Calculate the ac resistance of the coil.

Solution

First, calculate X_L.

$$X_L = 6.28 \times 5 \times 10^6 \times 40 \times 10^{-6} = 1256 \ \Omega$$

Now, solve for r_s.

$$Q = \frac{X_L}{r_s}$$

$$r_s = \frac{X_L}{Q} = \frac{1256}{120} = 10.47$$

Variation of Q with Frequency

Because the ac resistance of a coil increases with frequency, as does X_L, it might be assumed that the circuit Q remains constant. However, this assumption is not exactly true. In practice, Q for an inductor reaches a relatively broad peak at some particular frequency. The design of coils is such that this peak of Q versus frequency occurs in the normal operating range for which it is designed.

At audio frequencies, the Q of an inductor will be substantially less than the value calculated by dividing the inductive reactance by the resistance. Losses in the iron core represent circuit resistance, just as though it occurred in the windings themselves.

Voltage Rise across L and C

In Example 19-2 and Table 19-1, we saw how the voltage across both L and C can rise to many times the supply. This rise is the result of the resonant rise of current and is equal to Q for the circuit. This phenomenon is sometimes called *circuit magnification* and can be expressed as follows:

$$V_L = V_C = QE_S \qquad \text{19-8}$$

In Figure 19-4, the supply voltage is 50 mV, and according to Table 19-1, V_L and V_C are 10,000 mV at resonance. From this value, we find the Q of the circuit to be the ratio of V_L (or V_C) to E_S, or 10,000 mV to 50 mV. This ratio is 200 to 1, which is the Q of the tuned circuit. A general formula for determining the Q of a circuit is as follows:

$$Q = \frac{E_{out}}{E_{in}} \qquad \text{19-9}$$

For a series-resonant circuit, measuring input and output voltages is a practical and accurate method of determining Q. The method suggested by Equation 19-7 is likely to be less accurate, because the presence of skin effect makes it difficult to determine an exact value for r_s.

Example 19-8

Calculate the Q of a series-resonant circuit that develops 375 mV across a 150 pF capacitor with 5 mV input.

Solution

$$Q = \frac{E_{out}}{E_{in}} = \frac{375 \text{ mV}}{5 \text{ mV}} = 75$$

Effect of *L/C* Ratio on *Q*

Resonance is determined by the product of LC, as indicated in Equation 19-1. A large L and a small C can produce the same product as a small L and a large C. However, the reactances at resonance are the same, because $X_L = X_C$. If a circuit is designed so that a relatively large L is required, the circuit's X_L will be greater than if a small L and large C were used. Therefore, using a large L/C ratio results in higher values of X_L for a given frequency. These higher X_L values will result in high circuit Q's, provided the design is good and r_s does not increase at the same rate as X_L.

19–5
Ideal Parallel-Resonant Circuit

> **Parallel resonance**   The frequency at which $X_L = X_C$ in a parallel
> LC circuit. Also, the condition that exists in a parallel LC cir-
> cuit when the current in the circuit is in phase with the applied
> voltage.

The phenomenon of resonance, which has been described in
terms of the series circuit, is applicable to the parallel LC circuit. However, **parallel resonance** reveals a different set of operating conditions.
To avoid confusion, and to keep the distinction between series-resonant
and parallel-resonant circuits clear, we define the *series-resonant circuit
as one in which the signal (voltage) originates within the resonant circuit* (for example, Figure 19–4). In contrast, the *parallel-resonant circuit
is one in which the signal originates outside the LC combination.* The
latter condition is shown in Figure 19–5a.

> **Tank circuit**   A parallel-resonant circuit capable of storing r–f
> energy over a band of frequencies but centered on the resonant
> frequency.

Frequently, the term **tank circuit** is used in connection with series-resonant and parallel-resonant circuits. The term derives from
*the ability of an LC combination to store electrostatic and electromag-
netic energy.*

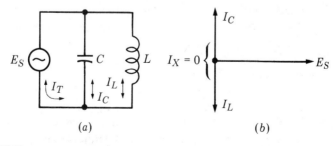

(a)   (b)

Figure 19–5

Parallel-resonant circuit: (a) Ideal circuit. (b) Phasor diagram.

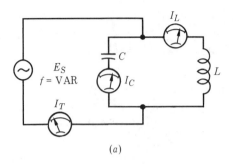

(a)

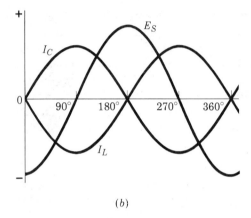

(b)

Figure 19–6

Ideal parallel-resonant tank circuit:
(a) Circuit. (b) Waveforms.

An ideal, or theoretically perfect, parallel-resonant circuit would have no resistance in either the L or C branch (see Figure 19–5a). An ac voltage, whose frequency is equal to f_r of the tank, will produce equal reactive currents I_C and I_L, as indicated in Figure 19–5b. These currents are 180° out of phase with each other and 90° out of phase with the supply voltage, which is the reference vector. In this hypothetical condition, the branch currents cancel and the line current I_T is zero. Thus, this ideal circuit presents an infinite impedance to the line and acts as an open circuit.

Figure 19–6a shows an ideal tank with three current meters installed. (Normally, current meters would never be installed in the L and C branches, because their internal resistances would seriously impair the Q of the circuit.) The frequency of the voltage source is variable. At frequencies much lower than f_r, $X_L \ll X_C$, and meter I_L will indicate a value almost equal to the line current I_T. At frequencies much higher than f_r, $X_C \ll X_L$, and meter I_C will indicate a value almost equal to I_T.

At f_r, $X_L = X_C$ and $I_C = I_L$, and the meter in the line, I_T, reads zero, as represented in Figure 19-6a. The graph in Figure 19-6b shows that, at any instant, the branch currents are equal and opposite and are 90° out of phase with the reference phasor E_S.

Even though the line current is assumed to be zero at resonance, the current in the capacitor and the inductor is the same current. These reactive currents, which are in phase in the tank itself, can be considerable. They circulate from capacitor, to inductor, back to capacitor, ad infinitum. Thus, the capacitor, once charged, delivers its energy (electrostatic) to the inductor, which stores it electromagnetically and then returns it to the capacitor.

> **Antiresonant frequency** The resonant frequency of a parallel *LC* circuit.

The frequency of resonance in a parallel *LC* circuit requires that $X_L = X_C$, as with the series-resonant circuit. Consequently, the resonant frequency is determined by the same formula, Equation 19-1:

$$f_r = \frac{1}{2\pi\sqrt{LC}}$$

The frequency of resonance in a parallel *LC* circuit is sometimes called the **antiresonant frequency**, to indicate that it is a parallel circuit.

19-6
Practical Parallel-Resonant Circuit

As a matter of practicality, all inductors have some resistance, particularly at radio frequencies. A practical parallel-resonant circuit must, therefore, include this circuit parameter, as shown in Figure 19-7. The inclusion of r_s in the inductive branch has a significant effect on the circuit's performance at resonance. The inductive and capacitive currents will no longer exactly cancel at f_r, because the impedance of the inductive branch will be slightly greater, due to the presence of r_s. Also, I_L will not lag exactly 90° from E_S but will lag a fraction of a degree less.

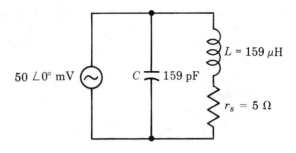

Figure 19–7

Practical parallel-resonant circuit

Determining I_τ

At a resonant frequency of 1 MHz, $X_L = X_C = 1000 \; \Omega$ in the circuit of Figure 19-7. The capacitive current is as follows:

$$I_C = \frac{50 \text{ mV}}{1000 \; \Omega} = 50 \; \mu\text{A}$$

Because of r_s, the impedance of the inductive branch will be slightly greater than 1000 Ω. Empirically, I_L will be approximately 49.75 μA. Thus, the line current I_T is as follows:

$$I_T = I_C - I_L = 50 - 49.75 = 0.25 \; \mu\text{A}$$

I_τ Minimum at Resonance

The previous calculations point out that the inductive and capacitive currents oppose each other and that their difference becomes I_T at resonance. Since I_L and I_C are nearly equal, their difference is very small, as confirmed by the preceding calculations.

A tabular presentation of the basic facts about a practical parallel-resonant circuit are shown in Table 19-2. Except for the calculation of I_T at resonance, the presence of the 5 Ω series resistance in the inductive branch has been ignored, since its effect is insignificant. The pronounced *dip* of line current at resonance is abundantly clear. The term *dip* is used to signify the minimum current reading in a parallel-resonant circuit.

Table 19-2 Calculations for Parallel-Resonant Circuit in Figure 19-7

Frequency (kHz)	X_L (Ω)	X_C (Ω)	I_L (μA)*	I_C (μA)	$I_L - I_C$ (μA)	I_T (μA)	Z_T (Ω)
600	600	1667	83	30	53	53	943
800	800	1250	63	40	23	23	2,174
1000	1000	1000	50	50	0	0.25*	200,000†
1200	1200	833	42	60	18	18	2,778
1400	1400	714	36	70	34	34	1,471

* I_L calculations made without r_s, except for I_T at f_r (see text).
† Z_T calculated by $Z_T = QX_L$.

Z_T Maximum at Resonance

Because the line current is minimum at resonance, we can expect the impedance at f_r to be maximum (see Table 19-2). Observe the rapid fall-off of impedance on either side of f_r. The characteristic of a parallel LC circuit to present a high impedance at one particular frequency makes it very useful in selecting a desired frequency within a given band of frequencies. The most common application of parallel-resonant circuits is a *load impedance* in the output circuit of radio frequency (RF) amplifiers. Not only is the circuit selective to its tuned frequency, but its high impedance causes the amplifier stage to provide highest amplification or gain.

Graphical Representation of Z_T and I_T

A graphical representation of the variation of circuit impedance with frequency is shown in Figure 19-8. This presentation is qualitative and not quantitative, since no particular circuit is being represented. The line current is represented by the dotted curve, which dips at f_r. The dip varies inversely with Z_T. An analysis of these curves reveals why the output current of an active device dips to its lowest value when its load impedance (a parallel-tuned circuit) is tuned to resonance.

At resonance, in the case of the circuit in Figure 19-7, the impedance reaches a value of 200 kΩ and is effectively a pure resistance. On the low-frequency side of resonance, X_L is less than X_C, and the circuit current favors the inductive branch. As Figure 19-8 indicates, I_L predominates in this region. The reverse is true for frequencies beyond resonance. In this case, X_C is less than X_L, and the circuit current favors the capacitive branch. The circuit power factor is unity at resonance; it is lagging below f_r and leading above f_r.

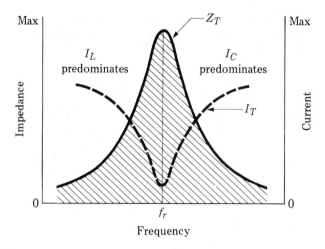

Figure 19–8

Variation of impedance (solid line) and current (dotted line) with frequency for parallel-resonant circuit

Circuit Q and Current Magnification

The determination of Q in the parallel-resonant circuit is done in the same manner as in the series-resonant circuit. Thus:

$$Q = \frac{X_L}{r_s}$$ 19–10

In the case of the circuit in Figure 19–7, the inductive reactance at resonance is 1000 Ω. With an ac resistance of 5 Ω, the circuit Q is as follows:

$$Q = \frac{1000}{5} = 200$$

At resonance, I_T is 0.25 μA, yet the circulating tank current is 50 μA in each branch, which is 200 times the line current and conforms to the circuit Q. This conformance is referred to as *current magnification*, and it is the counterpart of the voltage magnification experienced in the series-resonant circuit. Expressed mathematically, current magnification at f_r is as follows:

$$I_L = I_C = QI_T$$ 19–11

The circuit Q can also be expressed as follows:

$$Q = \frac{I_L}{I_T} = \frac{I_C}{I_T}$$ 19–12

Z Magnification

From Table 19-2, it is evident that the impedance at resonance is significantly greater than either X_L or X_C. There is a correlation between circuit Q and the reactances of the inductive and capacitive branches. The relationship at f_r is as follows:

$$Z_T = QX_L = QX_C \qquad \text{19-13}$$

This equation can be confirmed by inspecting the data in Table 19-2. Note that, at resonance, $X_L = X_C = 1000\ \Omega$. Since the circuit Q is 200, it follows that Z_T will be $200 \times 1000 = 200\ k\Omega$.

Example 19-9

The Q of a parallel-resonant tank in the output circuit of an RF amplifier is 15. Calculate the tank's impedance if the inductive reactance is 1650 Ω.

Solution

$$Z_T = QX_L = 15 \times 1650 = 24{,}750\ \Omega$$

19-7
Analysis of Parallel-Resonant Circuits

Table 19-2 indicates that X_L and X_C are equal at resonance. However, if the circuit Q is low, the presence of series r_s in the inductive branch causes its impedance to be slightly greater than that of the capacitive branch. Hence, the branch currents are not exactly equal.

High-Q Circuits

In the case of high-Q circuits, the impedance can be found by using Equation 18-4:

$$Z_T = \frac{Z_1 Z_2}{Z_1 + Z_2}$$

For the circuit of Figure 19–7 and its data, the circuit impedance is as follows ($f_r = 1$ MHz):

$$Z_T = \frac{-j\,1000 \times (j\,1000 + 5)}{-j\,1000 + (j\,1000 + 5)} = \frac{-j^2\,1 \times 10^6 - j\,1000}{5}$$

$$= -j^2\,0.2 \times 10^6 - j\,2000 \cong 200,000\ \angle 0°\ \Omega$$

This result compares with the impedance shown in Table 19–2.

Low-Q Circuits

Consider the same basic circuit (Figure 19–7) but with an r_s of 200 Ω. The Q is then $1000/200 = 5$. Any Q less than 10 is considered low. In this case, the impedance of the inductive branch will be greater than that of the capacitive branch, and its current will be less. The total impedance is as follows:

$$Z_T = \frac{-j\,1000 \times (j\,1000 + 200)}{-j\,1000 + (j\,1000 + 200)} = \frac{-j^2\,1 \times 10^6 - j\,2 \times 10^5}{200}$$

$$= -j^2\,5000 - j\,1000 = 5099\ \angle -11.31°\ \Omega$$

From this expression, we see that the circuit impedance is substantially reduced with the lower-Q circuit.

A parallel-resonant, high-Q circuit has the following characteristics:

1. Resonant frequency $f_r = 1/2\pi\,\sqrt{LC}$.
2. Impedance Z_T is maximum and I_{line} is minimum.
3. The power factor is unity (zero phase angle).

However, in a low-Q, parallel-resonant circuit, the above characteristics do not occur at the same frequency. When $X_L = X_C$, the line current will not be at its exact minimum. When we tune low-Q circuits, L or C is adjusted for minimum line current (as with tank circuits in RF power amplifiers). This phenomenon occurs at a slightly different frequency than $f_r = 1/2\pi\,\sqrt{LC}$, and it is called the antiresonant frequency. When Q is less than 10, the following formula can be used.

$$f_{ar} = \frac{1}{2\pi\sqrt{LC}}\,\sqrt{\frac{1}{1 + (1/Q^2)}} \qquad \text{19–14}$$

where $\qquad f_{ar}$ = antiresonant frequency

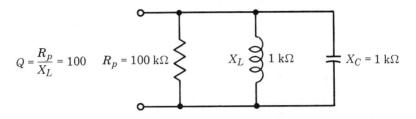

Figure 19–9

Parallel-resonant circuit with damping resistor R_p

19–8
Dampening a Parallel-Resonant Circuit

The effects of a series r_s on a parallel resonant circuit have been investigated. Now, we must consider the effect a parallel resistance, or load, will have on circuit performance. Resistor R_p in the circuit of Figure 19–9 may represent the resistance of the driving source, or it can be an actual resistance added to increase the bandwidth by lowering the circuit Q. The effect is to *dampen the resonant circuit*, which results in a more *symmetrical response curve* than is obtained by increasing r_s.

Calculating Q for R_p Present

Varying R_p produces the opposite effect from varying r_s. For example, reducing R_p *lowers* the circuit Q and selectivity. If r_s is reduced, the circuit Q *increases* along with selectivity. For a parallel circuit, Q is determined by the following equation:

$$Q = \frac{R_p}{X_L} \qquad \text{19–15}$$

Here, r_s is considered negligible. For example, in Figure 19–9, $Q = 100$ kΩ/1 k$\Omega = 100$.

From a more practical standpoint, the effects of r_s must be considered along with the effects of R_p. For the circuit of Figure 19–10, Q is determined as follows:

$$Q = \frac{X_L}{r_s + (X_L^2/R_p)} \qquad \text{19–16}$$

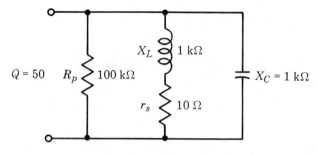

Figure 19-10

Parallel-resonant circuit with R_p and r_s

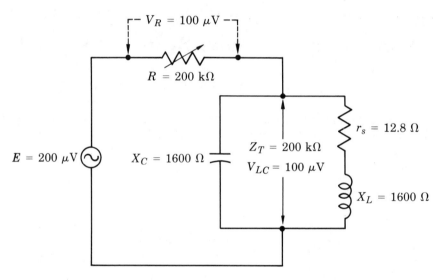

Figure 19-11

Determining impedance of parallel-resonant circuit

Calculating Q for the circuit of Figure 19-10, using the data supplied, we obtain the following result:

$$Q = \frac{1000}{10 + (1{,}000{,}000/100{,}000)} = \frac{1000}{10 + 10} = 50$$

Since the circuit has both r_s and R_p, it has a lower Q than if only one or the other were present.

Measuring Z_T of a Parallel-Resonant Circuit

An effective method of measuring Z_T of a parallel circuit is shown in Figure 19-11. Assuming the circuit has been tuned to resonance, adjust R so that $V_R = V_{Z_T}$. If the input voltage is 200 μV, then $V_R = V_{Z_T} = 100$ μV. Carefully remove the potentiometer R from the circuit, and measure its resistance. This resistance will equal Z_T since $V_R = V_{Z_T}$. From Equation 19-13, we can determine the circuit's Q:

$$Z_T = QX_L$$

$$Q = \frac{Z_T}{X_L} = \frac{200{,}000\ \Omega}{1600\ \Omega} = 125$$

19-9
Resonance Curves

> **Selectivity**   The ability of a tuned circuit, or receiver, to accept the frequency to which it is tuned and reject all others.

The effect of series resistance on circuit Q for a series-resonant circuit is shown in Figure 19-12. The graphs are plots of line current versus varying frequency with supply voltage and circuit resistance

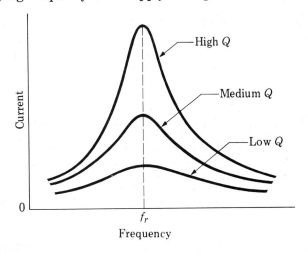

Figure 19-12

Series-resonant curves for circuits with different Q's

constant. The steepness of the curve for the high-Q circuit is very evident. As the circuit resistance increases, the response curves become flatter and broader about the resonant frequency. If the resistance becomes too large, the circuit loses its **selectivity**, and very little discrimination is made between different signal currents at resonance and off-resonance.

19-10
Bandwidth

Concept of Half-Power Points

The concept of half-power points on a resonant circuit's response curve can be explained by means of the series-resonant circuit in Figure 19-13a and the graph in Figure 19-13b. With the circuit at resonance, $X_L = X_C = 1000$ Ω. The impedance is as follows:

$$Z = R + j(X_L - X_C) = 10 + j(1000 - 1000) = 10 + j\,0 = 10 \angle 0° \ \Omega$$

With an $E_S = 10 \angle 0°$ mV, the circuit current is as follows:

$$I_T = \frac{10 \angle 0° \text{ mV}}{10 \angle 0° \ \Omega} = 1 \angle 0° \text{ mA}$$

This result represents the maximum current in the circuit and is shown as the highest point on the response curve in Figure 19-13b. Now, if the frequency of E_S is decreased by an amount $1/2 Q$ times f_r, the circuit current decreases to 0.707 of its resonant value and leads the E_S by 45°.

Example 19-10

Calculate the new frequency of E_S when it decreases from f_r by an amount equal to $1/2Q$ ($Q = 100$ for the circuit of Figure 19-13a).

Solution

$$f_{\text{dec}} = \frac{1}{2Q} \times f_r = \frac{1}{2 \times 100} \times 1000 = 5 \text{ kHz}$$

Then $\quad f_{\text{new}} = f_r - f_{\text{dec}} = 1000 - 5 = 995 \text{ kHz}$

At this new frequency, $X_L = 995 \angle 90°$ Ω and $X_C = 1005 \angle -90°$ Ω. The circuit impedance is as follows:

$$Z = R + j(X_L - X_C) = 10 + j(995 - 1005)$$
$$= 10 - j\,10\ \Omega = 14.14 \angle -45°\ \Omega$$

At 995 kHz, the circuit current is as follows:

$$I_T = \frac{10\ \angle 0°\ \text{mV}}{14.14\ \angle -45°\ \Omega} = 0.707\ \angle 45°\ \text{mA}$$

Now, calculate the power at resonance:

$$P_r = EI = 10\ \text{mV} \times 1\ \text{mA} = 10\ \mu\text{W}$$

The power at the new frequency of 995 kHz is as follows:

$$P = EI \cos \Theta = 10\ \text{mV} \times 0.707\ \text{mA} \times \cos 45° = 5\ \mu\text{W}$$

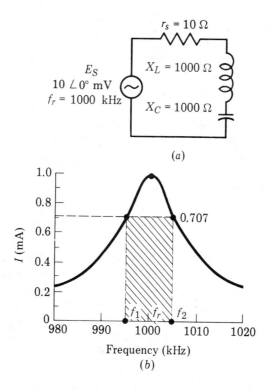

(a)

(b)

Figure 19–13

Half-power points: (a) Series-resonant circuit. (b) Response curve showing half-power points.

Plainly, when the current drops to 0.707 of the peak value, the power in the circuit is reduced to 50% of the maximum value. Therefore, the 0.707 points identified on the response curve in Figure 19–13*b* represent the frequencies (995 and 1005 kHz) at which the power drops to 0.5 of full value. It can also be shown that the power in a circuit is reduced 50% if the applied voltage is decreased by 70.7%, or 0.707 of full value. The response characteristics of filters, tuned circuits, and so on, are frequently identified in this manner.

Bandwidth   The range of frequencies between the half-power, or -3 dB (decibel), points of a tuned circuit.

While the effects of resonance are most pronounced at f_r, *adjacent frequencies* on either side of resonance can also produce effective results. For example, frequencies that are close to and on either side of f_r can produce nearly the same total circuit impedance in a parallel *LC* circuit as the resonant frequency. The same can be said of the series-resonant circuit; adjacent frequencies can produce almost the same value of total current as f_r. From these parallels, we see that, for every resonant circuit, there is a band of frequencies that provides the same characteristic resonant effects. The width of this band is called the **bandwidth** (sometimes referred to as *bandpass*) of the resonant circuit. The width of this band, centered on f_r, is determined by the circuit Q.

Calculating Bandwidth

One-half of the bandwidth is below f_r. The lower limit of this part of the bandpass is the frequency that causes the line current in a series *LC* circuit to drop to 70.7% of the maximum value at f_r. This lower-side frequency is identified as f_1 in Figure 19–13*b* and corresponds to 995 kHz. The upper limit of the bandpass is the frequency that also causes the line current to drop to 70.7% of I_T. This upper-side frequency is identified as f_2 on the graph and corresponds to 1005 kHz. The shaded portion of the figure represents the total bandwidth of the circuit in Figure 19–13*a*. Because the line current is reduced to 0.707 of I_T at f_1 and f_2, these points are referred to as the *half-power points*. (*Note:* Squaring a current that has been reduced to 0.707 of its former value results in one-half power; $0.707^2 = 0.5$.)

In the case of the parallel-resonant circuit, the 70.7% points would identify those frequencies where the circuit impedance dropped

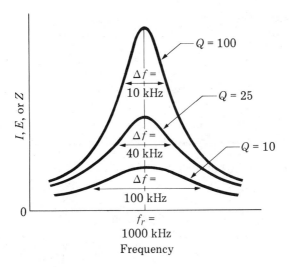

Figure 19–14

Effects of Q on bandwidth

to 0.707 of Z_T. If voltage measurements were made across a parallel-resonant circuit, the frequency at which the readings dropped to 70.7% of E_{max} would identify the half-power points.

The bandwidth of a tuned circuit is expressed by the following equation:

$$BW = f_2 - f_1 \qquad\qquad \textbf{19–17}$$

where BW = bandwidth in hertz

f_1 = lower frequency where response drops to 0.707 of maximum

f_2 = upper frequency where response drops to 0.707 of maximum

Example 19–11

Determine the bandwidth of the circuit in Figure 19–13a.

Solution

$$BW = f_2 - f_1 = 1005 \text{ kHz} - 995 \text{ kHz} = 10 \text{ kHz}$$

The bandwidth is dependent on circuit Q. High-Q circuits have smaller bandwidths than low-Q circuits. Observe that the sides, or skirts, of the curves become steeper. Figure 19–14 shows the Q's of three different circuits. The relationship between bandwidth and other tuned circuit parameters can be expressed as follows:

$$BW = f_2 - f_1 = \frac{f_r}{Q} \qquad\qquad \text{19-18}$$

Example 19–12

Calculate the bandwidth of a tuned circuit where f_r is 1000 kHz and $Q = 25$.

Solution

$$BW = \frac{f_r}{Q} = \frac{1000 \text{ kHz}}{25} = 40 \text{ kHz}$$

This condition is represented by the center curve of the graph in Figure 19–14.

There is yet another way that a circuit's Q can be determined. It is related to the L/C ratio and the series resistance.

$$Q = \frac{1}{R}\sqrt{\frac{L}{C}} \qquad\qquad \text{19-19}$$

Example 19–13

A series-resonant circuit has the following characteristics: $r_s = 5\ \Omega$, $C = 160$ pF, and $L = 40\ \mu$H. Determine the circuit Q.

Solution

$$Q = \frac{1}{R}\sqrt{\frac{L}{C}} = \frac{1}{5}\sqrt{\frac{40 \times 10^{-6}}{160 \times 10^{-12}}} = \frac{1}{5} \times 500 = 100$$

The *side frequencies* of the bandwidth curve are separated from f_r by one-half the bandwidth. Thus, to determine the side frequencies of the curve $Q = 25$ in Figure 19–14, use the following equations (Δ represents change):

$$f_1 = f_r - \frac{\Delta f}{2} = 1000 - 20 = 980 \text{ kHz}$$

$$f_2 = f_r + \frac{\Delta f}{2} = 1000 + 20 = 1020 \text{ kHz}$$

Example 19–14 ──────────────────────────────

Determine the bandwidth and side frequencies of a parallel-resonant circuit, where $f_r = 456$ kHz and $Q = 60$.

Solution

$$BW = \frac{f_r}{Q} = \frac{456 \text{ kHz}}{60} = 7.6 \text{ kHz}$$

Then $\qquad f_1 = f_r - \frac{\Delta f}{2} = 456 - \frac{7.6}{2} = 452.2 \text{ kHz}$

$$f_2 = f_r + \frac{\Delta f}{2} = 456 + \frac{7.6}{2} = 459.8 \text{ kHz}$$

19–11
Tuning

An analysis of the LC circuit of a typical RF amplifier stage of a radio receiver will help us understand its tuning action. Such a circuit, in somewhat simplified form, is represented in Figure 19–15. Induced into

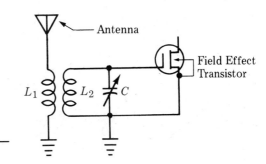

Figure 19–15

Tuned series RF circuit

the antenna is a multiplicity of voltages produced by the numerous passing radio waves. Each voltage has a particular frequency, identified by the station transmitting the wave. These voltages develop currents in winding L_1 of the RF input transformer. The completed circuits for these currents are via the distributed capacity between antenna and ground (not shown). The currents induce voltages into winding L_2.

Let us assume that the inductance of L_2 is 232 μH, and that tuning capacitor C has a range of 40–360 pF. Incidentally, these values are the approximate values encountered in a standard broadcast receiver covering the 535–1605 kHz band. Therefore, our hypothetical receiver will be able to tune over the broadcast band. Now, suppose we adjust capacitor C to 348 pF (the rotor and stator plates will be nearly fully meshed). Which of the many frequencies present in the tuned series circuit L_2C will appear "amplified" Q times across the tuning capacitor and impressed on the input to the transistor? Because $L_2 = 232$ μH, we must calculate the frequency at which it will resonate when tuned with a 348 pF capacitor:

$$f = \frac{0.159}{\sqrt{LC}} = \frac{0.159}{\sqrt{232 \times 10^{-6} \times 348 \times 10^{-12}}} = 560 \text{ kHz}$$

Therefore, if a station is transmitting on 560 kHz, its frequency will be the one that is selected, or tuned in, and will appear at the input to the transistor. Assuming the circuit has good selectivity (high Q), all other frequencies present in the tank will be discriminated against.

If the tuning capacitor is now adjusted to 68.7 pF, the circuit will resonate at 1260 kHz. This frequency only will be fed to the transistor; all other frequencies will be rejected (assuming they fall outside the bandpass of the circuit). Thus, by adjusting the tuning capacitor, we can select any of the frequencies induced into L_2 within the tuning range of L_2C. If the received signals are too weak, the circuit will not have sufficient sensitivity (ability to amplify weak signals) to select them.

In practice, a pointer and a scale for frequency band are employed. The pointer positions the tuning capacitor to the desired frequency, providing the exact capacity required for resonance with its associated inductance. No significant rise in current will occur with frequencies outside the bandpass of the circuit.

Equation 19–1 indicates that *a change of resonant frequency is inversely proportional to the square root of the change in L or C.* For example, earlier in this section, reference was made to the fact that the tuning capacitor of a broadcast AM receiver varies from 40 to 360 pF, representing a change of 9:1. The square root of this ratio is 3:1, which corresponds to the tuning ratio of the receiver that covers the band from 535 to 1605 kHz. Therefore, if C is decreased by a factor of 4, the frequency doubles.

Example 19–15 ───────────────────────────────────

A tuning capacitor when set to 160 pF resonates an inductor at 800 kHz. At what frequency will the circuit resonate when the capacitor is adjusted to one-fourth of this capacity?

Solution

$$\frac{160 \text{ pF}}{4} = 40 \text{ pF}$$

Therefore, the capacitance is reduced by a 4:1 ratio, or 1/4 of its former value. Now $\sqrt{1/4} = 1/2$; hence, the frequency will double, or equal 1600 kHz.

19–12
Bandspreading

Tuning capacitors generally have a 180° rotation. When they are connected across an appropriate inductor, such as those used in the broadcast band, the amount of frequency change for each degree of rotation will be as follows:

$$\frac{1605 - 535}{180} \cong 6 \text{ kHz per degree rotation}$$

With 10 kHz minimum frequency difference between adjacent stations, there is no problem in tuning the desired station.

On shortwave bands, where there is often considerable frequency congestion, tuning to a specific station is more difficult. This problem can be resolved by spreading the amount of kilohertz change over a larger part of the dial, a procedure called *bandspreading*. There are two ways to do so: mechanical and electrical. With the mechanical method, a small gear train is used so that a relatively large dial movement is required for a small tuning capacitor movement. The electrical method uses a *trimmer capacitor* to obtain bandspreading. The trimmer is a small, variable capacitor of approximately 10–30 pF connected across the main tuning capacitor, as shown in Figure 19-16. A rela-

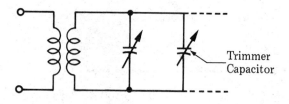

Figure 19–16

Trimmer capacitor used with main tuning capacitor for bandspreading

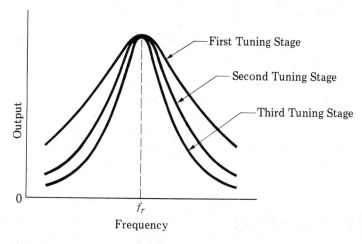

Figure 19–17

Graph illustrating how selectivity of a radio receiver increases with additional tuned stages

tively large angular movement of the trimmer causes only a small change in capacity. Therefore, a given number of kilohertz can be spread over a large movement of the trimmer dial.

19–13
Multiple Tuned Stages

Selectivity Related to Number of Tuned Stages

Radio receivers rarely use a single stage of tuning because it would not provide adequate selectivity. Additional tuned stages are required to achieve the overall selectivity needed for proper performance. The effect of adding extra tuned stages to a receiver is shown in Figure 19-17. Each stage is assumed to be tuned to a *common resonant frequency*. The characteristics of each tuned stage are assumed to be the same. The progressively steeper slopes of the skirts of the overall response curves are clearly evident as the effects of the first, second, and third tuned stages are added.

Limit to Number of Tuned Stages

With selectivity increasing as additional stages are added, it would seem that extreme selectivity could be accomplished by adding a number of stages. However, there are practical limits to the number of stages that can be added. For example, if the selectivity is too great, the bandpass characteristics of the receiver will be reduced so that the half-power points will be less than 10 kHz apart. As a result, the fidelity of the received signal will be reduced.

⌐—**Summary**

Resonance is a condition where $X_L = X_C$.

The values of L and C determine the specific resonant frequency.

As either L or C increases, the resonant frequency decreases.

At series resonance, the circuit impedance is determined by the amount of series resistance and is normally very low.

The resonant frequency of an LC combination can be calculated by the following formula:

$$f_r = \frac{1}{2\pi\sqrt{LC}}$$

Line current is maximum in a series-resonant circuit and minimum in a parallel-resonant circuit.

Maximum voltage occurs across L and C at series resonance and can be many times the supply voltage.

Voltages E_L and E_C in a series-resonant circuit equal QE_S.

The value of inductance required to resonate with a given capacitor can be calculated by $L = 0.0253/f_r^2 C$.

The value of capacitance required to resonate with a given inductor can be calculated by $C = 0.0253/f_r^2 L$.

Circuit Q is equal to X_L/r_s.

The Q of a circuit can usually be increased by using a large L/C ratio.

A parallel LC produces maximum impedance at resonance.

A parallel-resonant circuit is often referred to as a tank because of its ability to store energy.

A series LC circuit is capacitive below resonance and inductive above resonance.

A parallel LC circuit is inductive below resonance and capacitive above resonance.

The impedance of a parallel-resonant circuit equals QX_L or QX_C.

The selectivity of a resonant circuit increases as the circuit Q increases.

The bandwidth of a tuned circuit is equal to the frequency difference between the upper and lower half-power points on its response curve:

$$BW = f_2 - f_1 = \frac{f_r}{Q}$$

The overall selectivity of a receiver can be increased by adding several tuned stages.

┌──Progress Test

The bracketed number after each question indicates the section of this chapter where the answer can be found.

1. For a series or parallel LC circuit, resonance occurs when: (a) X_L is 10 times X_C or more. (b) X_C is 10 times X_L or more. (c) $X_L = X_C$. (d) the phase angle of the circuit is 90°. [19–2]

2. When either L or C is increased, the resonant frequency of the LC circuit: (a) increases. (b) decreases. (c) remains the same. (d) is determined by the shunt resistance. [19–3]

3. The resonant frequency of an LC circuit is 1000 kHz. If L is doubled but C is reduced to one-eighth of its original value, the resonant frequency in kilohertz is: (a) 250. (b) 500. (c) 1000. (d) 2000. [19–3]

4.   A coil has a 1000 Ω X_L and a 5 Ω internal resistance. Its Q equals: (a) 0.005. (b) 5.0. (c) 200. (d) 1000.   [19–4]

5.   In a parallel LC circuit, at the resonant frequency, the: (a) line current is maximum. (b) inductive branch current is minimum. (c) total impedance is minimum. (d) total impedance is maximum.   [19–5 and 19–6]

6.   In a series LC circuit, at the resonant frequency, the: (a) current is minimum. (b) voltage across C is minimum. (c) impedance is maximum. (d) current is maximum.   [19–2]

7.   A series LC circuit has a Q of 100 at resonance. When 5 mV is applied at the resonant frequency, the voltage across C in millivolts is: (a) 5. (b) 20. (c) 100. (d) 500.   [19–4]

8.   An LC circuit resonant at 1000 kHz has a Q of 100. The bandwidth equals: (a) 10 kHz between 995 and 1005 kHz. (b) 10 kHz between 1000 and 1010 kHz. (c) 5 kHz between 995 and 1000 kHz. (d) 200 kHz between 900 and 1100 kHz.   [19–10]

9.   The resonant frequency of a 150 μH inductor and a 180 pF capacitor is: (a) 0.97 MHz. (b) 10.7 MHz. (c) 993.8 kHz. (d) 455 kHz.   [19–3]

10.   The impedance of a tuned circuit at resonance is equal to: (a) X_L^2. (b) QX_L. (c) $X_L - X_C$. (d) X_C/Q.   [19–6]

⎡──Problems

Answers to odd-numbered problems are at the back of the book.

1.   Calculate the resonant frequency of a 270 pF capacitor in series with a 33 μH inductor.

2.   A 47 pF capacitor is in parallel with a 68 μH inductor. What is the resonant frequency of the combination?

3.   What value of capacitance is needed to provide parallel resonance with a 250 μH inductor at 1430 kHz?

4.   Calculate the required capacitance to resonate a 1.8 μH inductor at 144 MHz.

5.   Determine the value of inductance needed for series resonance with a 320 pF capacitor at 680 kHz.

6.   How much inductance is required to resonate with 10 pF of capacitance at 188 MHz?

7. A series circuit resonates at 456 kHz. The inductor is 125 μH. Calculate its ac resistance if it has a Q of 85.

8. Consider a 240 μH inductor having an ac resistance of 25 Ω. (a) At what frequency will the inductor have an X_L of 1500 Ω? (b) What value capacitance is required to have the same reactance? (c) Determine the Q of the circuit. (d) What is the resonant frequency of the combination?

9. Refer to Problem 8. Suppose 80 mV is applied to the parallel combination of LC. Determine the following values: (a) Z_T (b) I_T (c) Θ at f_r.

10. A 25 V source is connected to a series circuit, where $L = 40$ mH, $C = 0.05$ μF, and the inductor's effective resistance is 30 Ω. Calculate: (a) f_r (b) Z_T (c) I_T (d) V_L (e) V_C.

11. A series LCR circuit has a Q of 150. The input voltage is 25 mV at f_r. Calculate the following: (a) V_R (b) V_L and V_C at resonance.

12. A series-resonant circuit having $L = 4$ μH and $C = 8$ pF is supplied from a 5 V source; $V_C = 185$ V. Calculate: (a) f_r (b) Q (c) I_T.

13. Determine the range of tuning capacitance needed to resonate a 0.15 μH inductor through the FM broadcast band (88–108 MHz).

14. The tank circuit of a transistorized RF amplifier is resonant at 27 MHz. Suppose $X_L = 1700$ Ω, the tank current is 50 mA, and the effective resistance of the circuit is 125 Ω. Calculate the following: (a) L (b) Q (c) I_T (d) V_L.

15. At what frequency will the power of the circuit of Figure 19–18 drop to 0.5 of the original frequency?

16. Refer to Problem 15 and Figure 19–18. (a) Calculate the circuit impedance at the new frequency determined in Problem 15. (b) What is the line current at this frequency?

17. What is the bandwidth of the circuit in Figure 19–18?

18. Determine Q for a series-resonant circuit with $C = 360$ pF, $L = 20$ μH, and $r_s = 7$ Ω.

19. Determine the side frequencies of the curve $Q = 100$ in Figure 19–14.

20. Calculate the bandwidth and side frequencies of a parallel-resonant circuit where $f_r = 44.5$ MHz and $Q = 70$.

21. Assuming that maximum Q will result when an inductor's reactance is 1000 Ω, what value of L and C should be used for resonance at 15 MHz to realize the highest Q?

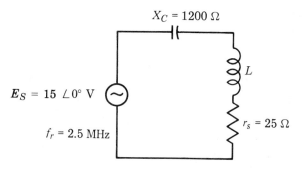

$X_C = 1200\ \Omega$

L

$E_S = 15\ \angle 0°\ V$

$r_s = 25\ \Omega$

$f_r = 2.5\ \text{MHz}$

Figure 19–18

Circuit for Problems 15–17

22. A series circuit using a coil of 250 μH resonates at what frequency when connected to a capacitor of 250 pF? What is the Q if the resistance of the coil is 5 Ω?

23. A series-resonant circuit with a resistance of 8 Ω has an X_L of 480 Ω. (a) What is the Q of the circuit? (b) What are the impedance and current at resonance?

24. In Problem 23, what are the voltages across R, C, and L when the source voltage is 24 V?

25. A 0.5 mH coil is to be used to tune a parallel circuit for 455 kHz. (a) What value capacitor is needed? (b) What is the impedance of the circuit at resonance if the r_s of the circuit is 30 Ω?

26. In Problem 25, what are the values of the line current and the capacitor and the inductor currents if $E_S = 60$ mV?

27. A parallel-resonant circuit will be used to reject a frequency of 1400 kHz. If the coil has a value of 150 μH and a resistance of 20 Ω, what will be the width of the rejected band?

28. Refer to Problem 27. (a) What is the impedance of the circuit at resonance? (b) What is the impedance of the circuit to a frequency at the lowest frequency in the bandpass and to the highest frequency in the bandpass?

29. A selectivity curve for a resonant circuit at 10 MHz has a bandwidth of 66.7 kHz. If the Q of the circuit is halved, what would be the upper and lower band edge frequencies?

30. Recall that resonant frequency varies inversely with the square root of the inductance or capacitance. What is the new resonant frequency if a 50 MHz tuned circuit has its inductance increased 25 times and its capacity decreased 4 times?

20

Filters

Objectives

After studying this chapter, you will be able to:

Recognize the basic characteristics of the principal filters used in industry.

Calculate the required component values for these basic filters.

Measure filter performance in decibels.

598

Introduction

The reactive characteristics of inductors and capacitors make it possible for these devices to be *frequency-selective*. By using various combinations of resistors, capacitors, and inductors, we can make circuits that will pass either low or high frequencies or bands of frequencies. These frequency-selective devices are called *filters*. Simple filters can be rather easily designed, while the more elaborate ones are usually obtained from manufacturers specializing in their design.

Filter performance is frequently expressed in terms of how much *attenuation* a band of frequencies experiences. Measurements are made in units called *decibels* (abbreviated dB).

Applications

A familiar application of filter circuits is in audio systems, many of which have already been mentioned. Bandpass filters select frequency ranges corresponding to desired radio or television stations. High-pass filters shunt carrier wave frequencies to ground, while keeping the lower audio frequency component of a signal in the amplifier circuit.

Low-pass filters eliminate undesirable hum in dc power supplies. The tone controls in some audio amplifiers act as simple filters to accept or reject frequencies above or below a given value. By varying a resistor in the filter circuit, we can change the range of bass or treble frequencies affected.

Loudspeakers are not equally efficient over the entire audible range of frequencies. Typical high-fidelity loudspeaker systems use filters to direct sounds of the appropriate frequency range to each of two or three speakers making up the system. High-fidelity audio amplifiers frequently use a combination of low-pass, bandpass, and high-pass filters, called a *crossover network*, in their output circuit. These filters

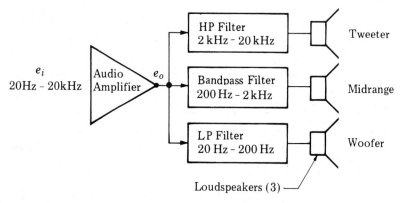

Figure 20–1

Block diagram showing use of filters, or crossover network

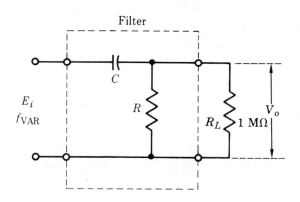

Figure 20–2

Simple *RC* high-pass filter

separate the low, midrange, and high frequencies and feed them to individual loudspeakers best able to reproduce them. A block diagram of this arrangement is shown in Figure 20-1. The amplifier output contains frequencies from 20 Hz to 20 kHz. Since no one loudspeaker can faithfully reproduce such a broad range of frequencies, filters are used to separate them into bands. The low-pass filter permits frequencies from about 20 to 200 Hz to pass through to the woofer, which is designed to handle these low frequencies. Similarly, the midrange and high-pass filters allow the bands of frequencies indicated in the block diagram to pass to the midrange and tweeter speakers.

20-1
High-Pass Filters

Filter   A selective network of inductors, resistors, or capacitors that offers little opposition to certain frequencies, or direct current, while attenuating or blocking other frequencies.

High-pass (HP) filter   A filter network that passes all frequencies above a specified frequency with little or no attenuation but discriminates or rejects all lower frequencies.

A **filter** will pass either low or high frequencies. A **high-pass (HP) filter** permits all currents having frequencies above a certain value to pass through to the load, and it discriminates against all frequencies below this value.

RC High-Pass Filter

A simple circuit exhibiting the characteristics of a high-pass filter is shown in Figure 20-2. Observe that it is a typical *RC* circuit, similar to those studied in previous chapters.

Lower frequencies experience considerable reactance by the capacitor in the circuit of Figure 20-2 and are not easily passed. Higher frequencies encounter little reactance and are easily passed. Those frequencies passing through the filter appear as an output voltage (V_o) that is connected to load R_L. To prevent any appreciable loading of the filter, we will assume that R_L is on the order of 1 MΩ, as indicated in Figure 20–2.

> **Cutoff frequency (f_{co})**   The point in a filter's frequency response characteristic where the output has dropped to 0.707, or -3 dB, of the input; also called the *half-power point*.

Let us assume that $R = 1$ kΩ and $C = 0.159$ μF, and that the input signal E_i can be varied from dc to some very high frequency. At dc and low frequencies, there will be no output voltage because the reactance of capacitor C will be extremely high. As the input frequency is increased, we arrive at a point where some output occurs. This point is represented in Figure 20–3 where the curve begins to leave the zero axis. As the input frequency continues to increase, we reach a frequency where the output voltage V_o has increased to 0.707 (or 70.7%) of E_i. This frequency is referred to as the **cutoff frequency** and has been identified as f_{co} in Figure 20–3. Frequencies below f_{co} are considered to be *rejected* by the filter, while those above are *passed*. In Section 19–10, we saw that, when an output voltage is 70.7% of the input, the power output is 50% of the input power. This condition exists at f_{co} because $X_C = R = 1000$ Ω. It is often called the *half-power point* (frequency).

The shaded area to the right of the curve represents the range of

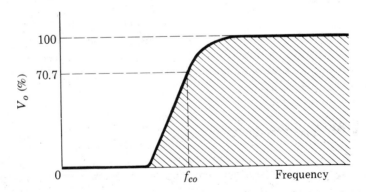

Figure 20–3

Pass characteristics of simple HP filter showing cutoff frequency and half-power point ($V_o = 0.707E_i$); shaded area represents passed frequencies

frequencies the filter passes. Those to the left are effectively blocked. If the value of C or R, or both, is changed (see Figure 20–2), the position of the curve will change. For example, if either of these filter parameters is reduced, the curve will be shifted to the right. An increase in the value of either will shift the curve to the left and cause the cutoff frequency to be decreased.

Other HP Filters

Other HP filter circuits exist besides the one shown in Figure 20–2. Typical examples of other simple types are represented in Figure 20–4. Figure 20–4a consists of inductance and capacitance connected in what is known as an *L type filter* (the capacitor and inductor form an upside down L in filters of this type; hence the name L type filter). Low frequencies encounter a high impedance at the capacitor and therefore are bypassed around the load through the low impedance of the inductor. As the input frequencies increase, the impedance of C diminishes while L increases, permitting the high frequencies to pass to the load.

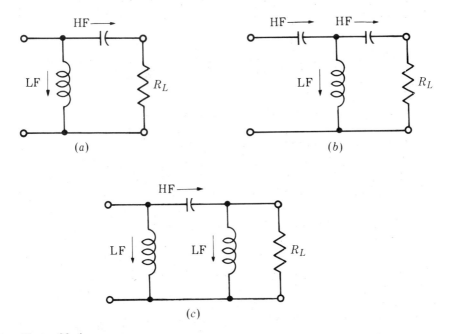

Figure 20–4

Variations of HP filters: (*a*) L type filter. (*b*) T type filter. (*c*) π type filter.

Filter action is essentially the same in the circuits of Figures 20-4b and 20-4c, except that the roll off, or attenuation, of frequencies below cutoff will be greater. The more elements (C and L) a filter has, the *steeper* is the slope of its bandpass. The circuit in Figure 20-4b is referred to as a *T type filter*, while the circuit in Figure 20-4c is a *π type filter*.

Calculation of Cutoff Frequency

The values of C and L of a single-section, high-pass filter, such as shown in Figure 20-4a, can be easily calculated from the following formulas:

$$L = \frac{R}{2\pi f_{co}} \qquad\qquad \text{20-1}$$

$$C_{\mu F} = \frac{1 \times 10^6}{2\pi f_{co} R} \qquad\qquad \text{20-2}$$

where
L = inductance in henrys
C = capacitance in microfarads
f_{co} = cutoff frequency in hertz
R = impedance of both source and load in ohms

Thus, if a particular cutoff frequency is desired, and the input and output impedances are known, the necessary values of inductance and capacitance can be determined. The following example illustrates the calculations.

Example 20-1 ──────────────────

Determine the necessary values of C and L for a high-pass filter having 1000 Ω of input and output impedance to cut off at 1 kHz.

Solution

Use Equations 20-1 and 20-2 to solve for L and C.

$$L = \frac{R}{2\pi f_{co}} = \frac{1000}{6.28 \times 1000} = 0.159 \text{ H}$$

$$C_{\mu F} = \frac{1 \times 10^6}{2\pi f_{co} R} = \frac{1 \times 10^6}{6.28 \times 10^3 \times 10^3} = 0.159 \ \mu F$$

Capacitive Coupling

Figure 20–5*a* represents a simplified equivalent circuit of the *capacitive-coupling arrangement* between two typical audio amplifier stages. The battery-generator combination represents the voltages present in the output of one stage, and R_i is the input resistance of the next. This arrangement is basically a high-pass filter designed to allow all frequencies above about 20 Hz to pass while rejecting those below 20 Hz. Customarily, the dc component must be blocked from the input to an ac amplifier so as not to upset the stage's bias.

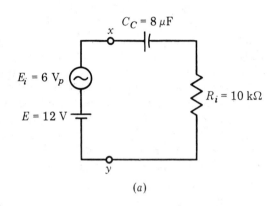

(*a*)

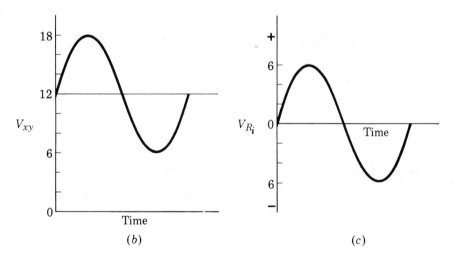

(*b*)   (*c*)

Figure 20–5

RC coupling for audio frequencies: (*a*) Circuit. (*b*) Voltages appearing at terminals *xy*, showing dc with ac superimposed. (*c*) Voltage across R_i, showing only ac present.

The voltages present at xy, which represents the output terminals of the amplifier, are shown in Figure 20-5b and consist of 12 V dc upon which a 6 V_p ac signal has been superimposed. Capacitor C_C will charge to the 12 V dc level, thus effectively blocking any dc voltage across the input resistance R_i. However, the ac signal is coupled through C_C and is developed across R_i. With the dc blocked, note that the ac signal across R_i swings alternately above and below the zero axis, as indicated in Figure 20-5c.

Capacitor C_C and resistor R_i have a relatively long time constant for the normal ac signals passing through. In the circuit of Figure 20-5a, this time constant is 80 ms. Input signal E_i tries to charge C_C to 6 V above and below its steady-state 12 V dc charge. Before it receives any appreciable charge from the ac source, the polarity changes. What little charge was received is returned to the source. On the next half cycle, the capacitor starts to charge in the reverse direction. Because of the long time constant, it receives only a slight charge before the next half cycle arrives. The relatively large charging and discharging currents produce an ac voltage across the series resistance R_i. This voltage has the same frequency as E_i and approximately the same amplitude. In typical RC coupling circuits, the reactance of C_C should be approximately $0.1R_i$ at the lowest frequency to be passed, to ensure negligible loss of signal voltage across the coupling capacitor.

Octaves and Decades of Frequency

A filter's performance is usually expressed in terms of the number of decibels (dB) the signal is increased, or attenuated, per *frequency octave* or *frequency decade*. An octave is a doubling or halving of a frequency; a decade represents a tenfold increase, or decrease, in frequency.

Example 20-2

A frequency of 500 Hz is increased three octaves. What is the new frequency?

Solution

$$f = [(500 \times 2) \times 2] \times 2 = (1000 \times 2) \times 2$$
$$= 2000 \times 2 = 4000 \text{ Hz}$$

Example 20-3

An HP filter has a cutoff frequency of 2.5 kHz. Changing certain filter elements increases f_{co} by one frequency decade. What is the new f_{co}?

Solution

$$f_{new} = 10f_{co}$$
$$f_{co} = 10 \times 2.5 \text{ kHz} = 25 \text{ kHz}$$

Decibels

The logarithm of the ratio of two voltages, currents, or power levels is customarily measured in a unit called the *bel*. However, it is too large a unit for most applications, so the *decibel* (1/10 bel) is used. To calculate the difference between two power levels in decibels, use the following formula:

$$\text{dB} = 10 \log_{10} \frac{P_o}{P_i} \qquad \text{20-3}$$

where

P_o = power output
P_i = power input
$\log_{10}$ = logarithm of the number to the base 10

Since P_o/P_i is a ratio, the formula can be written simply as follows:

$$\text{dB} = 10 \log_{10} PR$$

Here, PR is the symbol for power ratio. (*Note*: Since all calculations with decibels are to the base 10, it is redundant to write the subscript 10. Hereafter, it will be omitted.)

Example 20-4

As the input frequency to a certain HP filter increases from 5 to 10 kHz, the output power rises from 25 to 50 mW. What is the dB increase in power?

Solution

$$\text{dB} = 10 \log PR = 10 \log \frac{50 \text{ mW}}{25 \text{ mW}} = 10 \log 2$$

$$= 10 \times 0.3 = 3$$

From Example 20–4, we see that a doubling of power results in a 3 dB increase of power. If the power were reduced to one-half, there would be a *loss* of 3 dB. The input and output impedances need not be the same when calculating dB power gains or losses. Power depends on E^2/R or I^2R and not on resistance alone.

When we are dealing with voltage or current ratios, however, it is *essential* that the input and output impedances be equal. Thus, if we are to determine the dB change in a filter's output at different frequencies, we must be certain that Z_{in} and Z_{out} are equal. The formula used to calculate dB change in voltage or current output is as follows:

$$dB = 20 \log \frac{V_o}{E_i} \qquad \qquad \textbf{20–4}$$

$$dB = 20 \log \frac{I_o}{I_i} \qquad \qquad \textbf{20–5}$$

The following example illustrates the use of Equation 20–4.

Example 20–5

The output of an HP filter increases from 35 to 240 mV as the frequency increases from 10 to 50 kHz (input level remains constant). What is the dB increase?

Solution

$$dB = 20 \log \frac{V_o}{E_i} = 20 \log \frac{240}{35} = 20 \log 6.86$$
$$= 20 \times 0.84 = 16.72$$

20–2
Low-Pass Filters

Low-pass (LP) filter   A filter network that passes all frequencies below a specified frequency with little or no attenuation but discriminates against all higher frequencies.

Any circuit arrangement that allows only low frequencies to pass may be classified as a **low-pass (LP) filter**. What is considered to be low pass is somewhat relative to the band of frequencies encountered. For example, a low-pass audio filter may pass all frequencies from dc to 400 Hz before it begins to *roll off* (that is, attenuate frequencies beyond 400 Hz). An RF filter may be considered low pass and yet function from dc to 4 MHz before rolling off.

RC Low-Pass Filter

Perhaps the simplest LP filter is one consisting of a resistor and capacitor, as shown in Figure 20–6. At dc, the reactance of C is infinite, and no attenuation of the incoming signal E_i occurs, regardless of the value of R. Let us assume that $R = 1$ kΩ, that $C = 0.159$ μF, and that the input signal E_i can be varied from dc to some very high frequency. At dc and very low frequencies, the output voltage V_o will be equal to E_i. However, as the input frequency is continuously increased, we will arrive at a point where the output begins to fall off. Thus, we are approaching the filter's *cutoff frequency*, a point where V_o has dropped to 0.707 of E_i.

As we mentioned earlier, the cutoff frequency is the half-power point. At this frequency, $X_C = R = 1000$ Ω. Since R_L is 1 MΩ, its loading effect is nil, and V_o will be down 0.707 of E_i. In Figure 20–7, the shaded area under the curve represents the frequencies that are passed by the filter.

When we are determining the *attenuation* of a filter in decibels,

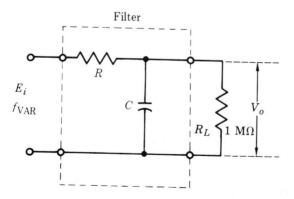

Figure 20–6

Simple *RC* low-pass filter

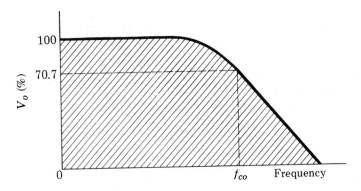

Figure 20–7

Pass characteristics of simple LP filter showing cutoff frequency and half-power point ($V_o = 0.707E_i$; shaded area represents passed frequencies

the calculation can be simplified by dividing the input voltage by the output voltage and placing a minus sign in front of dB. Thus:

$$-\text{dB} = 20 \log \frac{E_i}{V_o} \qquad\qquad \textbf{20–6}$$

The minus sign indicates a loss or attenuation of signal.

Example 20–6

An LP filter has an output of 2.5 V at f_{co}. One octave beyond cutoff, the output drops to 0.85 V. What is the dB roll off per octave?

Solution

$$-\text{dB} = 20 \log \frac{2.5}{0.85} = 20 \log 2.94 = 20 \times 0.47 = 9.37$$

Example 20–7

An LP filter has a cutoff frequency of 1000 Hz. If it rolls off at 20 dB per frequency decade from f_{co}, at what frequency will it be down 40 dB?

Solution

One frequency decade beyond 1000 Hz is 10,000 Hz, or -20 dB. A second frequency decade beyond 10 kHz is 100 kHz, which is -40 dB.

RL Low-Pass Filter

A filter with a 6 dB per octave roll off begins rolling off before cutoff. The same can be said for one that attenuates at 10 dB per frequency decade. An *RL* low-pass filter is shown in Figure 20–8, with its attenuation characteristics. The solid curve shows the actual response. At the cutoff frequency, in this case 1000 Hz, the output is down 3 dB. Attenuation begins at approximately 200 Hz and does not drop off linearly until a little below 6 dB. The theoretical 6 dB per octave roll off is shown by the dotted portion of the curve.

The data from which the curve in Figure 20–8 was drawn are listed in Table 20–1. At 1 kHz, the reactance of L is equal to R. Therefore, the voltages across each of these components will be equal and will be 70.7% of the input signal. At 1 kHz, the output will be down 3 dB, as shown by the following calculation:

$$-\text{dB} = 20 \log \frac{E_i}{V_R} = 20 \log \frac{10 \text{ V}}{7.07 \text{ V}} = 20 \log 1.414 = 20 \times 0.15 = 3$$

Attenuation begins gradually at approximately 300 Hz, which is identified as point A on the graph of Figure 20–8. Line CB is the tangent to the attenuation curve AB and has a constant slope of 6 dB per octave (or 20 dB per decade), thus representing the *ideal curve* of the filter. Its intercept with the zero dB coordinate at point C corresponds to the theoretical cutoff frequency of 1 kHz. For ease in calculation, the straight-line approximate characteristic ACB is sometimes used instead of the actual attenuation characteristic. The resulting error is of the order of 3 dB.

Correlation between octave and decade is indicated in the following data:

$$-6 \text{ dB per octave} = -20 \text{ dB per decade}$$
$$-12 \text{ dB per octave} = -40 \text{ dB per decade}$$
$$-18 \text{ dB per octave} = -60 \text{ dB per decade}$$

Table 20–1 Attenuation Characteristics of LP Filter in Figure 20–8

f (kHz)	V_R (V)*	V_L (V)*	$-$dB
1	7.07	7.07	3
2	4.47	8.94	6.99
4	2.43	9.70	12.29
8	1.24	9.92	18.13
16	0.62	9.98	24.15
32	0.31	$\cong 10.0$	30.17

*Based on input voltage of 10 V.

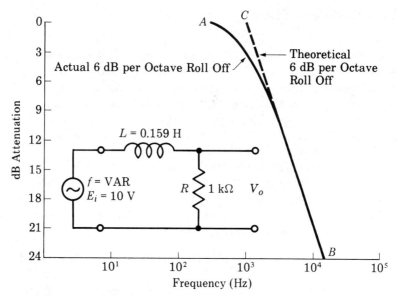

Figure 20–8

Roll-off characteristics of simple *RL* low-pass circuit (shown in inset), with attenuation (V_o/E_i) of 6 dB per octave

Other LP Filters

A number of different combinations of *RCL* can be arranged to provide an LP filter. Some are shown in Figure 20–9. Variations of the L type filter are illustrated in Figures 20–9*a* and 20–9*b*. Sharper attenuation results in the filters shown in Figures 20–9*c* and 20–9*d* because of the additional reactive elements. Attenuation increases 6 dB per octave with each reactive element in the filter. Thus, the circuits in Figures 20–9*c* and 20–9*d* each provide 18 dB per octave attenuation of frequencies beyond cutoff.

20–3
Bandpass Filters

Bandpass filter   A resonant circuit that is tuned to pass a single band of frequencies and to attenuate all other signals.

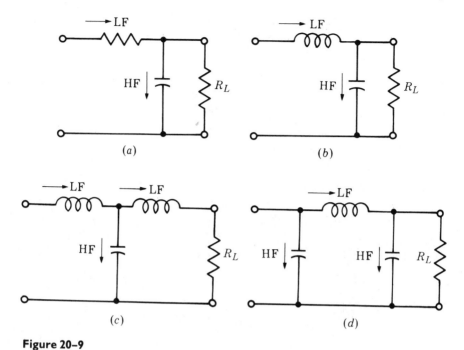

Figure 20–9

Variations of LP filters: (a) L type filter with resistor. (b) L type filter with inductor. (c) T type filter. (d) π type filter.

Frequently, *tuned circuits are used in electronic systems to separate a band of frequencies from a number of available signals,* as in the tuning circuits of radios and TV receivers. The band of frequencies to be passed may be narrow or broad. One of the more elementary **bandpass filters** consists of a series LC circuit resonated to the particular band of frequencies to be passed. The specific arrangement referred to is illustrated in Figure 20–10a. At resonance, $X_L = X_C$, and the output voltage is equal to the input signal, since the series-resonant circuit has minimal impedance at resonance. The response of such a circuit is shown in Figure 20–10b. Actually, the width of the bandpass will be determined by the Q of the resonant circuit. If Q is high, the filter will pass only a narrow band of frequencies.

The output voltage of the filter in Figure 20–10a can be determined by using the voltage divider formula:

$$V_o = E_i \frac{R}{R + j(X_L - X_C)} \qquad \textbf{20–7}$$

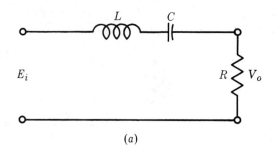

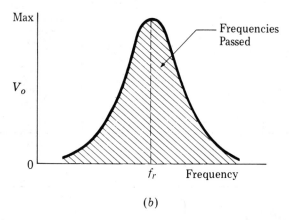

Figure 20–10

Simple bandpass circuit: (a) Filter circuit. (b) Filter response.

At resonance, the output voltage is as follows:

$$V_o = E_i \frac{R}{R} = E_i$$

Thus, $V_o = E_i$ at f_r. Because the filter passes signals at the resonant frequency and attenuates those on either side of f_r, it is called a bandpass filter.

Example 20–8

Calculate the bandpass frequency and bandwidth of the filter circuit in Figure 20–10a if $R = 100\ \Omega$, $C = 8\ \text{pF}$, and $L = 50\ \mu\text{H}$.

Solution

The resonant frequency is as follows:

$$f_r = \frac{1}{2\pi\sqrt{LC}} = \frac{0.159}{\sqrt{50 \times 10^{-6} \times 8 \times 10^{-12}}}$$

$$= \frac{0.159}{\sqrt{400 \times 10^{-18}}} = \frac{0.159}{20 \times 10^{-9}} = 7.95 \text{ MHz}$$

Calculate the circuit Q:

$$Q = \frac{\omega L}{R} = \frac{2.498 \times 10^3}{1 \times 10^2} \cong 25$$

The bandwidth, then, is as follows:

$$BW = \frac{f_r}{Q} = \frac{7.95 \times 10^6}{25} = 318 \text{ kHz}$$

Example 20–9

Refer to Example 20–8. What would the output voltage V_o be if $E_i = 5$ V at 4 MHz?

Solution

Use Equation 20–7, but first calculate ω, X_L, and X_C.

$$\omega = 2\pi f = 6.28 \times 4 \times 10^6 = 2.51 \times 10^7$$

$$X_L = \omega L = 2.51 \times 10^7 \times 50 \times 10^{-6} = 1.256 \times 10^3$$

$$X_C = \frac{1}{\omega C} = \frac{1}{2 \times 10^{-4}} = 4.976 \times 10^3$$

$$V_o = 5.0 \angle 0° \text{ V} \frac{100}{100 + (1.256 \times 10^3 - 4.976 \times 10^3)}$$

$$= 5.0 \angle 0° \text{ V} \frac{100}{100 - j\,3.72 \times 10^3} = \frac{500 \angle 0°}{3.721 \times 10^3 \angle -88.5°}$$

$$= 0.134 \angle 88.5° \text{ V}$$

Figure 20–11 shows a variation of a typical bandpass filter. The steepness of the slopes of the bandpass characteristic curve is a function of the Q's of both the series and the parallel sections.

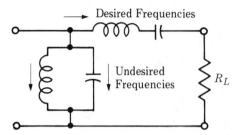

Figure 20–11

Another type of bandpass filter

20–4
Band Rejection Filters

Band rejection filter   A network that attenuates a single band of frequencies and allows those on either side to pass; sometimes called *band stop filter.*

Many electronic circuits require the use of special filters or *wave traps* to remove unwanted frequencies. These filters are called **band rejection filters**, or *band stop filters.* For example, an interfering signal can seriously affect the performance of radio receivers. Wave traps can be installed to eliminate this interference.

Basic Band Rejection Filter

A simple band rejection filter is shown in Figure 20–12a. Observe that an antiresonant circuit is placed in one leg of the filter. Because its impedance is very high at resonance, the unwanted signal will not be allowed to pass through to load resistor R_L. Essentially, this configuration amounts to a signal divider network where the Z of the LC circuit is in series with R_L. Frequencies at and near f_r will experience this high impedance, with the result that they will be effectively blocked. The frequency response curve for this kind of filter is shown in Figure 20–12b. Note that the output is reduced to its lowest level at f_r, with the shaded area representing the band of discriminated frequencies. As can be expected, the overall shape of the curve is a function of the circuit Q.

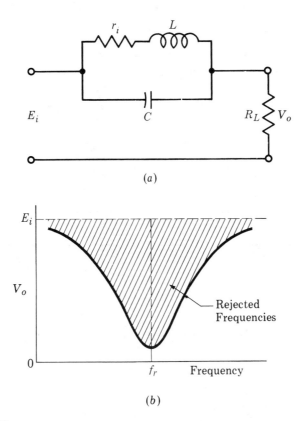

(a)

(b)

Figure 20–12

Band rejection filter: (a) Typical circuit. (b) Response characteristic.

Other Filter Types

Improved performance can be obtained by adding one or more series LC sections across the band rejection filter's input or output. Such an arrangement is shown in Figure 20–13. If the series LC element is designed to resonate at the antiresonant frequency of the parallel LC combination, then unwanted frequencies will be shorted around the load.

Other types of band rejection filters exist. One of the simplest is shown in Figure 20–14. At the resonant frequency, the series-resonant characteristics of LC provide a near-short across R_L, with the result that the unwanted frequencies are bypassed. By controlling the Q of the circuit, this arrangement allows the width of the band stop to be altered.

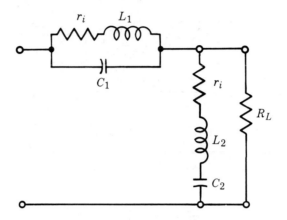

Figure 20–13

Band rejection filter with improved rejection characteristics

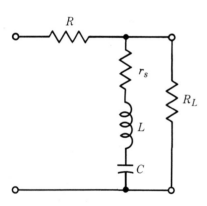

Figure 20–14

Simple form of band rejection filter

Example 20–10

Calculate the band stop center frequency of the circuit in Figure 20–12a if $r_i = 5 \ \Omega$, $L = 5$ mH, $C = 0.05 \ \mu F$, and $R_L = 1$ kΩ.

Solution

$$f_r = \frac{1}{2\pi\sqrt{LC}} = \frac{0.159}{\sqrt{5 \times 10^{-3} \times 0.05 \times 10^{-6}}} = 10.071 \text{ kHz}$$

Example 20–11

Calculate the bandwidth of the filter in Figure 20–12a.

Solution

In Example 20–10, we determined the resonant frequency to be 10.071 kHz. The circuit Q is then as follows:

$$Q = \frac{\omega L}{r_i} = \frac{6.28 \times 10.071 \times 10^3 \times 5 \times 10^{-3}}{5} = 63.25$$

Therefore $$BW = \frac{f_r}{Q} = \frac{10.071 \text{ kHz}}{63.25} = 159.2 \text{ Hz}$$

This result is the bandwidth at the filter's half-power points.

20–5
Constant-*k* Filters

Basic Components

All filters are made up of elementary sections called *L sections*, as shown in Figure 20–15. Here, Z_s represents the series part(s) of the filter, while Z_p represents the parallel part(s).

By combining series and parallel elements, we can make T or π networks, as desired. For example, a T type filter would be made up of two L sections, where each Z_p would be one-half the total Z_p required, as indicated in Figures 20–16a and 20–16b. A π filter can be represented in the same manner, as shown in Figures 20–17a and 20–17b. Both types of filters may therefore be divided in two to form half sections.

Basic Constant-*k* Filters

Filters of the type shown in Figure 20–18 are known as *constant-k filters*. They are so named because the product of X_L and X_C is constant at all frequencies. For example, if $X_L = 500 \ \Omega$ at a certain frequency, X_C

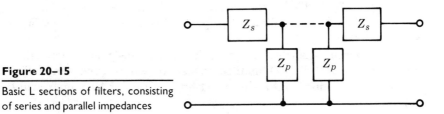

Figure 20–15

Basic L sections of filters, consisting of series and parallel impedances

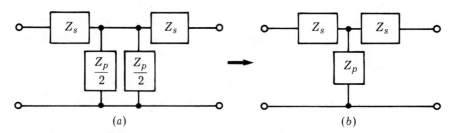

Figure 20–16

Combining L sections: (a) T filter. (b) T network.

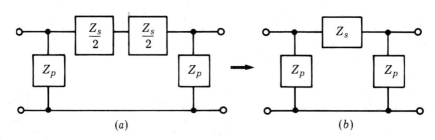

Figure 20–17

Combining L sections: (a) π filter. (b) π network.

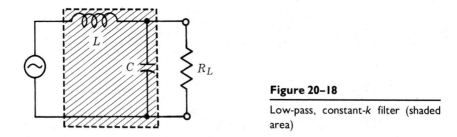

Figure 20–18

Low-pass, constant-k filter (shaded area)

may equal 1000 Ω. Their product is 500 kΩ. If the frequency is doubled, $X_L = 1000$ Ω and $X_C = 500$ Ω. Their product is unchanged, and k is constant at 500 kΩ.

If circuit elements L and C in Figure 20–18 are reversed, we have a simple HP filter. This filter is also known as a constant-k type, because the product of X_C and X_L at any particular frequency will be constant.

Simple filter performance is based on the assumption that the input and output impedances of the filter are equal and that they match their respective loads. When the impedances are not matched, the filter's characteristics may be altered.

20–6
m-Derived Filters

m-derived filter A filter employing either series or parallel resonant combinations, the attenuation of which is greater than can be achieved by a constant-*k* filter; *m* represents the ratio of the cutoff frequency to the frequency of infinite attenuation.

Constant-*k* filters are limited in the sharpness of their response. If greater attenuation of undesired frequencies is required, then *tuned circuit elements* must be employed, either in series- or parallel-resonant combinations, as shown in Figures 20–13 and 20–14. These elements are known as *m*-**derived filters**. The *m* represents a ratio of the cutoff frequency to the frequency of infinite attenuation. The constant-*k* section is the reference or prototype for the *m*-derived filter, which has sharper cutoff characteristics. Refer to Figure 20–19. There, the characteristics of

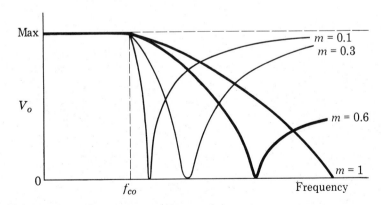

Figure 20–19

Effect of *m* on sharpness of f_{co} in LP filter

the constant-k filter have a value of $m = 1$. The lower the value of m, the sharper is the cutoff characteristic. Thus, a filter with $m = 0.1$ has greater attenuation past cutoff than one with $m = 0.3$. The sharpness of cutoff, therefore, increases as m approaches zero. In practice, an $m = 0.6$ value is a good compromise between sharp cutoff ($m = 0.1$) and uniform, but gradual, attenuation throughout the attenuated band, such as the constant-k type where $m = 1$.

20–7
Phase Shift Networks

Simple Phase Shift Circuit

An important aspect of a reactive circuit is its ability to *shift the phase of an incoming signal*. This phase may or may not be desirable, depending on the function of the circuit. Figure 20–20 shows the simplest kind of phase shift network. The number of degrees by which the output voltage leads the input voltage is determined by the relative magnitudes of C and R and the frequency of the applied signal. For example, in previous chapters, we saw that when $X_C = R$ at a particular frequency, the phase is shifted $45°$ and the output reduced to 0.707 of the input. By proper choice of C and R at a specific frequency, phase shifts up to about $90°$ can be achieved.

Multisection Phase Shift Network

When more than $90°$ of phase shift is required, additional sections can be added, such as shown in Figure 20–20. By using Thévenin's theorem, we can calculate the output voltage and phase shift.

Since the output voltage is across R_2, it will be designated as the load resistor. Figure 20–21 shows the network with R_2 open so that the Thévenin voltage can be computed.

$$E_{TH} = \frac{E_i R_1}{R_1 - jX_C} = \frac{(50 \angle 0°)(50 \angle 0°)}{50 - j\,87} = \frac{2500 \angle 0°}{50 - j\,87} = 25 \angle 60.1° \text{ V}$$

The source is now replaced with a short, as shown in Figure 20–22. Impedance Z_{TH} is composed of X_{C_2} in series with the equivalent impedance of X_{C_1} and R_1.

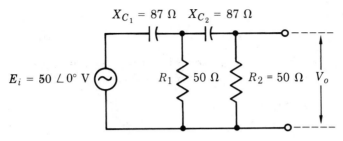

Figure 20-20

Two-section phase shift network

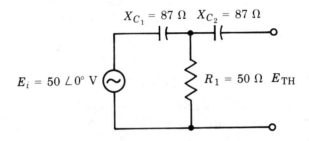

Figure 20-21

Thévenized circuit

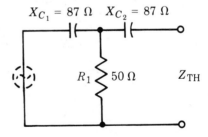

Figure 20-22

Calculating Z_{TH}

$$Z_{TH} = -jX_{C_2} + \frac{(R_1)(-jX_{C_1})}{R_1 + (-jX_{C_1})} = -j\,87 + \frac{(50 \angle 0°)(87 \angle -90°)}{50 + (-j\,87)}$$

$$= -j\,87 + 43.5 \angle -29.9° \ \Omega$$

Converting the second term to rectangular form so that it can be added to the first term gives the following:

$43.5 \angle -29.9° = 37.7 - j\, 21.7\, \Omega$

Therefore $\quad Z_{TH} = -j\, 87 + 37.7 - j\, 21.7\, \Omega = 37.7 - j\, 108.7\, \Omega$

The Thévenin equivalent circuit is shown in Figure 20–23. The output voltage V_{RL} is calculated by using the voltage divider formula:

$$V_{RL} = \frac{E_{TH}R_L}{Z_{TH} + R_L} = \frac{(25\ \angle 60.1°)\,(50\ \angle 0°)}{37.7 - j\, 108.7 + 50} = \frac{1250\ \angle 60.1°}{87.7 - j\, 108.7}$$

$$= \frac{1250\ \angle 60.1°}{139.7\ \angle -51.1°} = 8.95\ \angle 111.2°\ V$$

From these calculations, we see that the total phase shift of the two sections is $111.2°$ and that the voltage across R_2 is 8.95 V. By adding more sections, we can obtain additional phase shift — *at the expense of reduced output voltage.*

Network for 90° Phase Shift

In the special case where a phase shift of exactly 90° is desired, a network such as depicted in Figure 20–24 is used. For simplicity, a frequency has been assumed that provides an $X_{C_1} = X_{C_2} = 5\, \Omega$. Since $R_1 = R_2 = 5\, \Omega$, the impedance of section 1 and section 2 is as follows:

$$Z_1 = Z_2 = 5 - j\, 5\, \Omega = 7.07\ \angle -45°\ \Omega$$

Use the voltage divider equation to find V_{R_1}:

$$V_{R_1} = \frac{E_i R_1}{R_1 - jX_{C_1}} = \frac{(10\ \angle 0°)\,(5\ \angle 0°)}{7.07\ \angle -45°} = 7.07\ \angle 45°\ V$$

Assuming V_{R_1} to be the input to the second section, we calculate V_{R_2}:

$$V_{R_2} = \frac{V_{R_1} R_2}{R_2 - jX_{C_2}} = \frac{(7.07\ \angle 45°)\,(5\ \angle 0°)}{7.07\ \angle -45°} = 5\ \angle 90°\ V$$

The results indicate that the output voltage is leading the input voltage by 90°. The answer obtained for the output voltage is correct with regard to phase shift, but it is *only an approximation of the magnitude of V_o* (actual voltage is 3.3 V). The discrepancy arises from the shunting effect of X_{C_2} and R_2 on R_1, which lowers the input voltage to the second section of the network. This simplified technique of finding the total phase shift is valid only when the impedance of all the components in the network is equal.

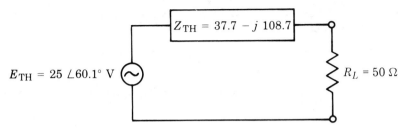

Figure 20–23

Thévenin equivalent circuit

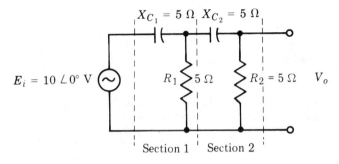

Figure 20–24

Network for 90° phase shift

Summary

Low-pass filters pass all frequencies from dc to cutoff.

High-pass filters permit frequencies above a cutoff value to pass while discriminating against frequencies below f_{co}.

The cutoff frequency of a filter is that frequency where the output is 0.707 of the input. It is also known as the half-power point.

Filters may be classified as low-pass, high-pass, bandpass, and band rejection. They may also be considered as L type, π type, or T type filters.

An octave is a doubling or halving of a given frequency.

The term *frequency decade* refers to a tenfold increase or decrease in frequency.

Filter performance is customarily measured in decibels (dB) per octave or frequency decade.

When filter performance is measured in voltage or current ratios, the following formulas are used:

$$dB = 20 \log \frac{V_o}{E_i} \quad \text{or} \quad dB = 20 \log \frac{I_o}{I_i}$$

If a loss occurs through the filter (as is usually the case), the formulas can be modified as follows:

$$-dB = 20 \log \frac{E_i}{V_o} \quad \text{or} \quad -dB = 20 \log \frac{I_i}{I_o}$$

Attenuation occurs at a rate of 6 dB per octave for each reactive element in a filter.

A constant-k filter derives its name from the fact that the product of X_L and X_C is constant at all frequencies.

In comparison with constant-k filters, m-derived filters have sharper responses.

The m-derived filters use tuned circuits in either series or parallel combinations, or both.

The m represents the ratio of the cutoff frequency to the frequency of infinite attenuation and varies between zero and one. Lower values indicate sharper cutoff characteristics.

A simple phase shift circuit consists of a series RC combination. One section can provide up to 90° phase shift. If more than 90° is required, additional RC combinations can be placed in series with each other.

Progress Test

The bracketed number after each question indicates the section of this chapter where the answer can be found.

    1. A signal source supplies a voltage of constant amplitude, from dc to 20 kHz, to a low-pass filter. The load resistor across the filter

output will experience greatest voltage at which of the following frequencies? (*a*) dc (*b*) 500 Hz (*c*) 5 kHz (*d*) 20 kHz   [20–2]

2.   A constant-amplitude signal source supplies a voltage from dc to 20 kHz to the input of a high-pass filter. The terminating load resistor will have the greatest voltage at which of the following frequencies? (*a*) dc (*b*) 500 kHz (*c*) 5 kHz (*d*) 20 kHz   [20–1]

3.   Which of the following filter configurations would make the best LP filter? (*a*) π type, with series C and shunt L (*b*) L type, with series C and shunt R (*c*) π type, with series L and shunt C (*d*) T type, with series R and shunt L [20–2]

4.   For maximum signal transfer, which of the following filter types would serve best as an HP filter? (*a*) π type, with shunt C and series L (*b*) T type, with series C and shunt L (*c*) L type, with shunt C and series L (*d*) L type, with series R and shunt C [20–1]

5.   In an LP filter, the cutoff frequency is the point where the output: (*a*) voltage is reduced to 50% of the input. (*b*) power is reduced to 70.7% of the input power. (*c*) voltage is reduced to 70.7% of the input voltage. (*d*) current is reduced to 70.7% of the input current.   [20–2]

6.   The output of a filter drops from 5.0 to 2.5 V as the frequency is increased from 1 to 2 kHz. The dB change is: (*a*) -3 dB per octave. (*b*) -6 dB per octave. (*c*) 6 dB per octave. (*d*) -3 dB per decade.   [20–1]

7.   A band rejection filter is characterized by: (*a*) a high-Z series pass element and a low-Z shunt element. (*b*) a low-Z series pass element and a low-Z shunt element. (*c*) a low-Z series pass element and a high-Z shunt element. (*d*) rejecting all frequencies except a certain band.   [20–4]

8.   In a constant-k filter: (*a*) $X_L = X_C$. (*b*) $X_L^2 = X_C^2$. (*c*) sharper response is obtained than with m-derived filters. (*d*) the product of X_L and X_C is constant at all frequencies.   [20–5]

9.   Which of the following would be the best bandpass filter? (*a*) a parallel-resonant LC circuit in series with the load and a series-resonant LC circuit in parallel (*b*) a parallel-resonant LC circuit in series with the load and a parallel-resonant LC circuit in parallel (*c*) a series-resonant LC circuit in series with R_L and a parallel-resonant LC circuit across R_L (*d*) a capacitor in series with R_L and a parallel-resonant LC circuit across the load   [20–3]

10.   Refer to Figure 20–5a. If source E_i were varied between 20 Hz and 20 kHz (constant amplitude), an ac voltmeter across C_C would: (*a*) read zero. (*b*) show a decreasing value of voltage. (*c*) indicate an increasing value of voltage. (*d*) indicate the maximum value of E_i.   [20–1]

┌─── Problems

Answers to odd-numbered problems are at the back of the book.

1.   Assume the RC coupling network of Figure 20-5a is designed to pass 20 Hz to 20 kHz. What must be the value of C_C to pass 20 Hz using the value of R_i indicated?

2.   The output power of a filter rises from 115 to 530 mW as the frequency increases from 50 to 400 Hz. (a) What is the increase in dB? (b) How many octaves are involved?

3.   What is the dB increase of a filter's output if it changes from 60 μW to 250 mW?

4.   An HP filter's output drops from 1.7 V to 35 mV as the frequency is changed from 1.5 MHz to 375 kHz. (a) What is the attenuation in decibels? (b) How many octaves are involved?

5.   An attenuator reduces the input voltage by -27 dB. (a) What is the ratio of output to input voltage? (b) If the input voltage is 3.55 V, what is the output voltage?

6.   Assume the circuit parameters in Figure 20-10a are $R = 60\ \Omega$, $C = 12$ pF, and $L = 35\ \mu$H. (a) Calculate the bandpass frequency. (b) What is the bandwidth?

7.   Refer to Problem 6. What will the output voltage of the filter be if the input frequency is changed to 5 MHz and $E_i = 10$ V?

8.   In the circuit of Figure 20-10a, suppose $C = 180$ pF, $L = 0.677$ mH, and $R = 600\ \Omega$. (a) Calculate the bandpass frequency. (b) Calculate the bandwidth.

9.   Determine the output voltage of the circuit described in Problem 8 if the input frequency is changed to 475 kHz and the input is 1.35 V.

10.   A band stop filter, similar to the one in Figure 20-12a, has the following parameters: $L = 220\ \mu$H, $C = 35$ pF, $r_i = 20\ \Omega$, and $R_L = 200\ \Omega$. (a) Calculate the band stop frequency. (b) Calculate the bandwidth.

11.   For a band rejection filter, such as the one in Figure 20-12a, calculate V_o if the source potential is 85 mV and the circuit parameters are $X_L = 1750\ \Omega$, $r_i = 21\ \Omega$, and $R_L = 1$ kΩ.

12.   A simple band stop filter, of the type indicated in Figure 20-14, has the following values: $R_L = 500\ \Omega$, $L = 67\ \mu$H, $f_r = 3.85$

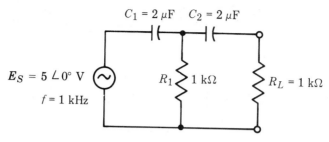

Figure 20–25

Circuit for Problems 14 and 15

MHz, and $R = 150 \ \Omega$. If the resistance in series with L is $7 \ \Omega$ and the filter input voltage is 9.5 V, what is V_{RL}?

13.   An LP filter, of the type shown in Figure 20–18, has an inductance of 4 H and a capacitance of 0.04 μF. What is the value of k at 400 Hz? At 800 Hz?

14.   For the circuit of Figure 20–25, determine the output voltage and phase shift across R_L.

15.   Refer to the circuit in Figure 20–25. If the input frequency is changed to 30 Hz (all other parameters unchanged), what will the output voltage and phase shift be?

16.   A low-pass filter with a cutoff frequency of 3 MHz has a voltage of 250 mV when 300 MHz is applied. How many decades separate the two frequencies? If the filter consists of a single coil and capacitor, what would the output be in decibels?

17.   In Problem 16, what is the output in volts?

18.   In a capacitive-coupling circuit designed to pass 300–3000 Hz, using a 1 μF coupling capacitor, what is the value of the input resistor R_i?

19.   A line amplifier with input and output impedances of 600 Ω has an input voltage of 0.02 V and a gain of 200. (a) What is the output voltage? (b) What is the voltage gain in decibels? (c) What is the power gain in decibels?

20.   In a simple RC high-pass filter section with $R = 5 \ \text{k}\Omega$, what is the cutoff frequency if a 0.01 μF capacitor is used?

21.   In a simple LR low-pass filter, what is the value of L if $R = 1500 \ \Omega$ and $f_{co} = 1000$ Hz? What is the output in decibels at 4000 Hz?

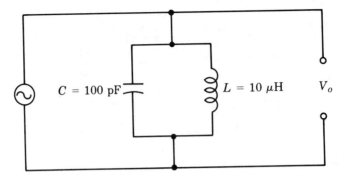

Figure 20–26

Circuit for Problem 24

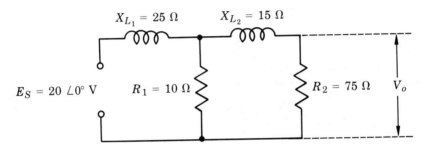

Figure 20–27

Circuit for Problem 25

22. A constant-k high-pass filter section has a k value of 1,000,000. At a frequency at which $X_L = 2.5$ kΩ, what will X_C be?

23. In Problem 22, if $C = 0.1$ μF, at what frequency would the reactances be equal? What would the cutoff frequency be if $R_L = 1$ kΩ?

24. In the circuit of Figure 20–26, what is the bandpass center frequency? If the circuit Q is 100, what are the upper and lower limits of the band passed?

25. What is the phase shift between E_S and V_o in the circuit of Figure 20–27?

21

Transformers

Objectives

After studying this chapter, you will be able to:

Calculate voltage, current, and turns ratios.

Use a transformer to match a loudspeaker to an amplifier and perform similar impedance-matching problems.

Recognize the characteristics of RF transformers.

Introduction

A *transformer* is a device that transfers electric energy from one circuit to another by *electromagnetic induction*. The energy is transferred without any change in frequency, but the transfer usually involves a change in voltage and in current levels. Typically, a transformer consists of two coils, a primary and a secondary, so mounted that magnetic flux produced by the primary links with the secondary. Transformers require practically no care and maintenance because of their rugged, simple, and durable construction.

Applications

Transformers can vary in size from the large stationary devices, which handle commercial power, to miniature components using only a few windings and with air as the core material. All are used in one way or another to transfer power between two different circuits, to change a voltage level, or to match two dissimilar impedances.

Transformers are frequently used to electrically isolate electronic equipment from the ac power source. This precautionary measure can eliminate the possibility of a person receiving a lethal electric shock from the chassis of an electric device back to the "hot" side of the power source. These transformers are called *isolation transformers*.

Most transistorized or integrated circuit devices require dc voltage levels of 5 to 20 V for their operation. When these devices operate on ordinary household current, a step-down transformer is used as part of the power supply to reduce the voltage level from 120 V prior to rectification. The ignition coil of an automobile ignition system is a step-up transformer, converting a transient potential of 12 V to several thousand volts.

Impedance-matching transformers are used to optimize the performance of amplifiers by matching the output circuit to the load. The output terminals of audio amplifiers marked 4, 8, and 16 Ω are connected to separate transformer windings and indicate the impedance of the speaker that should be connected to each terminal.

21–1
Transformer Types

A number of different types of transformers are available to meet the diverse needs of industry. In spite of their different sizes, weights, and shapes, they all have some things in common. The typical transformer

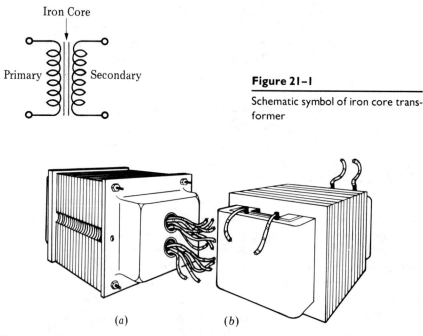

Figure 21–1

Schematic symbol of iron core transformer

Figure 21–2

Basic transformers: (a) Fully shielded. (b) Unshielded.

has two windings, a *primary* and a *secondary*. They are wound on a common magnetic core and electrically insulated from each other. The schematic symbol of a transformer is shown in Figure 21-1. Four parts make up the basic transformer:

1. A soft-iron core that provides a magnetic circuit of low reluctance.
2. A primary winding that receives energy from an ac source.
3. A secondary winding that receives energy from the primary by mutual induction and delivers it to the load.
4. An enclosure (optional).

Transformers are primarily classified according to their use. Thus, we have transformers such as *audio, power, radio frequency, modulation,* and *filament.* Transformers are also classed as *shielded* or *unshielded,* depending on whether or not they are enclosed in a soft-iron case to minimize stray magnetic flux. Figure 21-2a shows a fully shielded type, and Figure 21-2b shows an unshielded type.

Single-phase transformers are used for small-power applications, such as power supplies for audio amplifiers, tape recorder/reproducers, and small transmitters. For larger industrial applications, transformers are designed to operate on three-phase power and are called three-phase, or polyphase, transformers. Examples of polyphase transformers are step-down transformers used to reduce high voltages from utility companies to 440 or 220 V for distribution throughout a manufacturing facility, and step-up and step-down transformers used for large TV and radio station transmitter power supplies.

Lastly, transformers can be classified according to the type of core material used and the shape of the core. For example, RF transformers can have powdered-iron, ferrite, or air cores. Audio and power transformers use soft iron.

21-2
Transformer Principles

Basic Concepts

Primary winding The winding of a transformer that normally connects to a power source and causes current to flow in one or more secondary windings.

Transformers depend on the principle of electromagnetic induction. Basically, they consist of two or more inductors, electrically
isolated from each other and sharing a common core. A changing electromagnetic field set up by an alternating current in one induces an
alternating voltage in the other. Thus, *mutual inductance* exists between them, and they are said to be *inductively coupled*. A pictorial
representation of this coupling is shown in Figure 21–3, where the
transformer is connected between an ac source and a load. The coil N_p
connected to the source is called the **primary winding**, and the coil N_s
connected to the load is the **secondary winding**. The coils are magneti-
cally coupled by the flux flowing in the core. Energy delivered to the
primary establishes a magnetic flux that passes through the core and
induces a voltage in the secondary. This voltage gives rise to current
in R_L and, thus, completes the transformation of power from source to
load.

For maximum transfer of power, the flux linkage must be complete; that is, all the flux established by the primary must link the sec-

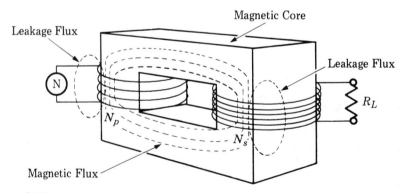

Figure 21–3

Pictorial representation of simple transformer

ondary. For this reason, the secondary is often wound *directly on the primary* with only protective insulation separating the two coils. Even with the use of high-permeability cores, a few of the flux lines fail to link the secondary and constitute a **leakage flux** that prevents the device from being 100% efficient. Nevertheless, a well-designed iron core transformer may have a coefficient of coupling k of about 0.98. While there are other losses (to be discussed), the efficiency runs around 98% for a well-designed transformer. A cross-sectional view of a transformer is shown in Figure 21–4. Observe the placement of insulating materials between turns and windings.

Theory of Operation

For maximum efficiency, the inductance of each winding and k should be as high as possible. These conditions mean that less magnetizing current will be needed to set up the required flux linkage.

The importance of the reactance of the primary, and the effect of the secondary winding on the primary, may be understood by the following analysis. Refer to Figure 21–5a. The secondary is left open so that it has no effect on the first phase of our analysis. The primary, then, is nothing but a simple iron core inductor with a current of $I_p = E/X_{L_p}$. Because of the large reactance, the current and voltage will be 90° out of phase. During one half cycle of I_p, a flux will be built up in the core,

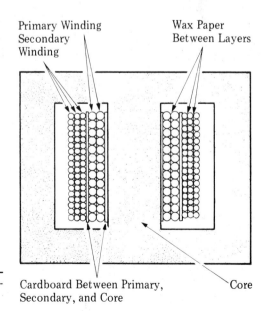

Primary Winding
Secondary
Winding

Wax Paper
Between Layers

Figure 21–4

Cross section of an iron core transformer

Cardboard Between Primary, Secondary, and Core

Core

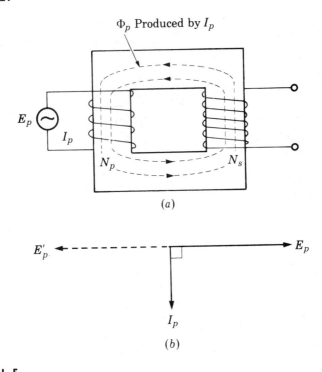

Φ_p Produced by I_p

(a)

(b)

Figure 21-5

Transformer operation: (a) Direction of flux lines at a given instant produced by I_p. (b) Phasor diagram of primary voltages and current.

the direction of which is shown in the figure by the dotted lines and arrowheads. At this same instant, a cemf E'_p is set up in the primary that is 180° out of phase with the applied voltage E_p and 90° out of phase with I_p (see Figure 21–5b).

In Figure 21–6a, the same transformer is shown with the action of the secondary indicated. It, too, acts like a simple inductance into which an emf has been induced by the changing flux created by the primary. This emf has the same direction as the cemf in the primary and gives rise to I_s. A phasor diagram of the secondary voltages and current is shown in Figure 21–6b. Observe that the reactive voltage E'_s is 90° out of phase and leading I_s.

At the same instant assumed for our analysis of the primary (Figure 21–5), the secondary current and flux lines are in the direction indicated in Figure 21–6. Note that the flux lines set up by I_s are opposite to those set up by I_p. Thus, the secondary current I_s effectively *decreases* the impedance of the primary circuit by opposing the flux

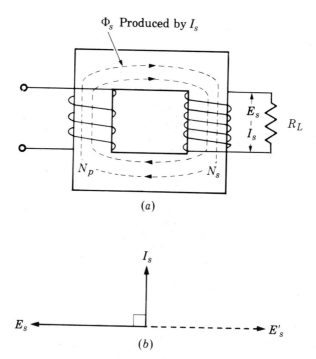

Φ_s Produced by I_s

N_p N_s

E_s

I_s

R_L

(a)

I_s

E_s E'_s

(b)

Figure 21–6

Transformer operation:(a) Direction of flux lines at a given instant produced by I_s. (b) Phasor diagram of secondary voltages and current.

set up by I_p. Therefore, I_p must increase to maintain the flux linkage between the windings. If the secondary load increases, I_s increases and causes I_p to increase. This opposition of flux lines in a transformer is in accordance with Lenz's law.

The total action of a transformer may now be considered in terms of phasor diagrams. With a small current in the secondary of the transformer in Figure 21–7a, the phasor diagram of voltages and current is as shown in Figure 21–7b. Voltages across the transformer are 180° out of phase, as are the currents. In both windings, the current is lagging the voltage by 90°.

If R_L is decreased, I_s increases and the circuit becomes more resistive. That is, the current and voltage in the secondary approach an in-phase condition, and the phase angle Θ_s approaches 0°. Because the primary current increases with an increase in secondary current, the X_L of the primary is reduced and the circuit appears more *resistive*. Therefore, voltage and current in the primary approach an in-phase condition, and Θ_p approaches 0°, as represented in Figure 21–7c.

Chapter 21

An ideal transformer would show the current and voltage phasors superimposed (Θ_s and Θ_p both $0°$); that is, voltage and current in each winding would be in phase. In this case, complete transfer of energy would occur, and the source would see the load as an equivalent

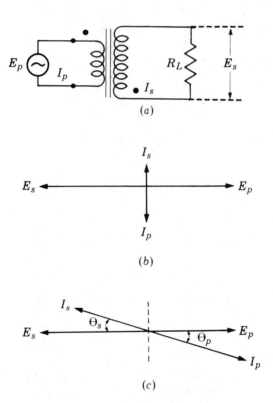

(a)

(b)

(c)

Figure 21–7

Total action of a transformer: (a) Circuit. (b) Phase relationships for small I_s. (c) Phase relationships when R_L is decreased.

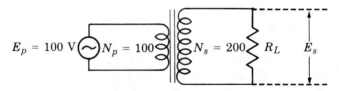

Figure 21–8

Circuit to explain voltage ratios

pure resistance (as though the transformer were not there) but with the voltage and current shifted 180°. In some circuits, *phasing* is an important consideration. So, *dots are placed beside the leads* of the primary and secondary that have the same instantaneous polarities. Thus, in Figure 21–7a, the dots indicate that the top of the primary has the same instantaneous polarity as the bottom of the secondary.

21–3
Voltage and Current Ratios

Voltage Ratios

From Faraday's law, we learned that the induced voltage in a coil equals the number of turns (N) multiplied by the rate of change of flux:

$$e = -N\frac{d\Phi}{dt} \qquad\qquad 21\text{--}1$$

In the circuit of Figure 12–8, a 100 V source is shown connected to a transformer whose primary has 100 turns. Because the primary cemf is approximately equal to the supply, the primary flux must be changing at a rate such as to produce 100 V across the primary. Since $N_p = 100$, the rate of change of flux is as follows:

$$e_p = N\frac{d\Phi}{dt}$$

$$100\text{ V} = 100 \times \frac{d\Phi}{dt}$$

or $$\frac{d\Phi}{dt} = \frac{100}{100} = 1\text{ Wb/s}$$

Assuming a perfect condition (no loss), $\Phi_p = \Phi_s$ and the flux linking the secondary winding must be changing at the same rate. Now, since the secondary has 200 turns, we can calculate E_s as follows:

$$E_s = N_s\frac{d\Phi}{dt} = 200 \times 1\text{ Wb/s} = 200\text{ V}$$

From this equation, we see that the voltage in the secondary is related to the voltage in the primary by the ratio of N_s to N_p. Expressed mathematically, the voltage ratio is as follows:

$$\frac{E_p}{E_s} = \frac{N_p}{N_s}$$

21-2

This equation may be rewritten as follows:

$$E_p N_s = E_s N_p \quad \text{or} \quad E_s = \frac{E_p N_s}{N_p}$$

21-3

Example 21-1

If the number of turns on the secondary of the transformer in Figure 21-8 is increased to 600, what will the value of E_s be?

Solution

$$\frac{E_p}{E_s} = \frac{N_p}{N_s}$$

$$\frac{100}{E_s} = \frac{100}{600}$$

$$100E_s = 60,000$$

$$E_s = 600 \text{ V}$$

Turns ratio   The ratio of the turns in the primary to those in the secondary of a transformer; expressed as N_p/N_s.

The expression N_s/N_p or N_p/N_s is called the **turns ratio**. Thus, the transformer shown in Figure 21-8 has the following turns ratio:

$$\frac{N_s}{N_p} = \frac{200}{100} = 2{:}1$$

Such a transformer is called a *step-up transformer*. If the ratio N_s/N_p is less than one, the secondary has fewer turns than the primary. In this case, the secondary voltage is less than the primary, and it is called a *step-down transformer*.

It must be noted that the terms *step-up* and *step-down* always refer to *voltage levels*, not current levels. Turns ratio never applies to power, because a well-designed transformer has essentially the same amount of power in the secondary as the primary.

Example 21–2

A transformer has 250 turns on its primary and 75 turns on its secondary. If the primary voltage is 60 V, what is the value of voltage developed across the secondary?

Solution

$$E_s = \frac{E_p N_s}{N_p} = \frac{60 \times 75}{250} = 18 \text{ V}$$

Volts-per-Turn Ratio

Since the voltage induced in the secondary is directly proportional to the turns ratio, the voltage ratio E_s/E_p may be found by determining the number of volts per turn (V/t). Therefore, if 120 V ac is applied to a primary of 250 turns, the volts-per-turn ratio is as follows:

$$\frac{\text{volts}}{\text{turn}} = \frac{120}{250} = 0.48$$

Because the V/t is constant for any given transformer, the secondary voltage E_s may be determined by multiplying this constant by the number of turns in the secondary. The next example illustrates this calculation.

Example 21–3

Using the data in Example 21–2, calculate the secondary voltage using the V/t method.

Solution

$$\frac{V}{t} = \frac{60}{250} = 0.24$$

Therefore $\qquad E_s = 0.24 \times 75 = 18 \text{ V}$

When a transformer has several secondary windings, the V/t method provides an expedient way of determining the value of the several secondary voltages. For example, Figure 21–9 shows the sche-

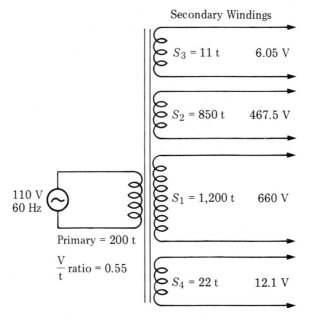

Secondary Windings

$S_3 = 11$ t   6.05 V

$S_2 = 850$ t   467.5 V

110 V
60 Hz

$S_1 = 1,200$ t   660 V

Primary = 200 t

$\dfrac{V}{t}$ ratio = 0.55

$S_4 = 22$ t   12.1 V

Figure 21–9

Multiwinding transformer showing secondary voltages calculated by the V/t method

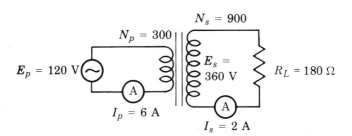

$N_s = 900$

$N_p = 300$

$E_p = 120$ V

$E_s = 360$ V

$R_L = 180$ Ω

$I_p = 6$ A

$I_s = 2$ A

Figure 21–10

Step-up transformer circuit for Example 21–4

matic diagram of a transformer with four secondary windings. The number of turns on each winding is as indicated. The V/t ratio can be determined from data relative to the primary winding:

$$\frac{V}{t} = \frac{110}{200} = 0.55$$

Therefore, the four secondary voltages are as follows:

$$E_{S_1} = 1200 \text{ t} \times 0.55 = 660 \text{ V}$$

$$E_{S_2} = 850 \text{ t} \times 0.55 = 467.5 \text{ V}$$

$$E_{S_3} = 11 \text{ t} \times 0.55 = 6.05 \text{ V}$$

$$E_{S_4} = 22 \text{ t} \times 0.55 = 12.1 \text{ V}$$

Current Ratios

In the ideal transformer, the load appears as a pure resistance to the source and the power factor is unity. Then, the power delivered to the primary equals the power consumed by the load. Expressed mathematically, the power is as follows:

$$P_p = E_p I_p$$

and

$$P_s = E_s I_s$$

Then

$$E_p I_p = E_s I_s$$

Rewritten

$$\frac{I_p}{I_s} = \frac{E_s}{E_p} \qquad \text{21-4}$$

However, the ratio of E_s to E_p is equal to the turns ratio:

$$\frac{E_s}{E_p} = \frac{N_s}{N_p}$$

Therefore, the ratio of current in the primary to current in the secondary is equal to the inverse of the turns ratio:

$$\frac{I_p}{I_s} = \frac{N_s}{N_p}$$

or

$$I_p = I_s \left(\frac{N_s}{N_p} \right) \qquad \text{21-5}$$

Example 21-4

The transformer in Figure 21-10 has a 1:3 step-up ratio. From the data supplied in the drawing, determine the current ratio of the transformer.

Solution

$$I_s = \frac{E_s}{R_L} = \frac{360}{180} = 2 \text{ A}$$

$$I_p = \frac{I_s N_s}{N_p} = \frac{2 \times 900}{300} = 6 \text{ A}$$

With $I_p = 6$ A and $I_s = 2$ A, it is apparent that the current ratio is 3:1. Therefore, the current ratio is the inverse of either the turns or voltage ratio, which can be verified from the fact that in an ideal transformer, $P_p = P_s$. Thus, the power in the primary is $P_p = 120 \text{ V} \times 6 \text{ A} = 720$ W and the power in the secondary is $P_s = 360 \text{ V} \times 2 \text{ A} = 720$ W.

21–4
Transformer Losses

Our ideal transformer has been assumed to be 100% efficient. Transformers actually in use are more nearly 95% to 98% efficient. The several possible loss conditions are outlined in this section.

Hysteresis

Inductor losses as a result of *hysteresis* were fully explained in Section 12–12. So that these losses in transformers are minimized, the core is made of soft silicon steel.

Saturation

For maximum efficiency from a transformer, the flux density in the core should be maintained near the *knee of the saturation curve*. Increasing the winding current so that the generated flux goes beyond saturation represents a loss in efficiency, because the same magnetizing effect could be obtained with a smaller current if the core were not saturated. Saturation can be prevented in a given condition, but at the price of a larger, more expensive core.

Eddy Currents

The ferromagnetic material of which a core is made may not be a good conductor, but it does have the ability to conduct current. Alternating currents in the primary induce currents not only into the windings but also into the core material. This induced voltage causes *eddy currents* to circulate in the core. Eddy currents are electric currents that circulate within a magnetic material as a result of induced voltages created by a varying induction. Thus, I^2R losses occur, which result in heating of the core.

So that these losses are minimized, the cores are *laminated*: Thin stampings of soft iron are bolted together to form the core. Each lamination is coated with a thin film of varnish or other insulating material. Because the thin insulated laminations do not provide an easy path for current, eddy current losses are substantially reduced. Figure 21-11a shows the large eddy current loss of a solid-iron core; Figure 21-11b shows the greatly reduced losses with a laminated core.

Copper Losses

Currents in the primary and secondary windings encounter the dc resistance of the wires. As a result, I^2R losses occur. Transformers carrying considerable power require conductors of large cross section to minimize this heat loss. Since a high percentage of flux linkage requires large inductance (particularly at power line frequencies), some compromise between the size of the core and the number of turns must be made. For most applications, the dc resistance of the windings may be ignored if the ratio of its X_L to R is 10 to 1 or greater.

Skin Effect and Distributed Capacitance

Skin effect, or the tendency of ac to travel near the surface of a conductor, also raises the resistance. However, it is not significant at power frequencies. At higher frequencies, it can result in noticeable losses. At

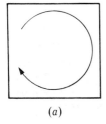
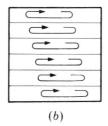

(a)   (b)

Figure 21-11

Eddy currents: (a) Large losses in solid-iron core. (b) Losses greatly reduced by laminating core.

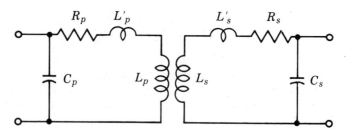

Figure 21–12

Approximate model of AF transformer

these same higher frequencies, the *distributed capacitance* between turns, windings, and windings and core can result in appreciable losses.

Another small loss occurs in transformers because not all of the flux lines produced by the primary link with all the turns on the windings. Stray flux linkages escape the core and are lost as far as inducing voltages are concerned.

An approximate model of a typical AF transformer is shown in Figure 21–12. Windings L_p and L_s represent the main primary and secondary coils. Leakage inductances are represented by L_p' and L_s'. Winding resistances are represented by R_p and R_s. Distributed capacitance between turns is represented by C_p and C_s.

21–5
Core Types

The principal types of magnetic cores used in transformers are illustrated in Figure 21–13. The core in Figure 21–13a is known as a *shell type*. Both primary and secondary windings are placed on the center leg. Frequently, the secondary winding will be wound directly on top of the primary, with a sheet of insulating material between them to minimize any possibility of arcing.

The manner in which the shell-type core is manufactured becomes evident by studying the somewhat exploded view in Figure 21–13c. The coils are wound on a thin cardboard or plastic form, exactly sized to slip over the center leg of the E laminations as the core is assembled. Through bolts (not shown) hold the finished assembly together.

Another popular shape is the *core type* illustrated in Figure 21–13b, obviously simpler in construction. Customarily, the primary winding is placed on one leg and the secondary on the other. This type of core is more susceptible to leakage flux than the shell-type core.

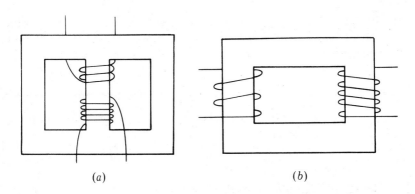

(a)   (b)

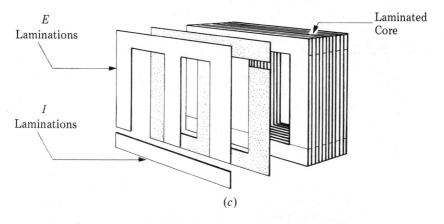

(c)

Figure 21–13

Types of cores for power and audio transformers: (a) Shell type. (b) Core type. (c) Construction of shell-type core.

21–6
Loading Effect

Resistive Load

When a load is placed across the secondary of a transformer, there will be a current as follows:

$$I_s = \frac{E_s}{Z_s}$$

If the load is resistive, the power factor will be approximately unity, and the secondary power will be $P_s = E_s^2/R_L$. In the ideal transformer, the primary power must therefore equal the secondary power.

Example 21–5

Calculate the primary current in the circuit of Figure 21–14 from the data supplied.

Solution

Calculate the secondary current:

$$I_s = \frac{E_s}{R_L} = \frac{600 \text{ V}}{12 \text{ k}\Omega} = 50 \text{ mA}$$

The power in the secondary is then as follows:

$$P_s = E_s I_s = 600 \text{ V} \times 0.05 \text{ A} = 30 \text{ W}$$

For the ideal transformer, $P_p = P_s$. Therefore:

$$P_p = E_p I_p$$

$$I_p = \frac{P_p}{E_p} = \frac{30 \text{ W}}{120 \text{ V}} = 0.25 \text{ A}$$

This equation may be confirmed from the turns ratio, which is 1:5 step-up. Use Equation 21–5:

$$I_p = I_s \left(\frac{N_s}{N_p}\right) = 0.05 \left(\frac{5}{1}\right) = 0.05 \times 5 = 0.25 \text{ A}$$

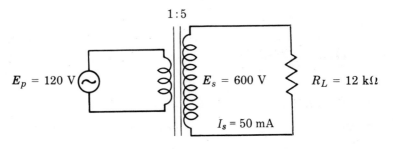

Figure 21–14

Circuit for Example 21–5

Reactive Load

When a reactive load is connected across the secondary, the voltage and current will no longer remain in phase. The following example illustrates the calculations involved.

Example 21-6

For the transformer circuit of Figure 21-15, calculate E_s, I_s, I_p, and the true power delivered by the source.

Solution

With a 1:3 step-up ratio, the secondary voltage and current are as follows:

$$E_s = 3 \times E_p = 3 \times 100 \angle 0° \text{ V} = 300 \angle 0° \text{ V}$$

$$I_s = \frac{E_s}{Z_L} = \frac{300 \angle 0° \text{ V}}{100 \angle 30° \text{ Ω}} = 3 \angle -30° \text{ A}$$

Since the transformer acts as a 3:1 step-down transformer for current, I_p is as follows:

$$I_p = 3 \times I_s = 3 \times 3 \angle -30° \text{ A} = 9 \angle -3 0° \text{ A}$$

The apparent power for Z_L is as follows:

$$P_A = E_s I_s = 300 \times 3 = 900 \text{ VA}$$

The true power for Z_L is as follows:

$$P_L = E_s I_s \cos \Theta = 900 \times 0.866 = 779.4 \text{ W}$$

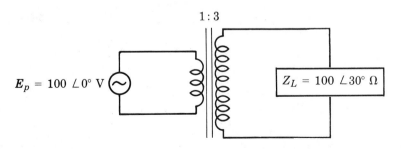

Figure 21-15

Reactive-load circuit for Example 21-6

With an ideal transformer, the load and input power are the same. Therefore:

$$P_{in} = P_L = 779.4 \text{ W}$$

21-7
Impedance Transformation

Impedance Ratio

Impedance ratio   The ratio of the impedance of the primary winding to the impedance of the secondary; expressed as Z_p/Z_s; equal to the square of the turns ratio.

In Figure 21-10, the secondary impedance Z_s is 180 Ω. The primary impedance must be as follows:

$$Z_p = \frac{120 \text{ V}}{6 \text{ A}} = 20 \text{ Ω}$$

The ratio of Z_p to Z_s is therefore as follows:

$$\frac{Z_p}{Z_s} = \frac{20}{180} = \frac{1}{9}$$

Consequently, the **impedance ratio** is not equal to the turns ratio of 1:3 but to the *square of the turns ratio*, $(1{:}3)^2$, or 1:9. Verification of this result is as follows. The voltage ratio equals the turns ratio:

$$\frac{E_p}{E_s} = \frac{N_p}{N_s}$$

And the current ratio is the reciprocal of the turns ratio:

$$\frac{I_p}{I_s} = \frac{N_s}{N_p}$$

Then, by dividing one by the other, we obtain the following ratio:

$$\frac{E_p/I_p}{E_s/I_s} = \frac{N_p^2}{N_s^2}$$

However, $E_p/I_p = Z_p$, the primary impedance, and $E_s/I_s = Z_s$, the secondary impedance. Therefore:

$$\frac{Z_p}{Z_s} = \frac{N_p^2}{N_s^2} \qquad\qquad \textbf{21-6}$$

From this derivation, we see that *the ratio of impedances across a transformer varies as the square of the turns ratio.* Therefore:

$$Z_p = Z_s \left(\frac{N_p}{N_s}\right)^2 \qquad\qquad \textbf{21-7}$$

$$Z_s = Z_p \left(\frac{N_s}{N_p}\right)^2 \qquad\qquad \textbf{21-8}$$

Example 21-7

A step-down transformer with a 12:1 turns ratio is connected to a load impedance of 2 Ω. Calculate the impedance of the primary.

Solution

$$Z_p = Z_s \left(\frac{N_p}{N_s}\right)^2 = 2 \left(\frac{12}{1}\right)^2 = 2 \times 144 = 288 \; \Omega$$

Example 21-8

A step-up transformer draws 3 A at 12.6 V ac in its primary. Calculate the secondary impedance if the turns ratio is 1:8.5.

Solution

Calculate the primary impedance:

$$Z_p = \frac{E_p}{I_p} = \frac{12.6}{3} = 4.2 \; \Omega$$

Then $\quad Z_s = Z_p \left(\frac{N_s}{N_p}\right)^2 = 4.2 \left(\frac{8.5}{1}\right)^2 = 4.2 \times 72.25 = 303.45 \; \Omega$

Impedance Matching

In the transfer of maximum power from a source to a load, *the load impedance must equal, or match, the internal impedance of the source.* With electronic circuits, it is frequently necessary to match a low-Z load to a high-Z source. For example, a loudspeaker with an 8 Ω voice coil has to be matched to a transistor with an output impedance of 150 Ω. An *impedance-matching transformer* is required. Its primary must match the source (transistor), and its secondary must match the loudspeaker. The turns ratio required to satisfy these requirements is as follows:

$$\frac{N_p}{N_s} = \sqrt{\frac{Z_p}{Z_s}} \qquad\qquad \textbf{21-9}$$

Then $$\frac{N_p}{N_s} = \sqrt{\frac{150}{8}} = \sqrt{18.75} = 4.33{:}1$$

21-8
Autotransformers

Basic Operation

> **Autotransformer** A transformer with a single electric winding, which can be used as a step-down or step-up device; a transformer characterized by no electrical isolation between primary and secondary.

Autotransformers are a special type of power transformer, designed for good voltage regulation *under varying loads.* A single tapped winding characterizes the autotransformer, as shown in Figure 21–16. When used as a step-up device, as in Figure 21–16a, all of the primary winding is part of the secondary. When used as a step-down device, all of the secondary winding is part of the primary, as shown in Figure 21–16b.

Voltages across the individual windings follow the turns ratios, as with conventional transformers. One difference, however, is that *part of the turns are common to both windings.* Generally, the winding is

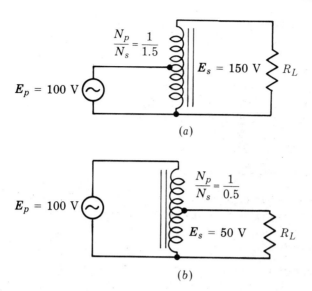

$$\frac{N_p}{N_s} = \frac{1}{1.5}$$

$E_s = 150$ V $\quad R_L$

$E_p = 100$ V

(a)

$E_p = 100$ V

$$\frac{N_p}{N_s} = \frac{1}{0.5}$$

$E_s = 50$ V $\quad R_L$

(b)

Figure 21–16

Autotransformer: (a) Step-up connection. (b) Step-down connection.

tapped so that the turns ratio can be varied. A disadvantage of this arrangement is that there is *no electrical isolation* between primary and secondary, as there is with a conventional transformer.

Voltage Regulation

In a conventional power transformer, any sudden change in load results in a large reactive voltage across the windings. In extreme cases, *arcing may occur* and short some of the winding, thus destroying the transformer. The chief advantage of the autotransformer is that its load may be varied without arcing and with little change in output voltage. This advantage results from the primary and secondary currents being 180° out of phase and *tending to cancel* in that part of the transformer common to both windings.

Consider the autotransformer in Figure 21–16a, which has a 1:1.5 step-up ratio. Assume that $R_L = 75 \ \Omega$. With $E_p = 100$ V, then $E_s = 150$ V and the secondary current is 2 A. By the turns ratio formula, I_p should equal 3 A. However, since I_p and I_s are 180° out of phase, the total primary current is the algebraic sum of the two, or 1 A. If the load current increases to 4 A, I_p increases to 6 − 4, or 2 A. If the secondary load is decreased to 1 A, I_p decreases to 1.5 − 1, or 0.5 A.

Now, consider a step-down autotransformer with a turns ratio of 1 to 0.5, as shown in Figure 21–16b. With the same applied voltage and $R_L = 25 \, \Omega$, $E_s = 50$ V and $I_s = 2$ A. The current in the part of the primary common to both is 1 A in the direction of the secondary current. Thus, $I_p = 1$ A may be conceived as flowing directly to the load and not passing through the entire primary winding. Any variation in the load causes a smaller change in I_p than is possible with the separate-winding power transformer.

Example 21–9 ──────────────────────────────────

An autotransformer has a 1:2 step-down ratio. Determine the current in the common winding when $E_p = 50$ V and $R_L = 50 \, \Omega$.

Solution

The circuit is essentially that of Figure 21–16b.

$$E_s = \frac{E_p}{2} = \frac{50}{2} = 25 \text{ V}$$

$$I_s = \frac{E_s}{R_L} = \frac{25 \text{ V}}{50 \, \Omega} = 0.5 \text{ A}$$

The current in the common winding is as follows:

$$I_s - I_p = 0.5 \text{ A} - 0.25 \text{ A} = 0.25 \text{ A}$$

21–9
Isolation Transformers

Some electronic equipment has one side of the ac power line connected to the chassis, which presents a *lethal hazard* to any operator. If the chassis happens to be connected to the grounded side of the ac line (50% chance), there is no particular problem. However, in the situation where the plug is inserted into the ac outlet in the other direction, the chassis is "hot." Anyone touching the chassis and an earth ground would suffer severe electric shock – possibly electrocution (see Figure 21–17a). Needless to say, this type of equipment is *not* approved by the Underwriters Laboratories.

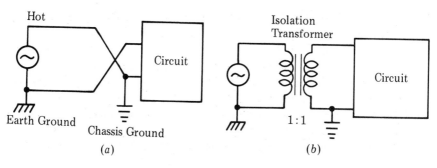

Figure 21-17

Power connections: (a) Possible connection resulting in "hot" chassis. (b) Use of isolation transformer to eliminate problem.

> **Isolation transformer**   An transformer that usually has a 1:1 turns ratio; designed to prevent electrostatic (capacitive) coupling between primary and secondary.

A simple solution to the power connection problem involves the use of an **isolation transformer**. This device is simply a conventional transformer with a 1:1 turns ratio that has a power-handling capability at least equal to that of the circuit to be isolated. Its use is illustrated in Figure 21-17b. Since there is no electric connection between primary and secondary, a grounded chassis will present no hazard.

21-10
Radio Frequency Transformers

Power and AF transformers are primarily power transfer or impedance-matching devices having high mutual coupling. At radio frequencies (RF), eddy current and hysteresis losses in a magnetic core become prohibitive, and therefore *RF transformers* are generally of powdered-iron or air core construction. Eddy current loss is the more important factor since it increases as the *square* of the frequency. Hysteresis loss, being constant for each cycle, increases *directly* with frequency. With air core construction, the tightest possible coupling results in a k equal to about 0.65, compared with the 0.98 of well-designed iron core transformers.

A large part of the primary and secondary windings of these partially coupled transformers constitutes a leakage reactance. Therefore, voltage, current, and impedance ratios cannot be calculated by the turns ratio equations, which are based on an assumption of almost perfect coupling. Hence, the turns ratios of RF transformers have no *exact significance.* In fact, the windings of these transformers are generally very loosely coupled (k may be as small as 0.005), and they function primarily as devices *coupling two circuits* rather than as voltage or current-level transformers.

At RF, the reactance of even a small inductance is very high, because the rate of change of the current is high. Therefore, the high mutual reactance between primary and secondary makes these transformers effective for impedance matching, for power transfer at low loss, and for high-voltage gain obtained by resonance.

The use of loose coupling at RF is illustrated by the intermediate-frequency (IF) transformer of a typical radio receiver, as shown in Figure 21–18. It is designed to pass 455 kHz with a 10 kHz bandwidth. Note that the primary is part of a parallel-resonant circuit, and the secondary is part of a series-resonant circuit. Loose coupling exists between the windings. Therefore, each is largely a leakage inductance acting mostly as the inductance of a tuned circuit. Since each is part of a tuned circuit, the overall voltage gain is determined by *the Q of the circuit* and not the turns ratio of the windings. Normally, the turns ratio is 1:1, and L and C for each winding are equal. Although Q should be high

FET

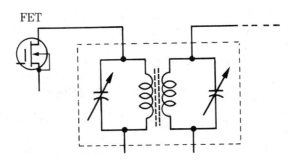

Figure 21–18

Transformer used in IF amplifier of typical receiver

for good selectivity, it must not be so high as to reduce the overall bandwidth below certain minimum requirements.

The bandpass characteristics of a *double-tuned stage* depend on several things, such as coefficient of coupling k, Q of the windings, and the mutual inductance. Figure 21–19 shows the effects of varying the coupling.

Construction details of two typical IF transformers are shown in Figure 21–20. The distance between the two coils is an indication of coupling. In Figure 21–20a, the windings are *permeability-tuned*, while in Figure 21–20b, they are resonated by small variable *capacitors* across each winding. The assembly is mounted inside a *small aluminum can*, which functions as a *shield* to prevent the electromagnetic fields of the coils from inducing voltages into nearby circuits. Small openings in either end of the can, or both, allow adjustments of the powdered-iron slugs, or adjustable capacitors.

The metal can represents a single-turn secondary into which is introduced an emf. The emf sets up a current counter to the original force, thus preventing flux lines from passing beyond the can. Because the can has very low resistance and is well grounded, the reactance it reflects back to the coils is the amount due to only a single turn.

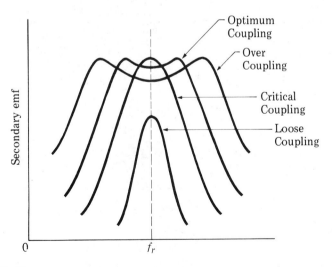

Figure 21–19

Effects of varying the coupling between primary and secondary of IF transformer

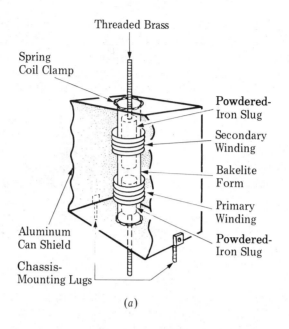

(a)

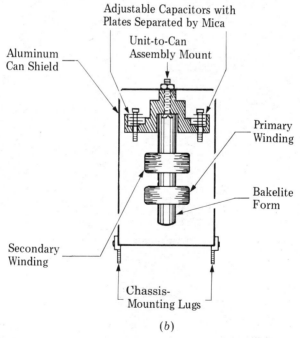

(b)

Figure 21–20

Details of IF transformer construction: (a) Slug-tuned. (b) Capacitor-tuned.

┌──Summary

A transformer is a device that transfers electric energy from one circuit to another by electromagnetic induction.

Audio and power transformers consist of a primary and one or more secondary windings on a soft-iron core.

The coefficient of coupling between primary and secondary is about 0.98 for AF and power transformers. Efficiency is about 95–98%.

Flux lines set up by I_s are opposite those set up by I_p. Therefore, the secondary current decreases the impedance of the primary and allows I_p to increase to maintain the proper flux linkages between the windings.

The voltage ratio of a transformer is as follows:

$$\frac{E_p}{E_s} = \frac{N_p}{N_s}$$

If the turns ratio is greater than one, it is a step-up transformer. If the turns ratio is less than one, it is a step-down transformer.

The volts-per-turn ratio is determined by dividing the voltage by the number of turns.

The current ratio of a transformer is as follows:

$$\frac{I_p}{I_s} = \frac{N_s}{N_p}$$

Transformer losses include hysteresis, copper (I^2R), eddy currents, leakage flux, skin effect, and distributed capacitance.

Transformer cores are usually shell- or core-shaped.

A purely resistive load on a secondary winding reflects a pure resistance back into the primary.

The impedance ratio is the square of the turns ratio; the square root of the impedance ratio is the turns ratio.

Transformers are frequently used as impedance-matching devices.

An autotransformer consists of one winding with a tap for either a secondary or a primary. They provide good voltage regulation.

A disadvantage of autotransformers is that no isolation exists between primary and secondary.

Isolation transformers provide magnetic coupling between two circuits without electric connections between them.

Radio frequency (RF) transformers have low k values so that each winding is largely a leakage inductance, acting primarily as the inductance of a tuned circuit.

The bandpass characteristics of an RF transformer depend on the coefficient of coupling, the Q of the windings, and the mutual inductance.

┌──Progress Test

The bracketed number after each question indicates the section of this chapter where the answer can be found.

1. A well-designed power transformer should have: (a) as few turns as possible. (b) a high k value. (c) $N_p = N_s$. (d) primary and secondary currents in phase. [21–2]

2. Which one of the following statements is correct concerning a power transformer with a lightly loaded secondary? (a) Secondary voltage is 90° out of phase with the applied voltage. (b) Secondary current leads the secondary voltage by 90°. (c) Secondary voltage leads the secondary current by 90°. (d) Secondary voltage and current are in phase. [21–2]

3. When an output transformer is loaded to design specifications, the : (a) Z of the primary looks nearly resistive. (b) X_L of the primary increases. (c) output and input currents are nearly in phase. (d) input emf and current are 90° out of phase. [21–2]

4. A step-up transformer with a turns ratio of 1:5 operates under the following conditions: $e_p = 120$ V, $Z_L = 600$ Ω, and 100% efficiency. Current I_p is: (a) 600 mA. (b) 5 A. (c) 0.2 A. (d) 10 A. [21–3]

5. The impedance of an output transformer varies: (a) directly as the turns ratio. (b) inversely as the square of the turns ratio. (c) inversely as the turns ratio. (d) directly as the square of the turns ratio. [21–7]

6. The turns ratio of an output transformer designed to

match a load resistance of 4.5 kΩ to an 8 Ω voice coil of a loudspeaker is: (a) 19.8:1. (b) 21.6:1. (c) 23.7:1. (d) 28.4:1.   [21–7]

7.   Iron cores are laminated to: (a) reduce hysteresis losses. (b) reduce eddy current losses. (c) provide better heat dissipation. (d) minimize distributed capacitance.   [21–4]

8.   Current I_p of an autotransformer with a 2.5:1 step-up ratio, 120 V ac connected to its primary, and a 470 Ω secondary load is: (a) 0.64 A. (b) 0.96 A. (c) 1.60 A. (d) 2.34 A.   [21–8]

9.   Which of the following statements is characteristic of an isolation transformer? (a) The secondary is electrically isolated from the primary. (b) They are ideally suited for RF applications. (c) They may be of either the step-up or step-down type. (d) They are usually of autotransformer design.   [21–9]

10.   Which of the following statements is characteristic of an IF transformer? (a) The primary winding is part of a series-resonant circuit. (b) Tight coupling exists between primary and secondary. (c) Maximum energy is transferred between primary and secondary. (d) Voltage gain is determined by the Q of the circuit.   [21–10]

Problems

Answers to odd-numbered problems are at the back of the book.

1.   A voltage of 120 V_{rms} is applied to the primary of a power transformer having a 1:4 step-up ratio. What is the secondary voltage?

2.   A transformer has $E_s = 750$ V and a turns ratio of 1:6.25. What is E_p?

3.   One of the secondary windings of a power transformer has 6.3 V across its terminals. If $E_p = 120$ V, what is the turns ratio?

4.   In Problem 3, what must the primary current be if the secondary current is 3 A?

5.   A 6:1 step-up transformer has a primary current of 2.5 A. What is I_s?

6.   If the primary current of a 4.5:1 step-up transformer is 50 mA, what is the secondary current?

7.   A step-down transformer delivers 5 V at 2 A. If the pri-

mary is connected to a 120 V ac source, what is the value of primary current?

8. A high-voltage transformer develops 1600 V across its secondary winding of 720 turns. The primary has 100 turns. (a) What is the turns ratio? (b) What is the primary voltage?

9. A transformer delivers 650 W to a resistive load. Calculate the primary current if the primary voltage is 117 V$_{rms}$.

10. A power transformer has a 3.2:1 step-up ratio and supplies 375 V to a resistive load. If the primary current is 2.25 A, what power is being furnished to the load?

11. For the circuit of Figure 21–21, calculate the following: (a) P_s (b) E_s (c) turns ratio (d) N_s.

12. From the data shown in Figure 21–22, calculate the following: (a) E_s (b) I_s (c) I_p (d) P_T.

13. For the step-down transformer circuit shown in Figure 21–23, calculate the following: (a) E_s (b) I_s (c) I_p (d) P_T.

14. An output transformer is connected to a 3.2 Ω load. If the step-down turns ratio is 37.5:1, what is Z_p?

15. An output transformer used with a solid-state amplifier

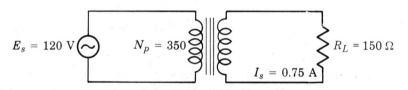

$E_s = 120$ V $N_p = 350$ $R_L = 150$ Ω $I_s = 0.75$ A

Figure 21–21

Circuit for Problem 11

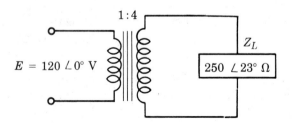

$1:4$

$E = 120 \angle 0°$ V Z_L $250 \angle 23°$ Ω

Figure 21–22

Circuit for Problem 12

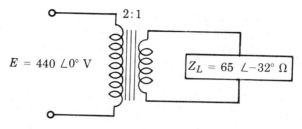

Figure 21–23

Circuit for Problem 13

has its secondary connected to an 8 Ω loudspeaker. If the step-down turns ratio is 3:1, what is Z_p?

16. Calculate the reflected impedance of a 6:1 step-up transformer with a 72 Ω load connected to its secondary.

17. Determine the reflected impedance in a transformer if the step-down turns ratio is 4:1 and $Z_L = 250$ Ω.

18. An impedance of 2200 Ω is reflected into the primary of a transformer. If $N_p = 550$ and $N_s = 110$, what is the load resistance?

19. A power supply transformer draws 2.1 A at 120 V_{rms} in its primary. Calculate Z_s if the step-up turns ratio is 1:6.

20. A primary winding of 400 turns has an impedance of 8 kΩ reflected into it from the secondary. If the load impedance is 480 kΩ, how many turns are there in the secondary?

21. An autotransformer has a 1:3 step-up ratio. If the secondary is connected to a 180 Ω resistive load and the primary to a 60 V ac source, what is the value of the primary current?

22. An autotransformer has 1:3 step-down ratio. Calculate the current in the common winding when the supply is 120 V ac and the load is 80 Ω resistive.

23. A step-down transformer with a turns ratio of 6.67:1 is connected to commercial 120 V ac power mains. What is the secondary voltage?

24. For the transformer described in Problem 23, solid-state circuitry in the secondary draws 120 mA. (a) What is I_p for an ideal transformer? (b) What is the primary impedance? (c) What is the secondary impedance?

25. A small computer utilizes a power transformer with the following data stamped on its case: 120 V ac, 60 Hz, 6.5 W, 10 V ac,

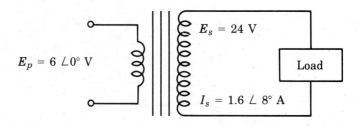

Figure 21-24

Circuit for Problem 29

0.65 A. (a) What is the probable secondary power? (b) What is the primary current? (c) What is the transformer turns ratio?

26.   For the transformer described in Problem 25, using only the turns ratio, calculate the impedance ratio. If $Z_s = 15.38$ Ω, calculate the primary impedance from the turns ratio.

27.   A multivoltage step-down transformer has 0.4 V per turn on its primary. (a) How many turns must it have on each of its secondaries if they produce 24, 18, and 6 V, respectively? (b) If the primary has 375 turns, what is the primary voltage?

28.   It is desired to match a 150 Ω microphone to a 600 Ω amplifier input. If the primary of the mike transformer has 225 turns, what number of turns should the secondary have to effect the proper match?

29.   For the circuit in Figure 21-24, calculate the following: (a) turns ratio (b) R_L (c) I_p (d) P_T.

30.   A step-up autotransformer with an actual 4.2 A in the secondary and an actual 4.2 A in the primary has what turns ratio?

22

Power Supplies

Objectives

After studying this chapter, you will be able to:

Describe the action of a PN junction.

Recognize the characteristics of half-wave and full-wave rectifiers.

Discuss the basic operation of a nonregulated and a regulated power supply.

Design a voltage divider network.

Calculate the regulation of a power supply.

Introduction

Most electronic equipment requires dc power for operation. Although this power may come from batteries, as in the case of portable equipment, it customarily comes from dc power supplies. Before a study of power supplies can be undertaken, it is necessary to consider the operation of diodes. There are a number of different kinds of diodes, but only the power rectifier diode will be studied in this chapter.

Applications

Power supplies provide the necessary voltage and current to operate electronic circuits. In some cases, the power requirements may be small, while in others, the power requirements may be large. A 100 W per channel stereo sound system will require more power from its power supply than would a small TV set.

The amount of dc power required to operate a particular electronic system dictates the kind of power transformer to be used and the types of rectifiers and capacitors needed. Low-power requirements can often be met with half-wave rectifiers; full-wave rectifiers are used when more power is needed.

22–1
Semiconductor Materials

> **Rectifiers/diodes**   Devices used to change ac to dc; they permit one-half of an ac wave to pass while blocking the other half.

Rectifiers and **diodes** are used in power supplies, in conjunction with transformers and filter circuits, to furnish the dc voltages and currents necessary to operate electronic equipment. They are also used in other applications in electronics.

Two basic types of diodes are used: *solid-state* and *vacuum tube*. Vacuum tubes are seldom used except for replacement parts in older equipment. A brief review of semiconductors is essential to our study of solid-state diodes since they are the materials of which diodes are made.

> **Semiconductors**   Materials that lie midway between conductors and insulators in their ability to pass an electric current.

The material used for **semiconductors** is either silicon (Si) or germanium (Ge). In their pure (intrinsic) state, these elements are insulators. They become semiconductors when extremely small, controlled amounts of other materials, called *impurities*, are added to them. The process is called *doping*.

A silicon atom has a charged nucleus of $+14$; the germanium atom has a charged nucleus of $+32$. In each case, the positive charges in the nucleus are equalized by a similar number of electrons orbiting the nucleus. Hence, the atoms are neutral under normal conditions. Each electron, in its relationship to its nucleus, has an energy value and a distinct energy level. Electrons that are closer to the nucleus are more tightly bound and require greater energy to break loose. Outer orbital electrons, frequently called *valence electrons*, are said to be stronger than those in the inner orbits because of their ability to break away from the parent atom. The outer orbit in which the valence electrons exist is called the *valence band* or *valence shell.*

Models of silicon and germanium atoms appear in Figure 22–1 in two-dimensional form. Figure 22–1*a* shows a silicon atom with its 14 orbital electrons and $+14$ charge in the nucleus. To the right is a simplified version of this atom with only the four outermost electrons shown and their corresponding $+4$ charges in the nucleus. These valence electrons are the ones involved in diode and transistor operation. The germanium model with its simplified version is shown in Figure 22–1*b*.

In the silicon and germanium atoms, the orbiting of one valence electron is associated with the orbiting of another valence electron in an adjacent atom. These electrons form an *electron pair bond*, or *valence bond*, and are represented symbolically by the four pairs of curved lines extending from each atom in Figure 22–2. The overall arrangement of nuclei and orbiting electrons is referred to as a *crystal lattice.* In a perfect crystal, there would be no free electrons to act as charge carriers and no current could flow; it would be a good insulator.

N type material   A semiconductor material with an extra electron (the N stands for "negative"); also called a donor.

P type material   A semiconductor material with a deficiency of one electron (the P stands for "positive"); also called an acceptor.

To make the material a semiconductor, the manufacturer dopes it by adding very small amounts of an element containing five electrons (*pentavalent element*) in the outermost orbit. Arsenic and antimony are typical dopants used. During fabrication, the pentavalent atoms disperse uniformly throughout the lattice. Figure 22–2*a* shows this condition with an atom having $+5$ located within the lattice. The fifth

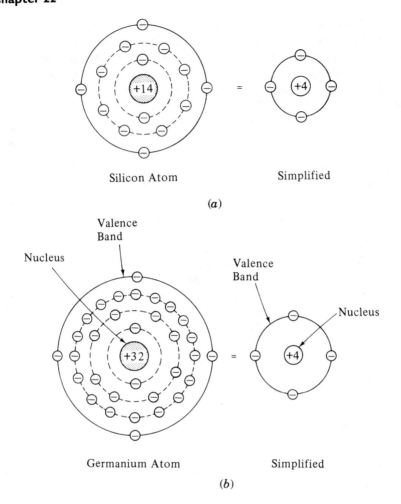

Figure 22–1

Two-dimensional models of atoms: (a) Silicon. (b) Germanium.

electron is not part of any covalent bond and is free to wander throughout the lattice; it becomes a *negative-charge carrier*. Such atoms are called *donor atoms* and produce **N type material**.

Silicon or germanium may be doped by *trivalent elements*, such as indium, boron, and gallium. Instead of having one excess electron, we now have a deficiency of one electron in the electron pair bond. A *hole* results, as shown in Figure 22–2*b*, which exhibits a *plus charge*. These trivalent elements are known as *acceptors*, and the crys-

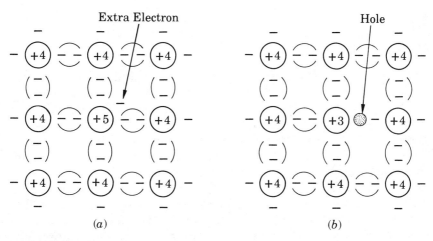

Figure 22–2

Crystal lattice of silicon: (a) N type. (b) P type.

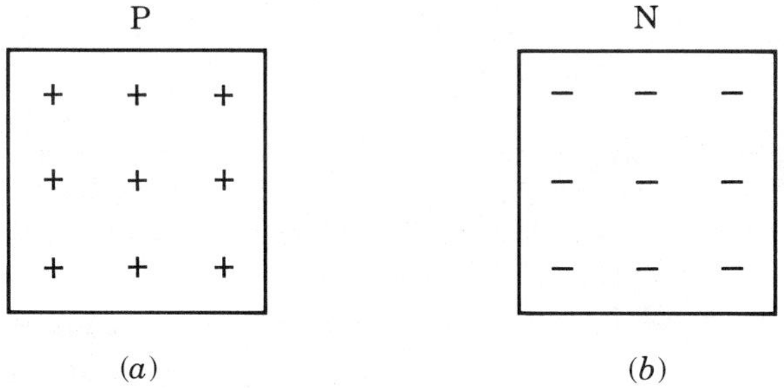

P

N

(a)

(b)

Figure 22–3

Semiconductors: (a) P type. (b) N type.

tals are known as **P type materials**. Figure 22–3 represents specimens of P type and N type semiconductor materials.

Electrons are the *majority carriers* in N type semiconductors. Because of manufacturing limitations, some holes are present in this type of semiconductor; they are known as *minority carriers*. Minority carriers are responsible for *leakage currents*. For P type material, holes are the majority carriers. Again, because of imperfections, electrons are present and become minority carriers responsible for leakage currents in this type of semiconductor.

22-2
PN Junction

Under controlled conditions, a manufacturer can produce a crystal with P material on one side and N material on the other, as shown in Figure 22-4a. At the instant the junction is formed, some of the free electrons in the N type material cross over the junction and combine with holes. Simultaneously, some holes cross over and combine with free electrons. Consequently, the atoms that have gained or lost electrons become *negative* and *positive ions,* as shown in Figure 22-4b.

   This action continues for a very brief time after the junction is formed and causes a *depletion region* to be formed. As this region builds up, a potential difference, called a *barrier voltage*, appears across the junction, as indicated in Figure 22-5. Soon, this potential difference becomes large enough to prevent further electron-hole movement across the junction. At ambient temperatures (about 25°C), this potential is approximately 0.3 V for germanium and 0.6 V for silicon.

22-3
Biasing the Junction

A *PN junction diode* is a device that permits carriers to flow in *only one direction.* If an external voltage source is connected across the device, as in Figure 22-5, the majority carriers in each material will be

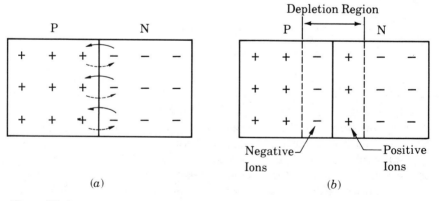

Depletion Region

(a)

(b)

Figure 22-4

PN junction: (a) At moment of formation. (b) With depletion region after junction is formed.

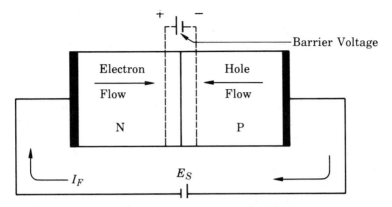

Figure 22–5

Forward-biased PN junction

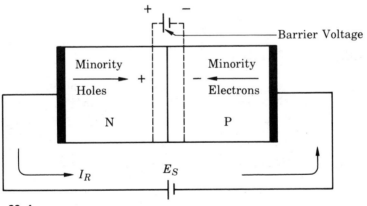

Figure 22–6

Reverse-biased PN junction

forced toward the junction. If the potential of E_S is greater than the barrier voltage, these carriers will cross the junction in large numbers and combine. The voltage source continues to inject new electrons at the left and produce new holes at the right. Consequently, a continuous flow of majority carriers flows toward the junction. The diode is said to be *forward-biased*, and a large forward current I_F results, indicated by the arrows.

Now, consider what happens when the potential of E_S is reversed, as in Figure 22–6. The majority carriers in each material are attracted toward the source. The source potential aids the barrier volt-

Cathode  Anode

N   P

Figure 22–7

Schematic symbol for diode

age and prevents majority carriers from crossing the junction. This condition is known as *reverse bias*. Actually, a very small amount of reverse current I_R will flow under these conditions, but it is so small that it is normally neglected. Reverse current results from minority carriers and surface leakage across the crystal.

A PN junction diode is schematically represented in Figure 22–7. The direction of the arrow represents conventional current flow. However, when we are analyzing electronic circuits, current is assumed to flow from negative to positive, or against the diode arrow.

22–4
Diode *I-V* Curves

A better understanding of a diode's performance can be obtained if we connect the device in a circuit, as in Figure 22–8a, and vary the potential of E_S. A plot of current versus voltage ($I-V$) will result in the curve shown in Figure 22–8b when the diode (made of silicon) is forward-biased. Between $E_S = 0$ V and $E_S \cong 0.6$ V (knee voltage V_K in Figure 22–8b), there is virtually no current. As V_K is approached, I_F begins to

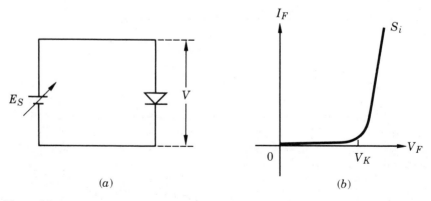

(a)   (b)

Figure 22–8

Forward-biased diode: (a) Circuit. (b) *I–V* characteristic curve.

flow, increasing rapidly as the supply voltage is slightly increased beyond V_K. The I–V curve for a germanium diode has the same shape, except that $V_K = 0.3$ V. When the voltage across the diode is increased more than a few tenths beyond V_K, current increases rapidly. A typical diode, such as the 1N4004, can handle a maximum of 1 A continuously without being damaged.

The dc resistance of a diode varies widely over different parts of the I–V curve. For example, at $V_F = 0.55$ V, I_F may be 0.05 mA, which gives the following diode resistance:

$$R_D = \frac{0.55 \text{ V}}{0.05 \text{ mA}} = 11 \text{ k}\Omega$$

The same diode with 0.61 V applied may pass a current of 8 mA. At this operating point, the resistance is as follows:

$$R_D = \frac{0.61 \text{ V}}{8 \text{ mA}} = 76.25 \text{ }\Omega$$

Plainly, the diode's resistance varies over a wide range.

The ac resistance of a diode is different from its dc resistance. For example, if the applied voltage changes from 0.61 to 0.63 V and the diode current varies from 8 to 9.5 mA, its resistance is as follows:

$$r_{\text{ac}} = \frac{\Delta V}{\Delta I} = \frac{0.63 - 0.61}{9.5 - 8} = 13.33 \text{ }\Omega$$

A diode is reverse-biased when connected as in Figure 22–9a. Very little current flows until *breakdown voltage* V_{BD} is reached. At this point, reverse current I_R increases sharply, as shown in Figure

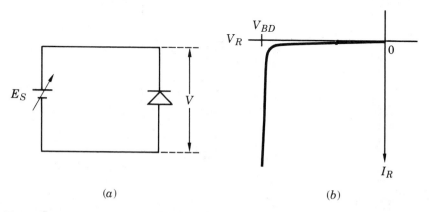

(a)   (b)

Figure 22–9

Reverse-biased diode: (a) Circuit. (b) I–V characteristic curve.

22-9b. If it is allowed to increase too much, the diode will be destroyed. Breakdown is caused by bound electrons being knocked out of their valence shells and becoming free electrons. This phenomenon is a result of either *avalanche* or *zener breakdown*. A commonly used diode such as the 1N4004 has a V_{BD} of 400 V.

22–5
Zener Diodes

Zener diode   A diode whose breakdown region is very sharp and almost vertical.

If the manufacturer increases the doping of a PN junction, the breakdown region can be made very sharp and almost vertical. Diodes with these characteristics are called **zener diodes**. The depletion region

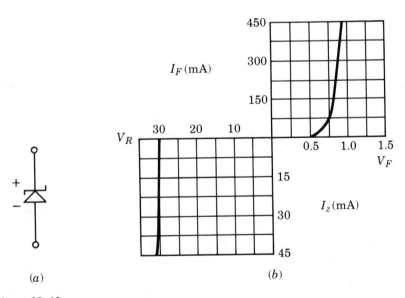

(a)   (b)

Figure 22–10

Zener diode: (a) Schematic symbol. (b) Characteristic curves.

is very thin, and the electric field across the junction is sufficient to break electron pair bonds. A large reverse current results.

Zener diodes fall into two categories: *voltage regulators* and *voltage reference devices*. When used in power supplies as regulators, they provide a nearly constant dc output voltage, even though the input voltage and load change. As a reference device, the zener utilizes the exact voltage drop across its reverse-biased junction for a specified current. Zener diodes are available in voltage ratings from about 2 to more than 200 V and in 1/4 to 50 W ratings.

Figure 22–10a shows the schematic symbol for a zener diode. Characteristic curves of a 30 V zener diode are shown in Figure 22–10b. The device is designed to operate in the reverse-bias region slightly below the knee of the curve.

The circuit in Figure 22–11 shows how the zener can be connected as a voltage regulator. Placing a load resistor across a zener di-

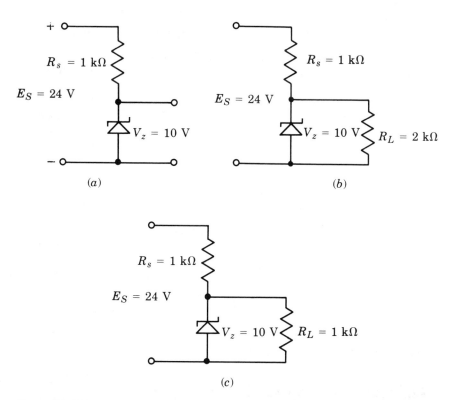

(a)

(b)

(c)

Figure 22–11

Zener diode (10 V) operation: (a) No load. (b) 2 kΩ load. (c) 1 kΩ load.

ode does not change the voltage drop across the zener if enough current is available to supply both the zener diode and the load resistor. The reason is that the zener needs only enough current through it to remain below the knee of the curve. For a typical zener, the current for maintaining zener operation is from 1 to 5 mA.

Suppose a 24 V dc source is connected to a 1 kΩ resistor in series with a 10 V zener diode. There will be 14 V dropped across the 1 kΩ resistor and 10 V across the zener diode. Now, if a load is placed across the zener, a portion of the available R_s current (14 mA) will flow to the load, and the remaining current will flow through the zener.

Figure 22–11 shows zener operation for no-load and loaded conditions. In Figure 22–11a, the zener current equals the current through R_s which is 14 mA (no load). In Figure 22–11b, the load current is 5 mA (10 V ÷ 2 kΩ). Therefore, the 14 mA of total current divides, with 5 mA to the load and 9 mA to the zener. In Figure 22–11c, the load current is 10 mA and the zener current is 4 mA.

Example 22–1

Refer to the basic schematics in Figure 22–11. Calculate the zener and load currents if $E_S = 30$ V, $R_s = 1$ kΩ, $V_z = 10$ V, and $R_L = 2$ kΩ.

Solution

$$V_{RS} = E_S - V_z = 30 - 10 = 20 \text{ V}$$

$$I_{R_s} = \frac{20 \text{ V}}{1 \text{ k}\Omega} = 20 \text{ mA}$$

$$I_{RL} = \frac{10 \text{ V}}{2 \text{ k}\Omega} = 5 \text{ mA}$$

Therefore $\quad I_z = I_{R_s} - I_{RL} = 20 - 5 = 15 \text{ mA}$

22–6
Rectifiers

Half-Wave Rectifiers

Since a diode passes current in only one direction, it is ideally suited for converting ac to dc. Figure 22–12a shows a simple *half-wave rectifier* circuit. During the half cycle, when the secondary voltage has the polarity

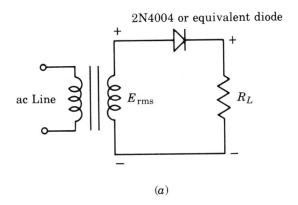

(a)

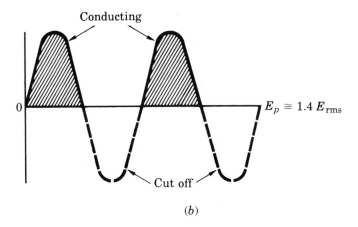

(b)

Figure 22–12

Half-wave rectifier: (a) Circuit. (b) Diode conducting and nonconducting periods (cutoff).

indicated, current flows since the diode is forward-biased. Appearing across load resistor R_L is a series of positive-going pulses whose peak value is approximately 1.4 E_{rms} (see Figure 22–12b). Peak voltage across the load will be 1.4 E_{rms} minus the 0.6 V barrier potential of the diode. The frequency of this voltage is equal to the line frequency.

On the negative half-cycle, the diode is reverse-biased, and its resistance is very high. Consequently, the secondary voltge of the transformer is nearly all across the diode. The average value of each half-wave pulse is 0.318 of the peak amplitude. Filters can be used to smooth out these variations and provide a higher average dc output voltage. (Filters are covered in following sections.) These devices are capable of supplying limited currents to a load.

Full-Wave Rectifiers with Center-Tapped Transformers

A *full-wave rectifier* uses two diodes arranged so that the load current flows in the same direction during each half of the ac cycle. Figure 22–13a is a schematic of such a circuit. During one half cycle, the polarity of the secondary voltages are as shown by the solid plus and minus signs. During this time interval, current leaves the center tap connection and flows through R_L (see solid arrows), diode D_1, and back to the top of the secondary winding. Diode D_2 cannot conduct since it is reverse-biased.

On the alternate half cycle, the secondary polarities are reversed, as shown by the dotted polarity signs. Current leaves the center-tapped connection and passes through R_L, D_2, and back to the bottom of the secondary winding (see dotted arrows). Diode D_1 will not

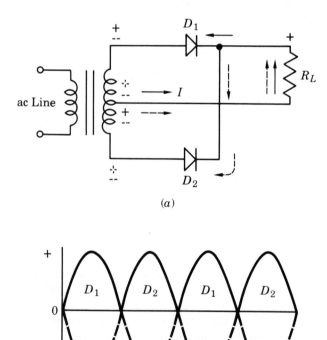

(a)

(b)

Figure 22–13

Full-wave, center-tapped rectifier: (a) Circuit. (b) Waveform appearing across R_L.

conduct since it is reverse-biased. Note that current pulses through R_L are *always in the same direction* regardless of which diode is conducting. Current pulses through R_L produce a series of positive-going waveforms, as shown in Figure 22–13*b* (solid curves). The frequency of these pulses, called *ripple frequency*, is twice the line frequency. The dotted curves represent the time intervals that the diodes are not conducting.

Average current through R_L is 0.638 of the peak current. Average output voltage is 0.9 times E_{rms} of one-half of the transformer secondary (twice the value of the half-wave circuit). Full-wave rectifiers can deliver twice the load current (for the same diodes and secondary voltages) as a half-wave circuit and are easier to filter.

Bridge Rectifiers

If four diodes are connected as shown in Figure 22–14*a*, the circuit is called a *bridge rectifier*. Input to the circuit is applied to diagonally opposite corners of the network, and the output is taken from the re-

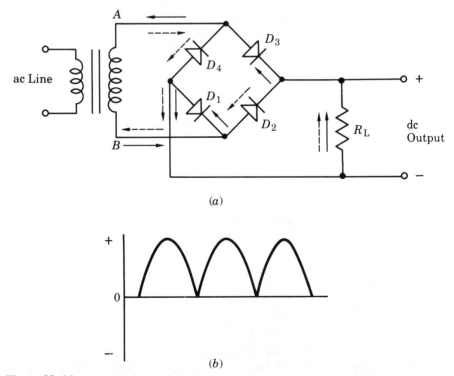

(*a*)

(*b*)

Figure 22–14

Bridge rectifier: (*a*) Circuit. (*b*) Output waveform.

maining two corners. During one half cycle, point A becomes positive with respect to point B by the amount of voltage induced into the secondary of the transformer. This voltage forward biases diodes D_1 and D_3, causing a pulse of current through them and R_L. The solid arrows indicate the direction of this rectified current.

One half cycle later, diodes D_2 and D_4 are forward-biased, and current flows in the direction indicated by the dashed arrows. During this interval, D_1 and D_3 are reverse-biased. Notice that the rectified currents from both pairs of diodes *always flow in the same direction* through the load. Ripple frequency of the output voltage is twice the line frequency. The output waveform for this device is shown in Figure 22-14b.

Bridge rectifiers have several advantages over full-wave, two-diode rectifiers:

1. For a given transformer, they have nearly twice the output voltage.
2. Their secondary windings need not be center-tapped.
3. They have a lower ratio of peak inverse voltage to average output voltage. (See Section 22–8 for a discussion of peak inverse voltage.)

22–7
Filters

> **Filter** A device located between the rectifier and the load that removes voltage fluctuations.

Before the output of a rectifier can be used, it must be filtered to remove any voltage fluctuation. A **filter** is required between the rectifier and load to provide this essentially constant dc voltage. Its design depends on such factors as the amount of output voltage and current to be supplied and the constancy of the output voltage under changing load conditions (known as *voltage regulation*). All power supply filters are low-pass filters since they are designed to pass dc but attenuate the ripple frequency.

Capacitive Input Filter

A simple method of smoothing out these fluctuations is to place a large capacitor across the load, as in the circuit of Figure 22–15a. As the output voltage rises, the capacitor begins to charge. It charges nearly as fast as the rate of rise of the voltage, owing to the limited reactance of the transformer secondary and low resistance of the diode.

When the rectifier output drops to zero, the voltage across the capacitor does not fall immediately. Instead, the energy stored in the capacitor is discharged through the load during the time that the rectifier is not conducting. If the load resistance is sufficiently high, the current drawn from the capacitor will be limited because of the relatively long RC discharge time. Thus, the amplitude of the ripple is reduced. Figure 22–15b graphically shows this action. The heavy line illustrates the voltage variation across the load.

If a full-wave rectifier were used, the output pulses would be doubled. Under these conditions, the capacitor would not be able to dis-

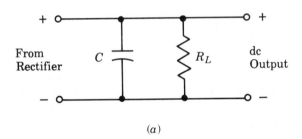

(a)

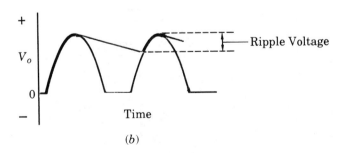

(b)

Figure 22–15

Half-wave rectifier with capacitor filter: (a) Circuit. (b) Output waveform showing ripple voltage.

charge as much as in the half-wave rectifier before the next pulse. Consequently, the amplitude of the ripple would be substantially reduced. This reduction is a decided advantage of the full-wave rectifier.

After the capacitor has been charged, it holds the cathode of the diode positive, and the rectifier will not conduct again until a positive-going secondary voltage at the anode exceeds the positive voltage at the cathode. Hence, it does not begin to conduct until the secondary voltage of the transformer exceeds the charge on the capacitor. Diode current flows until slightly after the peak of the sine wave. This current flow is of short duration but of sufficient amplitude to restore the charge on the filter capacitor. From this brief explanation, we see that only short pulses of current actually flow through the diode – and the transformer. The diode must have a current rating sufficient to safely handle these pulses.

Capacitor input filtering, so called because the capacitor is the first (and only, in this case) filter element the diode "sees," has limitations. The amount of ripple is a function of the size of the capacitor and the value of load resistance. Large values of capacitance (usually several hundred microfarads or more) will reduce the amplitude of the

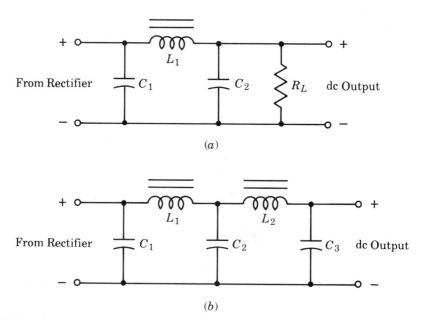

(a)

(b)

Figure 22–16

Capacitive input filters: (a) π type. (b) Double-section filter.

ripple to a low value if the load resistance is fairly high (low load). If the load is relatively large (low resistance), the capacitor discharges more rapidly, resulting in increased ripple amplitude. Therefore, this type of filtering can only be used successfully when the current demands are low.

Improved capacitive input filter systems are shown in Figure 22–16. The device indicated in Figure 22–16a is known as a π type filter and provides better filtering than the circuit of Figure 22–15a. Even better ripple reduction results when another LC section is added (called a double-section filter), as in Figure 22–16b. These filters are used in power supplies for higher output voltage, such as in vacuum tube circuits.

Inductive Input Filter

In large-current applications, inductance or choke input filters are used. A typical circuit is shown in Figure 22–17a; it is sometimes called an L section filter because the L and C filter components form an inverted L.

Because an inductor opposes any change of current, it will react

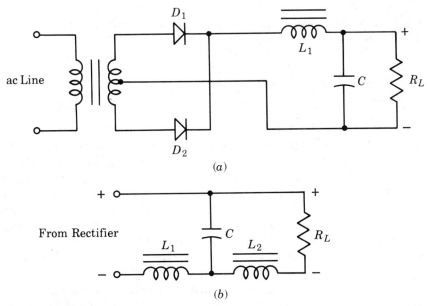

(a)

(b)

Figure 22–17

Inductive input filters: (a) Inverted L type. (b) Inverted T type with filter elements in the negative output lead.

against any change of load current. If the load current falls, the magnetic field about the inductor collapses and induces a voltage in the inductor, which opposes the current change. Likewise, as the load current increases, the magnetic field increases and produces a voltage in the inductor, which will again oppose the current change. Thus, the addition of the inductor prevents the current from building up or dying down quickly. If the inductor is large enough, the load current becomes nearly constant.

Because of the additional filtering action of the capacitance, the ripple amplitude across R_L in Figure 22–17a will be less than it would be if only the inductor were used. The large value of this electrolytic capacitor can also be thought of as bypassing any ripple frequency around R_L that may have passed through the choke. Typically, the reactance of C should be about one-tenth the value of R_L to be effective. Adding an additional inductor will further improve filtering.

For many years, it was customary to place the inductors in the positive side of the power supplies. However, in this arrangement, the insulation between the winding and its core (normally fastened to the chassis and grounded as a safety measure) must be sufficient to withstand the output voltage of the supply. This problem can be eliminated by placing the filter choke(s) in the negative lead of the supply, as in Figure 22–17b. These filters are more likely to be encountered in higher-voltage power supplies where the IR drop across the inductors can be tolerated.

Filter Chokes

There are a number of factors to consider in selecting a filter choke. Some of the more important ones are value of inductance required, dc resistance and current-carrying capacity of the winding, insulation, losses, and distributed capacitance. A choke coil must present a high reactance to the ripple frequency. Thus, the choke should consist of a large number of turns. While such a choke may provide the necessary inductance, it might also create a winding having a high dc resistance, resulting in excessive voltage losses across the coil. Of course, it is possible to use larger wire. However, for a given amount of inductance, larger wire will increase the physical size of the choke.

Choke coils are designed to provide a certain amount of inductance based on a given amount of dc current flowing through the winding. If the current falls below the specified value, the inductance increases. Conversely, if more current flows through the winding than it was designed for, saturation results, with a substantial loss of inductance. Saturation materially reduces the efficiency of the filter.

Laminating the core of the choke reduces eddy currents. Soft iron is used, which has low hysteresis losses. Encasing the transformer in a soft-iron case contains any stray magnetic fields.

Large reactive voltages may be built up across the choke; therefore, adequate insulation is needed. It is customary to have the insulation resistance able to withstand two or three times the normal values of reactive voltages appearing across the choke. For small-power applications, this voltage is about 1500 V dc. For transmitter power supplies, it will be much greater.

Often, a power supply must be capable of furnishing small and large values of current. So that the choke does not saturate under these extreme conditions, the core is generally constructed with a very small air gap. This gap may be only a few thousandths of an inch, but it is sufficient to prevent saturation.

22–8
Peak Inverse Voltage (PIV)

> **Peak inverse voltage (PIV)** The maximum voltage that a diode can experience, when it is reverse-biased, without breaking down.

One of the principal ratings of a diode is its **peak inverse voltage (PIV)**, sometimes called *peak reverse voltage (PRV)*. It represents the maximum voltage that the diode can experience, when reverse-biased, without breaking down.

Refer to Figure 22–18. Each half of the transformer secondary develops 300 V_{rms} = 424.2 V_p. Therefore, the peak voltage that the capacitor will charge to is 424.2 V_p − 0.6 V (barrier voltage of the diode) or 423.6 V_p. At this instant, the anode of D_2 has a negative 424.2 V_p applied by the lower half of the secondary winding. Therefore, D_2 has 424.2 V_p + 423.6 V_p = 847.8 V_p momentarily connected across it. On the alternate half cycle, D_1 is subjected to the same reverse potential. From a practical viewpoint, diodes D_1 and D_2 should have a PIV rating of 1000 V.

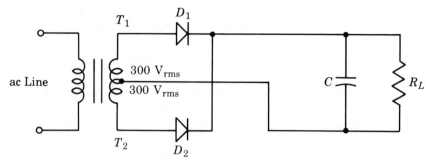

Figure 22–18

Circuit to illustrate PIV

Example 22–2

Determine the peak reverse voltage that the diodes in Figure 22–18 are subject to if each half of the secondary develops 500 V$_{rms}$.

Solution

The total peak voltage across the secondary is $500 \times 2 \times 1.414 = 1414$ V$_p$. Assume the voltage drop across the conducting diode is 0.6 V. Then:

$$PRV = 1414 \text{ V} - 0.6 \text{ V} = 1413.4 \text{ V}$$

22–9
Waveform Analysis

We have previously seen the unfiltered output of a half-wave rectifier (Figure 22–12b). With a capacitive input filter, as in Figure 22–19a, rectifier conduction is somewhat less than one half cycle. After the capacitor has charged to the peak rectified output voltage, diode conduction momentarily stops while the capacitor discharges through R_L. As the voltage rises on the next positive half cycle, a point is reached where the rectified voltage equals the voltage remaining on the capacitor. As the rectified voltage rises beyond that point, the diode begins to conduct and recharge the capacitor until the waveform reaches its peak, as shown. During the brief interval that the diode conducts, it must pass a

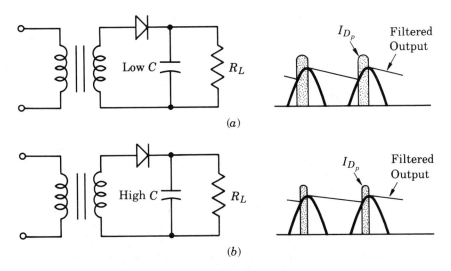

Figure 22–19

Half-wave rectifier circuits: (a) Low-C filter and associated waveform showing large conduction angle. (b) High-C filter and associated waveform showing narrow conduction angle but high pulse.

current equal to that used by the load. Consequently, the pulse of diode current I_{D_p} will be high.

If improved filtering is required (larger C needed) for a given load, the capacitor will not discharge as much as before. Therefore, the diode will conduct for an even shorter period of time, with a resulting larger current pulse. A study of the waveforms in Figures 22–19a and 22–19b reveals these changing conditions.

22–10
Diode Protection Techniques

From the preceding section, it is apparent that diodes must pass very short but powerful pulses of current. When the supply is first turned on, the discharged input capacitor *momentarily* looks like a short circuit, and the rectifier must pass a very large current. This current is known as I_{surge} and is usually rated for one cycle. Some form of protection is generally required to protect the diode(s) until the filter capacitor is charged. The dc resistance of the transformer windings provides some surge-current limiting but is often not enough. Series resistors can be added between the diodes and the transformer secondary winding, but they do not provide good regulation. For high-voltage supplies, as used with transmitter circuits, a surge-limiting resistor can be connected in series with the primary winding, thus eliminating the need for series resistors in the secondary circuit.

Rectifier diodes have relatively small junctions that often must carry many amperes. Considerable heat can be produced, which could destroy the junction. These diodes are usually designed with a stud mounting so that they can be fastened to the chassis by means of a thin insulating washer. This washer must be covered on both sides with a thin layer of thermal joint compound (a form of silicone grease) to promote good heat transfer. High-current rectifiers are usually mounted on special *heat sinks*, which radiate the heat and maintain a safe operating temperature.

Transient voltages can be a great source of trouble. These voltages are pulses of very short duration and high amplitude that cause the rectifiers to experience voltages much greater than normally supplied by the transformer. They usually come from circuits containing inductors that are turned on or off. Transients are often responsible for unexpected or unexplained diode failures.

An effective, yet simple, way to suppress transients is to connect 0.01 µF capacitors across the primary and secondary windings. At power line frequencies, their reactance is on the order of 265 kΩ; but to a voltage transient, that reactance will be an effective bypass. For high-voltage transformers (100 V or more), the capacitors should be 0.001 µF.

In circuits having high PRVs, beyond the rating of a single diode, several diodes of like kind may be connected in series. However, it is necessary to place a resistor and capacitor across each diode in the string, as shown in Figure 22–20a. These components equalize the PRV drops and protect against transients. In determining the ohmic value of these resistors, use the following rule of thumb: Multiply the PRV rating of the diode by 500 Ω. Thus, a 500 PRV diode should be shunted with 250 kΩ.

Some diodes take longer than others to switch from forward conduction to cutoff (high reverse resistance). Placing a 0.01 µF capacitor across each will protect the fast diodes until all are turned off (see Figure 22–20a).

Diodes may be parallel-connected (like kinds) to increase current-handling capacity. However, *equalizing resistors* should be added, as in Figure 22–20b, so that one diode will not hog the load. Choose resistors that will provide about 1 V drop at maximum current.

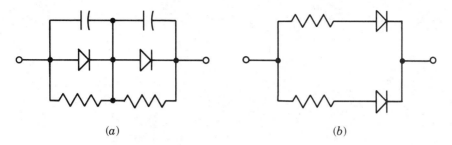

(a) (b)

Figure 22–20

Diode protection techniques: (a) Series string. (b) Parallel diodes.

22–11
Regulation

Bleeder Resistors and Voltage Dividers

A resistor is frequently placed across the output terminals of a high-voltage power supply. It is used to bleed off the charge on the filter capacitor(s) when the rectifier is turned off and to apply a fixed load to the filter and thus improve the *voltage regulation* of the power supply. In the latter case, the resistor is designed to draw at least 10% of the full-load current. Under these conditions, the change of output voltage between no-load and full-load will be substantially reduced. In both conditions, the resistor is called a *bleeder resistor*. If it is tapped to provide several voltages to different parts of a circuit, it is also called a *voltage divider*.

One side of the voltage divider is customarily grounded, as at D in Figure 22–21a. In some applications, it is necessary to have a negative voltage with respect to ground. An example is developing a bias voltage for vacuum tube circuits. This negative voltage can be achieved by grounding a point such as C in Figure 22–21b. The values of resistances R_{AB}, R_{BC}, and R_{CD} would be determined by the loads connected to the different voltage taps.

A voltage divider used to power several circuits from a common supply is shown in Figure 22–22. The following example shows how the different values of resistance can be calculated to furnish the required voltages and currents.

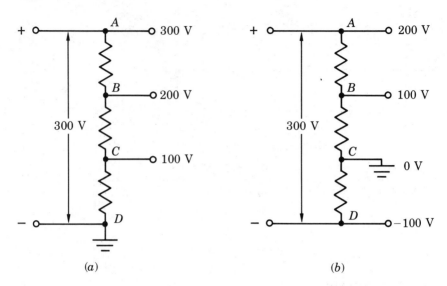

Figure 22–21

Combination bleeder resistor and voltage divider: (a) With ground at point D. (b) With ground at point C.

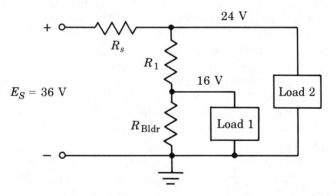

Figure 22–22

Voltage divider network with two loads

Example 22–3

Design a voltage divider circuit, as in Figure 22-22, to furnish the required voltages and currents if load 1 is 30 mA, load 2 is 70 mA, and the bleeder current is 10% of the load currents.

Solution

Calculate the bleeder resistance.

$$R_{\text{Bldr}} = \frac{16 \text{ V}}{10 \text{ mA}} = 1.6 \text{ k}\Omega$$

Then

$$R_1 = \frac{24 \text{ V} - 16 \text{ V}}{I_{\text{Bldr}} + I_{L_1}} = \frac{8 \text{ V}}{40 \text{ mA}} = 200 \ \Omega$$

$$R_s = \frac{36 \text{ V} - 24 \text{ V}}{I_{\text{Bldr}} + I_{L_1} + I_{L_2}} = \frac{12 \text{ V}}{110 \text{ mA}} = 109 \ \Omega$$

The wattage ratings of these resistors are as follows:

$$P_{R_{\text{Bldr}}} = EI = 16 \text{ V} \times 0.01 \text{ A} = 160 \text{ mW} \quad (\text{use } 0.5 \text{ W})$$
$$P_{R_1} = EI = 8 \text{ V} \times 0.04 \text{ A} = 320 \text{ mW} \quad (\text{use } 1 \text{ W})$$
$$P_{R_s} = EI = 12 \text{ V} \times 0.11 \text{ A} = 1.32 \text{ W} \quad (\text{use } 5 \text{ W})$$

Voltage Regulation

All power sources have some internal resistance. Current drawn from the supply produces a varying IR drop across this resistance as the load changes. Consequently, the output voltage fluctuates. Many electronic circuits require a stable operating voltage even though the current requirements vary. A simple voltage regulator can solve this problem by using a zener diode.

A typical zener regulator is shown in Figure 22–23a. (Zener diode characteristics were discussed in Section 22–5.) The ability of these diodes to stabilize a voltage depends on their varying impedance under different zener currents. Zener impedance may vary from an ohm or less in low-voltage, high-power units to a thousand ohms in high-voltage, low-power diodes.

The value of R_s in Figure 22–23a is determined by the load requirements. If R_s is too small, the diode dissipation rating may be exceeded at low values of load current. If it is too large, the diode will not regulate at large values of I_L (load current).

A recommended value for R_s may be calculated as follows (see Figure 22–23a):

$$R_s = \frac{E_{\text{supply (min)}} - V_z}{1.1 I_{L \text{ (max)}}}$$

22–1

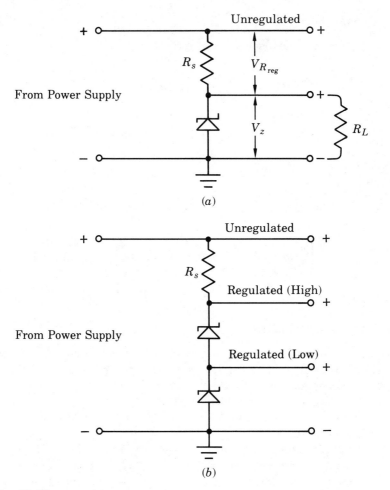

Figure 22–23

Zener diode voltage regulators: (a) Single regulated source. (b) High and low regulated source.

Example 22–4

Assume that the supply voltage is 30 V (min), a regulated 10 V output is required, and $I_{L_{(max)}} = 40$ mA. Calculate the value of R_s.

Solution

$$R_s = \frac{30 \text{ V} - 10 \text{ V}}{1.1 \times 40 \text{ mA}} = \frac{20 \text{ V}}{44 \text{ mA}} = 455 \ \Omega$$

The power dissipation (P_D) of the zener is as follows:

$$P_D = V_z I_z = 10 \times 4 = 40 \text{ mW}$$

By *cascading*, or connecting two zeners in series, we can obtain two separate, regulated output voltages, as in Figure 22–23b. They need not have the same breakdown voltages because the arrangement is self-equalizing. It is necessary, however, that their current-handling capacities be compatible. Calculation of R_s is as shown in Equation 22–1, except that V_z and I_L must be the sum of the zener voltages and load currents, respectively.

The regulation of a *power supply* is a measure of its ability to maintain a constant load voltage (or current) despite changes in line voltage or load impedance. Expressed as a formula, the percentage of regulation is as follows:

$$\% \text{ regulation} = \frac{E_{NL} - E_{FL}}{E_{FL}} \times 100 \qquad \textbf{22–2}$$

where
$$E_{NL} = \text{no-load voltage}$$
$$E_{FL} = \text{full-load voltage}$$

Example 22–5

What is the percentage of regulation of a power supply with a no-load voltage of 126.5 V and a full-load voltage of 115 V?

Solution

$$\% \text{ regulation} = \frac{126.5 - 115}{115} \times 100 = 10\%$$

Example 22–6

If a power supply has an output voltage of 140 V at no load and the regulation at full load is 15%, what is the output voltage at full load?

Solution

Start with the basic formula and transpose:

$$\% \text{ regulation} = \frac{E_{NL} - E_{FL}}{E_{FL}} \times 100$$

$$\% \text{ regulation} \times E_{FL} = 100E_{NL} - 100E_{FL}$$

$$\% \text{ regulation} \times E_{FL} + 100E_{FL} = 100E_{NL}$$

$$E_{FL} = \frac{100\,E_{NL}}{\% \text{ regulation} + 100} = \frac{14{,}000}{115} = 121.7 \text{ V}$$

Power supplies having poor regulation have relatively high internal resistance. A well-designed, solid-state, 12 V power supply might have an internal impedance of 0.012 Ω and a voltage regulation of 0.01% for a 100 mA load change.

Regulated Power Supplies

Many circuits require stable dc voltages even though the load fluctuates. This condition can be satisfied by using a *regulated power supply*. In past years, these supplies were made from discrete components (separate transistors, resistors, and so on). Today, it is more practical and economical to build a supply by using an *integrated circuit* (IC).

A typical regulator is the LM317 shown in Figure 22–24. It is a three-terminal, positive-voltage regulator connected in series with the output. It is capable of supplying in excess of 1.5 A over an output of 1.2 V to 37 V, depending on the kind of power transformer used. In the basic circuit of Figure 22–24, only two external resistors are required to set the output voltage. Extremely good load regulation (abil-

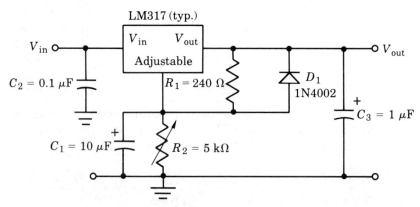

Figure 22–24

Basic adjustable voltage regulator with improved ripple rejection (Courtesy National Semiconductor)

ity to maintain a constant output voltage under changing load conditions) results from this simple circuit. Resistor R_1 is nominally 240 Ω and should be connected as close to the regulator as possible, preferably directly across terminals V_{out} and adjustable.

Input bypass capacitor C_2 is included in the circuit of Figure 22–24 to minimize the possibility of extraneous signals getting into the regulator circuit. A 0.1 μF disc or 1 μF solid-tantalum capacitor is suitable for this application. Likewise, C_3 should preferably be a tantalum capacitor and is needed to provide circuit stability. Solid-tantalum capacitors have low impedance even at high frequencies. Depending on capacitor construction, it takes about 25 μF of aluminum electrolyte to equal 1 μF solid tantalum at high frequencies.

It is necessary to bypass the adjustment terminal to ground to improve *ripple rejection*. Capacitor C_1 prevents any of the output voltage ripple from getting into the regulator and being amplified as the output voltage is increased. With a 10 μF bypass capacitor, 80 dB ripple rejection is obtainable at any output voltage.

When external capacitors are used with any IC regulator, it is a good idea to add protective diodes to prevent the capacitors from discharging through low-current points into the regulator. For example, most 10 μF capacitors have low enough internal series resistance to deliver 20 A spikes when shorted. This energy is sufficient to damage the IC. In Figure 22–24, D_1 will discharge C_1 if the output is shorted to ground.

A complete schematic of an adjustable voltage-regulated power supply is shown in Figure 22–25. For the kind of power trans-

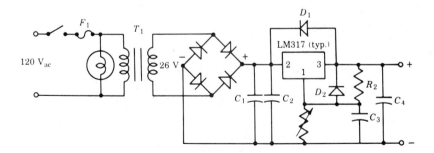

Figure 22–25

Complete power supply circuit with adjustable voltage regulator

former shown (T_1), the power supply will deliver approximately 22 V at 1 A. Up to 35 V dc at 1.5 A, output can be achieved by selecting a power transformer that delivers a proportionately higher secondary voltage.

The following parts list applies to the circuit in Figure 22–25:

Diode bridge	2 A at 100 PIV (min), Motorola HEPRO 853 or equivalent
C_1	4000 to 5000 μF at 40 W V dc
C_2 and C_4	1 μF at 40 W V dc, solid tantalum
C_3	10 μF at 40 W V dc
D_1 and D_2	1N4002 or equivalent
Fuse F_1	2 A, slo-blo
Potentiometer R_1	3 kΩ, 2 W, W. W.
Resistor R_2	240 Ω, 1 W

The LM317 must be mounted to a suitable heat sink and the case electrically insulated from the sink.

22–12
Voltage Multipliers

One way to obtain high voltage is to use step-up transformers. However, they are expensive and relatively large and heavy. An alternative method is to use a *voltage multiplier*. Depending on design, they can supply dc voltages that are several times that of the ac power line.

A typical *voltage doubler* circuit is shown in Figure 22–26a. Essentially, the circuit is a combination of two half-wave rectifiers. Assume that the instantaneous polarity of the secondary winding is as shown in Figure 22–26b. Current will flow through D_1 in the direction of the arrow and charge C_1 to $1.4E_{rms}$. When the ac reverses, as in Figure 22–26c, output capacitor C_2 is charged through D_2. The voltages available to charge this capacitor are the peak value of the ac plus the dc charge on C_1, which are in series at this moment. The actual current-charging paths for each half cycle are shown by the heavy lines in Figures 22–26b and 22–26c. Hence, C_2 charges to $2.8E_{rms}$.

The peak inverse voltage on the diodes is about equal to twice the value of the peak input voltage. Typical values of capacitors range from 40 to over 100 μF each. Their working voltage ratings must be at

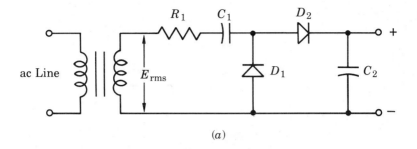

(a)

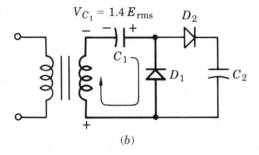

(b)

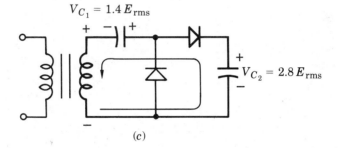

(c)

Figure 22–26

Typical voltage doubler: (a) Circuit. (b) Charging action on one half cycle. (c) Charging action on second half cycle.

least equal to the peak value of voltage appearing across them. Larger capacitors provide better voltage regulation and higher output voltages.

When large capacitors are used, the peak rectifier current may be very high. The surge is greatest when the supply is turned on. If the multiplier should be turned on at the instant the ac line is at its peak, the surge may damage the diodes. Adding a low-value, surge-limiting resistor (R_1 in Figure 22–26a) keeps the current surge at a safe value.

22–13
Three-Phase Power Supplies

Commercial transmitting stations use three-phase (3ϕ) power to operate their power supplies. Several advantages accrue through this technique:

1. A 60 Hz, 3ϕ, half-wave rectifier has a ripple frequency of 180 Hz, while a 60 Hz, full-wave rectifier has 360 Hz. Higher ripple frequencies are easier to filter (require smaller L and C components).
2. The rectifier current pulses overlap in 3ϕ circuits. Since the rectifier output current supplied to the filter never drops to zero, the filtering requirements are simplified.

A number of different circuit combinations are possible in polyphase systems, depending on the particular kind of transformer(s) used. They may be conveniently subdivided into systems giving ripple frequencies that are three and six times the line frequencies.

A three-phase, half-wave rectifier circuit is shown in Figure 22–27. Transformer T_2 is Δ to Y (delta-to-wye) connected, which provides the optimum step-up voltage ratio between primary and secondary windings for a given turns ratio. Switch SW1 is the filament power switch connecting single-phase power to the primary of the step-down filament transformer T_1. It should be turned on sufficiently ahead of high-voltage power switch SW2 to permit the rectifier tubes to come up to operating temperature. The ripple frequency in this circuit is three times the power line frequency.

A six-phase output wave is obtained with the Δ to Y connection shown in Figure 22–28. Here, the ripple is six times the supply frequency. Known as the six-phase, single-Y circuit, it makes the most efficient use of the transformer of any of the polyphase circuits and has the lowest inverse voltage.

Three-phase transformers may have their primary windings connected to delta or wye (sometimes called *star*). Similarly, their secondaries can be Δ or Y connected. If the transformers have a 1:1 turns ratio, there will be no step-up or step-down of voltage or current in either the Δ to Δ or Y to Y connections. However, if these same transformers have their primaries Δ connected and their secondaries Y connected, each output phase will have 1.73 times the primary phase voltages. If the transformers have 1:10 step-up ratios, the secondary voltages will be 17.3 times the primary voltage.

What happens if the primary is Y connected and the secondary

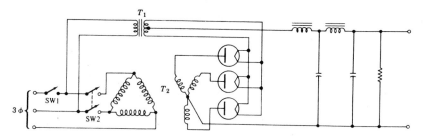

Figure 22–27

Three-phase, half-wave, delta-to-wye–connected power supply

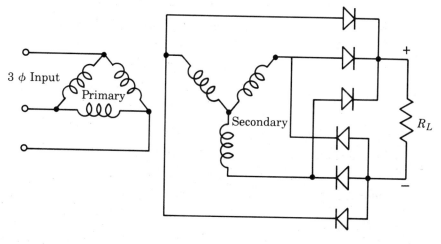

Figure 22–28

Six-phase, single-Y rectifier circuit using a delta-to-wye primary-to-secondary connection

is Δ connected? Each output phase will deliver a voltage equal to the reciprocal of 1.73, or 0.578, times the input voltage. However, the current in each phase of the output will be 1.73 times the current in each primary phase.

Example 22–7

Three single-phase transformers, each with a ratio of 220 to 2200 V, are connected across a 200 V, 3φ line, primaries in delta.

If the secondaries are wye-connected, what is the secondary line voltage?

Solution

In the Δ to Y connected, 3ϕ system, the secondary line voltage can be determined by $E_s = E_p \times$ turns ratio $\times$ 1.73. Therefore:

$$E_s = 220 \times 10 \times 1.73 = 3806 \text{ V}$$

Summary

Low-power requirements are often satisfied with half-wave rectifiers; full-wave rectifiers are used when more power is needed.

Rectifiers are used to change ac to dc.

Semiconductor material used for diodes is either silicon (Si) or germanium (Ge).

In N type semiconductors, the current majority carriers are electrons.

In P type semiconductors, the current majority carriers are holes.

A barrier voltage of approximately 0.6 V appears across a PN Si junction, and about 0.3 V appears across a Ge junction.

A PN junction is called a diode.

A diode is like a one-way check valve, allowing current to pass in one direction but not the other.

When a voltage of the correct polarity is applied to a PN junction, current flows as soon as the barrier (junction) voltage is overcome.

Zener diodes are used as voltage regulators and voltage reference devices.

Zener diodes operate in the reverse-bias region.

In a half-wave rectifier, the peak voltage across the load is approximately $1.4E_{rms}$.

The average value of each half-wave pulse is 0.318 of the peak amplitude. This average is equal to the dc component of the wave.

Full-wave rectifiers utilize both halves of the ac waveform.

In full-wave rectifiers, average current through the load is 0.638 of the peak current.

Bridge rectifiers involve four diodes and do not require center-tapped secondary windings on the power transformer.

The purpose of a filter is to remove any voltage fluctation in the power supply output.

Power supply filters are low-pass filters since they pass dc and attenuate the ripple frequency.

The simplest filter involves a large capacitor across the power supply output.

Peak inverse voltage represents the maximum voltage a diode can experience, when reverse-biased, without breaking down.

On larger diodes, heat sinks are required to radiate the heat generated.

A bleeder resistor is frequently connected across the output terminals of a power supply to bleed off the charge on the filter capacitor(s) when the rectifier is turned off and to improve voltage regulation.

The regulation of a power supply is a measure of its ability to maintain a constant output voltage despite changes in line voltage or load.

Power supplies with low internal resistance have good regulation.

A voltage doubler circuit is a combination of two half-wave rectifiers.

Three-phase power supplies are used in commercial transmitting stations.

Progress Test

The bracketed number after each question indicates the section of this chapter where the answer can be found.

1. Which of the following statements is true regarding diodes? (a) They change dc to ac. (b) They pass both halves of an ac wave. (c) They pass one-half of an ac wave. (d) They furnish ac voltages for different parts of an electronic circuit. [22–1]

2. Which statement is true regarding semiconductors? (a) They are made of either Si or Ge. (b) They are good insulators. (c) They are used in their intrinsic state. (d) They are good conductors. [22-1]

3. The element silicon becomes a P type semiconductor when: (a) it is doped with a pentavalent element. (b) controlled amounts of a trivalent element are added to it. (c) it forms a crystal lattice. (d) ac is passed through it. [22-1]

4. The majority carriers in N type semiconductors are: (a) electrons. (b) holes. (c) leakage currents. (d) ions. [22-1]

5. Which statement is correct regarding a silicon diode? (a) It consists of an NPN junction. (b) It has a barrier voltage of 0.3 V. (c) It has no depletion region. (d) It has a junction potential of 0.6 V. [22-2]

6. A PN junction: (a) is called an anode. (b) is called a diode. (c) passes current in both directions. (d) must be reverse-biased to pass current. [22-3]

7. What is the schematic symbol for a rectifier diode? [22-3]

8. Zener diodes: (a) are used as rectifiers. (b) have thick depletion regions. (c) pass current equally well in both directions. (d) are voltage reference devices. [22-5]

9. Which statement is correct regarding half-wave rectifiers? (a) They convert dc to ac. (b) Conduction takes place when the diode is reverse-biased. (c) They have an output frequency equal to twice the line frequency. (d) Peak voltage across the load will be approximately $1.4E_{rms}$ of the power transformer secondary. [22-6]

10. The average value of each half-wave pulse is what value of the peak amplitude? (a) 0.707 (b) 0.318 (c) $1.4E_{rms}$ (d) 0.9 [22-6]

11. The average value of the current through the load resistor in an unfiltered full-wave rectifier is: (a) 0.638 of peak current. (b) 0.9 of peak current. (c) 0.318 of primary current. (d) $1.4 E_{rms}$. [22-6]

12. The ripple frequency in a full-wave rectifier is: (a) 60 Hz. (b) 120 Hz. (c) 180 Hz. (d) 240 Hz. [22-6]

13. One advantage of a bridge rectifier over a full-wave, two-diode rectifier is that: (a) only one-half the dc output is needed. (b) it has a higher ratio of peak inverse voltage to average output voltage. (c) the secondary winding of the power transformer does not have to be center-tapped. (d) lower-amperage diodes can be used. [22-6]

14. A filter is needed in a power supply to: (a) provide an essentially constant dc voltage. (b) change some of the dc to ac. (c) act as a high-pass filter. (d) protect the diodes. [22-7]

15.   Because of the charge on the filter capacitor(s), diode current flows for: (a) a small conduction angle. (b) nearly 180° of the ac cycle. (c) more than 90° but less than 180° of the ac cycle. (d) more than 180° but less than 270°.   [22–7]

16.   Inductors can be used in filter circuits to: (a) oppose any changes in ripple voltage. (b) increase the current flow through the diodes. (c) eliminate the need for filter capacitors. (d) oppose changes in load current.   [22–7]

17.   Peak inverse voltage of a diode: (a) represents the maximum voltage across the diode when it is conducting. (b) is the maximum voltage a diode can withstand, when it is reverse-biased, without breaking down. (c) equals the rms voltage across the transformer secondary. (d) equals the peak voltage of the transformer secondary.   [22–8]

18.   Which of the following statements is correct regarding diode protection? (a) They require large-ohmic-value, current-limiting resistors. (b) High-value resistors should be connected across the diodes. (c) Diode heat sinks are required when large currents are involved. (d) Capacitors of 0.01 μF should be connected in series with the diodes to suppress voltage transients.   [22–10]

19.   Voltage regulation: (a) is improved when large series resistance is used. (b) is a measure of a power supply's ability to maintain a stable output under changing load. (c) provides output changes when line voltage fluctuates. (d) provides high output impedance of the power supply.   [22–11]

┌──Problems

Answers to odd-numbered problems are at the back of the book.

1.   What is the dc resistance of a diode when $V_F = 0.62$ V and $I_F = 2.75$ mA?

2.   If 0.015 A flows through a diode when 0.61 V is applied, what is its resistance?

3.   Calculate the static (dc) resistance of a diode when a forward voltage of 0.68 V produces 31.3 mA.

4.   What is the dynamic (ac) resistance of a diode when a forward voltage of 0.59 V causes 0.72 mA and 0.61 V causes 1.08 mA?

5. The following measurements were made on a silicon diode: $I_F = 26.2$ mA at $V_F = 0.67$ V and $I_F = 19.8$ mA at $V_F = 0.64$ V. Calculate its ac resistance.

6. What is the dc resistance of a reverse-biased diode when $V_R = 50$ V and $I_F = 32$ μA?

7. Calculate the reverse-bias resistance of a diode when the leakage current is 13.6 μA at $V_R = 600$ V.

8. Refer to the basic schematic in Figure 22-11b. Calculate the zener and load currents if $R_L = 1$ kΩ, $E_S = 28$ V dc, $V_z = 12$ V, and $R_s = 1$ kΩ.

9. Using the circuit in Figure 22-11b, calculate R_s, I_{R_L}, and I_z when $E_S = 36$ V, $R_L = 2.2$ kΩ, $I_T = 10.833$ mA, and $V_z = 16.5$ V.

10. The secondary voltage of a step-down transformer connected to a half-wave rectifier is 28 V. What is the approximate peak voltage across the load?

11. What is the average value of the half-wave pulse developed across the load described in Problem 10?

12. A step-up transformer has 360 V_{rms} across its secondary winding. This transformer is connected to a half-wave rectifier supplying a 1 kΩ load. (a) What is the peak value of the voltage? (b) What is the average value? (c) What is the peak current pulse delivered to the load?

13. A full-wave rectifier circuit is connected to a center-tapped transformer with 24 V_{rms} on each side of the center tap (see Figure 22-13a). Calculate the average voltage across 1 kΩ rectifier load.

14. A transformer secondary winding with 60 V on each side of the center tap is connected to two diodes in a full-wave rectifier configuration. The load resistance is 500 Ω. Calculate the following: (a) V_{R_L} (peak) (b) E_{R_L} (average) (c) I_{R_L} (peak) (d) I_{R_L} (average).

15. A full-wave rectifier circuit, similar to the circuit of Figure 22-18, has 80 V on each side of the center tap. What is the PIV across each diode?

16. A full-wave rectifier circuit is connected to a transformer with 360 V across each half of its secondary. Calculate the PIV across each diode (refer to the circuit in Figure 22-18).

17. Design a voltage divider similar to the circuit in Figure 22-22. The specifications are as follows: $E_S = 48$ V, load 2 is 42 V at 90 mA, and load 1 is 24 V at 16 mA. Bleeder current is 10% of I_{L_T}. Calculate the following: (a) R_s (b) R_1 (c) R_{Bldr}.

18.   Design a voltage divider similar to the divider of Figure 22–22, with $E_S = 80$ V, load 1 = 24 V at 20 mA, load 2 = 72 V at 150 mA, and $I_{Bldr} = 10\%$ of I_{L_T}. Calculate: (a) R_{Bldr} (b) R_1 (c) R_s.

19.   Calculate the value of R_s in a zener regulator where $V_z = 12$ V, $E_s = 30$ V, and maximum load current is 40 mA.

20.   What value of R_s is necessary in a zener regulator where $E_S = 36$ V, $V_z = 16$ V, and maximum load current is 55 mA?

21.   Calculate the percentage of regulation of a power supply with no-load voltage of 24 V and a full-load voltage of 23.85 V.

22.   A power supply has the following characteristics: $V_{FL} = 35.24$ V and $V_{NL} = 36$ V. What is its percentage of regulation?

23.   If a power supply has an output voltage of 120 V at no load and 3.5% regulation, what is its output voltage at full load?

24.   Three single-phase transformers, each having a turns ratio of 1:8, are delta-connected to a 240 V, 3ϕ power line. If the secondaries are wye-connected, what is the secondary line voltage?

Appendixes

A
Powers of 10, or Scientific Notation

A convenient method of writing very large or very small numbers is by means of scientific notation, or powers of 10. For example, $10 \times 10 \times 10$ may be written as 10^3. Any multiplication of equal factors may be expressed in this way. The exponent [3] tells how many times to use 10 as a factor. Thus, in the statement $10^3 = 1000$, 10 is the *base*, 1000 is the *power*, and [3] is the *exponent*. The number 10^3 is read *ten to the third power*, or simply *ten to the third*.

A number containing many zeros to the right or left of the decimal point can be expressed as in the following examples.

$$6{,}280{,}000 = 6.28 \times 10^6$$
$$0.0000037 = 3.7 \times 10^{-6}$$

Rule: To express a large number as a power of 10, move the decimal point to the *left* and count the number of places the decimal point is moved. The number of places counted indicates the power of 10.

Examples

$$735 = 7.35 \times 10^2$$
$$82{,}357 = 8.2357 \times 10^4$$

Note: The decimal point should be placed so that the number to the *left* of the decimal is between 1 and 10.

Rule: To express a decimal fraction as a whole number times a power of 10, move the decimal to the right and count the number of places it is moved. The number of places moved will be the proper negative power of 10.

Examples

$$0.00637 = 6.37 \times 10^{-3}$$
$$0.0000045 = 4.5 \times 10^{-6}$$

Multiplication with Powers of 10

When multiplying numbers written in powers of 10, simply multiply the base numbers and algebraically add the exponents.

Examples ──

$$10^3 \times 10^4 = 10^7$$
$$36,000 \times 4000 = 3.6 \times 10^4 \times 4 \times 10^3 = 1.44 \times 10^8$$

$$10^{-4} \times 10^{-7} = 10^{-11}$$
$$10^3 \times 10^{-5} = 10^{-2}$$

Division with Powers of 10

When dividing numbers written in powers of 10, simply subtract the exponent in the denominator from the exponent in the numerator.

Examples ──

$$\frac{10^4}{10^2} = 10^{4-2} = 10^2$$

$$\frac{64,000}{0.008} = \frac{6.4 \times 10^4}{8 \times 10^{-3}} = 0.8 \times 10^{4+3} = 8 \times 10^6$$

B
Greek Alphabet and Common Designations

Name	Capital	Lowercase	Used to Designate
Alpha	A	α	Angles, area, coefficients
Beta	B	β	Angles, flux density, coefficients
Gamma	Γ	γ	Conductivity, specific gravity
Delta	Δ	δ	Variation, density
Epsilon	E	ϵ	Base of natural logarithms
Zeta	Z	ζ	Impedance, coefficients, coordinates
Eta	H	η	Hysteresis coefficient, efficiency
Theta	Θ	θ	Temperature, phase angle
Iota	I	ι	
Kappa	K	κ	Dielectric constant, susceptibility
Lambda	Λ	λ	Wave length
Mu	M	μ	Micro, amplification factor, permeability
Nu	N	ν	Reluctivity
Xi	Ξ	ξ	
Omicron	O	o	
Pi	Π	π	Ratio of circumference to diameter $= 3.1416$
Rho	P	ρ	Resistivity
Sigma	Σ	σ	Sign of summation
Tau	T	τ	Time constant, time phase displacement
Upsilon	Υ	υ	
Phi	Φ	ϕ	Magnetic flux, angles
Chi	X	χ	
Psi	Ψ	ψ	Dielectric flux, phase difference
Omega	Ω	ω	Capital: ohms; lowercase: angular velocity

C
MILSPEC and Industrial-Grade Resistor Coding for Hot-Molded Fixed Resistors

EXPLANATION OF PART NUMBERS

All Allen-Bradley fixed composition resistors are identified by a Part Number which will provide information as to the type of resistor, resistance value, and tolerance. The Part Number is merely for identification on drawings, specifications, ordering, and other areas where it is convenient to use a Part Number to describe a particular resistor. The only markings that appear on the resistor are the Color Code bands.

INDUSTRIAL TYPE DESIGNATION ⟶ EB5145

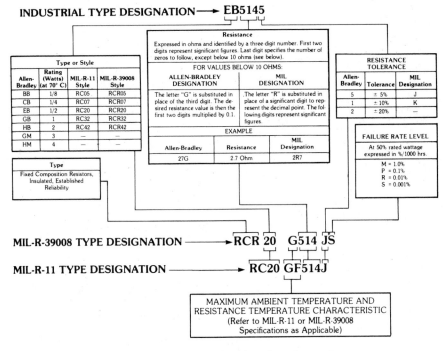

Type or Style			
Allen-Bradley	Rating (Watts) (at 70° C)	MIL-R-11 Style	MIL-R-39008 Style
BB	1/8	RC05	RCR05
CB	1/4	RC07	RCR07
EB	1/2	RC20	RCR20
GB	1	RC32	RCR32
HB	2	RC42	RCR42
GM	3	—	—
HM	4	—	—

Resistance

Expressed in ohms and identified by a three-digit number. First two digits represent significant figures. Last digit specifies the number of zeros to follow, except below 10 ohms (see below).

FOR VALUES BELOW 10 OHMS:

ALLEN-BRADLEY DESIGNATION	MIL DESIGNATION
The letter "G" is substituted in place of the third digit. The desired resistance value is then the first two digits multiplied by 0.1.	The letter "R" is substituted in place of a significant digit to represent the decimal point. The following digits represent significant figures.

EXAMPLE

Allen-Bradley	Resistance	MIL Designation
27G	2.7 Ohm	2R7

RESISTANCE TOLERANCE		
Allen-Bradley	Tolerance	MIL Designation
5	± 5%	J
1	± 10%	K
2	± 20%	—

FAILURE RATE LEVEL

At 50% rated wattage expressed in %/1000 hrs.

M = 1.0%
P = 0.1%
R = 0.01%
S = 0.001%

Type

Fixed Composition Resistors, Insulated, Established Reliability

MIL-R-39008 TYPE DESIGNATION ⟶ RCR 20 G514 JS

MIL-R-11 TYPE DESIGNATION ⟶ RC20 GF514J

MAXIMUM AMBIENT TEMPERATURE AND RESISTANCE TEMPERATURE CHARACTERISTIC
(Refer to MIL-R-11 or MIL-R-39008 Specifications as Applicable)

Standard color code and preferred number series

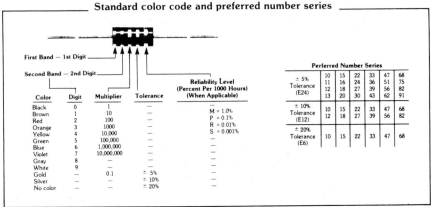

First Band — 1st Digit
Second Band — 2nd Digit

Color	Digit	Multiplier	Tolerance	Reliability Level (Percent Per 1000 Hours) (When Applicable)
Black	0	1	—	—
Brown	1	10	—	M = 1.0%
Red	2	100	—	P = 0.1%
Orange	3	1000	—	R = 0.01%
Yellow	4	10,000	—	S = 0.001%
Green	5	100,000	—	—
Blue	6	1,000,000	—	—
Violet	7	10,000,000	—	—
Gray	8	—	—	—
White	9	—	—	—
Gold	—	0.1	± 5%	—
Silver	—	—	± 10%	—
No color	—	—	± 20%	—

Perferred Number Series						
± 5% Tolerance (E24)	10	15	22	33	47	68
	11	16	24	36	51	75
	12	18	27	39	56	82
	13	20	30	43	62	91
± 10% Tolerance (E12)	10	15	22	33	47	68
	12	18	27	39	56	82
± 20% Tolerance (E6)	10	15	22	33	47	68

Courtesy of Allen-Bradley Co.

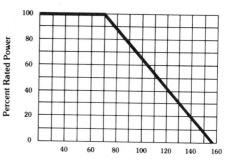

Ambient Temperature ° C

Marking

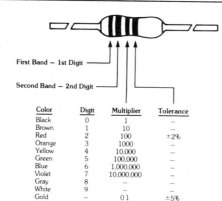

First Band — 1st Digit

Second Band — 2nd Digit

Color	Digit	Multiplier	Tolerance
Black	0	1	—
Brown	1	10	—
Red	2	100	±2%
Orange	3	1000	—
Yellow	4	10,000	—
Green	5	100,000	—
Blue	6	1,000,000	—
Violet	7	10,000,000	—
Gray	8	—	—
White	9	—	—
Gold	—	0.1	±5%

Standard Resistance Values
±5%, ±2%

10	15	22	33	47	68
11	16	24	36	51	75
12	18	27	39	56	82
13	20	30	43	62	91

Notes:
1. All resistors are marked with Standard EIA color code.

EXPLANATION OF PART NUMBERS

Part Numbering — Example: RND103J

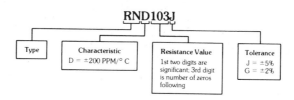

RND103J

| Type | Characteristic
D = ±200 PPM/° C | Resistance Value
1st two digits are significant; 3rd digit is number of zeros following | Tolerance
J = ±5%
G = ±2% |

Courtesy of Allen-Bradley Co.

Type **RNK**
Metal Film
Fixed
Resistor

Marking

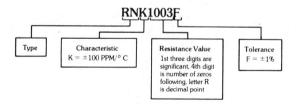

First Band —
   1st Digit

Second Band —
   2nd Digit

Third Band — 3rd Digit

Color	Digit	Multiplier	Tolerance
Black	0	1	—
Brown	1	10	±1%
Red	2	100	—
Orange	3	1000	—
Yellow	4	10.000	—
Green	5	100.000	—
Blue	6	1.000.000	—
Violet	7	10.000.000	—
Gray	8	—	—
White	9	—	—
Gold	—	0.1	—

STANDARD RESISTANCE VALUES — 1% (E96)

10.0	14.7	21.5	31.6	46.4	68.1
10.2	15.0	22.1	32.4	47.5	69.8
10.5	15.4	22.6	33.2	48.7	71.5
10.7	15.8	23.2	34.0	49.9	73.2
11.0	16.2	23.7	34.8	51.1	75.0
11.3	16.5	24.3	35.7	52.3	76.8
11.5	16.9	24.9	36.5	53.6	78.7
11.8	17.4	25.5	37.4	54.9	80.6
12.1	17.8	26.1	38.3	56.2	82.5
12.4	18.2	26.7	39.2	57.6	84.5
12.7	18.7	27.4	40.2	59.0	86.6
13.0	19.1	28.0	41.2	60.4	88.7
13.3	19.6	28.7	42.2	61.9	90.9
13.7	20.0	29.4	43.2	63.4	93.1
14.0	20.5	30.1	44.2	64.9	95.3
14.3	21.0	30.9	45.3	66.5	97.6

Notes:
1. All resistors are marked with Standard EIA color code.

EXPLANATION OF PART NUMBERS

Part Numbering — Example: RNK1003F

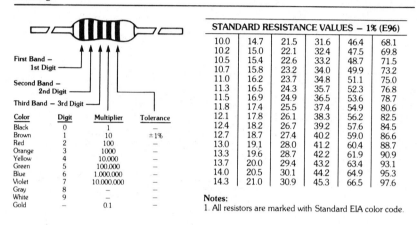

RNK1003F

Type	Characteristic	Resistance Value	Tolerance
	K = ±100 PPM/° C	1st three digits are significant, 4th digit is number of zeros following, letter R is decimal point	F = ±1%

D
Capacitor Codes

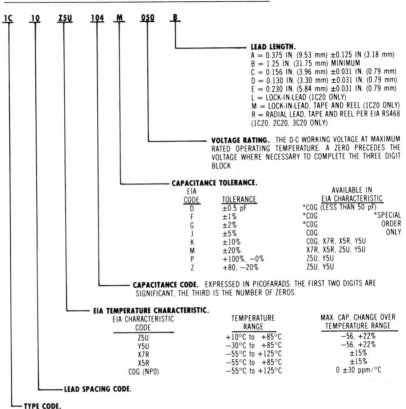

ALWAYS SPECIFY BY COMPLETE CATALOG NUMBER, SUCH AS:

1C   10   Z5U   104   M   050   B

LEAD LENGTH.
A = 0.375 IN. (9.53 mm) ±0.125 IN (3.18 mm)
B = 1.25 IN. (31.75 mm) MINIMUM
C = 0.156 IN. (3.96 mm) ±0.031 IN. (0.79 mm)
D = 0.130 IN. (3.30 mm) ±0.031 IN. (0.79 mm)
E = 0.230 IN. (5.84 mm) ±0.031 IN. (0.79 mm)
L = LOCK-IN-LEAD (1C20 ONLY)
M = LOCK-IN-LEAD, TAPE AND REEL (1C20 ONLY)
R = RADIAL LEAD, TAPE AND REEL PER EIA RS468
(1C20, 2C20, 3C20 ONLY)

VOLTAGE RATING. THE D-C WORKING VOLTAGE AT MAXIMUM RATED OPERATING TEMPERATURE. A ZERO PRECEDES THE VOLTAGE WHERE NECESSARY TO COMPLETE THE THREE DIGIT BLOCK

CAPACITANCE TOLERANCE.

EIA CODE	TOLERANCE	AVAILABLE IN EIA CHARACTERISTIC	
D	±0.5 pF	*COG (LESS THAN 50 pF)	
F	±1%	*COG	*SPECIAL
G	±2%	*COG	ORDER
J	±5%	COG	ONLY
K	±10%	COG, X7R, X5R, Y5U	
M	±20%	X7R, X5R, Z5U, Y5U	
P	+100%, −0%	Z5U, Y5U	
Z	+80, −20%	Z5U, Y5U	

CAPACITANCE CODE. EXPRESSED IN PICOFARADS. THE FIRST TWO DIGITS ARE SIGNIFICANT, THE THIRD IS THE NUMBER OF ZEROS.

EIA TEMPERATURE CHARACTERISTIC.

EIA CHARACTERISTIC CODE	TEMPERATURE RANGE	MAX. CAP. CHANGE OVER TEMPERATURE RANGE
Z5U	+10°C to +85°C	−56, +22%
Y5U	−30°C to +85°C	−56, +22%
X7R	−55°C to +125°C	±15%
X5R	−55°C to +85°C	±15%
COG (NP0)	−55°C to +125°C	0 ±30 ppm/°C

LEAD SPACING CODE.

TYPE CODE.

718
Appendix D

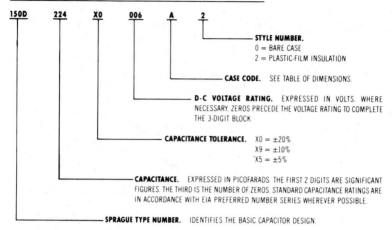

| 150D | 224 | X0 | 006 | A | 2 |

STYLE NUMBER.
0 = BARE CASE
2 = PLASTIC-FILM INSULATION

CASE CODE. SEE TABLE OF DIMENSIONS.

D-C VOLTAGE RATING. EXPRESSED IN VOLTS. WHERE NECESSARY, ZEROS PRECEDE THE VOLTAGE RATING TO COMPLETE THE 3-DIGIT BLOCK.

CAPACITANCE TOLERANCE.
X0 = ±20%
X9 = ±10%
X5 = ±5%

CAPACITANCE. EXPRESSED IN PICOFARADS. THE FIRST 2 DIGITS ARE SIGNIFICANT FIGURES. THE THIRD IS THE NUMBER OF ZEROS. STANDARD CAPACITANCE RATINGS ARE IN ACCORDANCE WITH EIA PREFERRED NUMBER SERIES WHEREVER POSSIBLE.

SPRAGUE TYPE NUMBER. IDENTIFIES THE BASIC CAPACITOR DESIGN.

Courtesy of Sprague Electric Co.

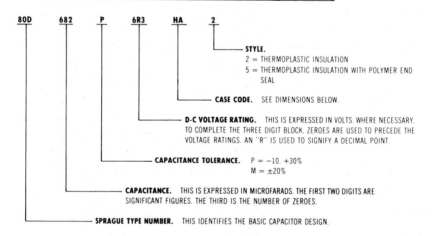

| 80D | 682 | P | 6R3 | HA | 2 |

STYLE.
2 = THERMOPLASTIC INSULATION
5 = THERMOPLASTIC INSULATION WITH POLYMER END SEAL

CASE CODE. SEE DIMENSIONS BELOW.

D-C VOLTAGE RATING. THIS IS EXPRESSED IN VOLTS. WHERE NECESSARY, TO COMPLETE THE THREE DIGIT BLOCK, ZEROES ARE USED TO PRECEDE THE VOLTAGE RATINGS. AN "R" IS USED TO SIGNIFY A DECIMAL POINT.

CAPACITANCE TOLERANCE.
P = −10, +30%
M = ±20%

CAPACITANCE. THIS IS EXPRESSED IN MICROFARADS. THE FIRST TWO DIGITS ARE SIGNIFICANT FIGURES. THE THIRD IS THE NUMBER OF ZEROES.

SPRAGUE TYPE NUMBER. THIS IDENTIFIES THE BASIC CAPACITOR DESIGN.

Courtesy of Sprague Electric Co.

E
Review of Trigonometric Functions

When two sides, or one side and the adjacent angle, of a right triangle are known, the other sides and angles can be determined with the aid of trigonometry. The sides and angles of a right triangle are related as indicated in the following list (refer to Figure E–1):

1. Ratio a/c is called the sine of the angle θ and is written as sin.
2. b/c is called the cosine of the angle θ and is written as cos.
3. a/b is called the tangent of the angle θ and is written as tan.

If, in Figure E–1, side $a = 5$ and $c = 10$, then sin of $\theta = 5/10 = 0.500$. Scanning for this number in a table of trigonometric functions, we find that it equals 30.0°. The cosine of this angle is $8.66/10 = 0.8660$, or 30.0°.

Suppose that a certain right triangle has side $a = 3.17$ and side $b = 5.58$. What is the tangent of the angle? We have tan $\theta = 3.17/5.58 = 0.5681$. What angle equals this tangent function? Searching the vertical columns for 0.5681, we find that the angle is 29.6°. Customarily, we would write this relationship as arctan $0.5681 = 29.6°$, but it can also be written as $\tan^{-1} 0.5681 = 29.6°$. The latter method is the one used in this text. In like manner, we can write arcsin or $\sin^{-1} 0.5000 = 30°$ and arccos or $\cos^{-1} 0.8660 = 30°$.

Figure E–1

Right triangle used to illustrate trigonometric functions

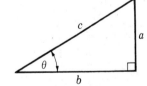

Selected Answers

Chapter 1

1. $-2\,C$
3. $0.00299\,C$ or $+3\,mC$
5. $3.5\,C$
7. $8\,s$

Chapter 2

1. $3900\,\Omega \pm 10\%$
3. $4.7\,\Omega \pm 20\%$
5. $0.0025\,S$
7. $66.67\,\Omega$
9. 455
11.

	Rated Value (Ω)	Tolerance (%)
1	22,000	5
2	470	10
3	100	20
4	56	5
5	15	2
6	3,900,000	10
7	6,800	5
8	2,700	5

13. (a) $4.7\,\Omega \pm 2\%$ (b) $2.2\,\Omega \pm 5\%$
    (c) $0.15\,\Omega \pm 5\%$ (d) $1.0\,\Omega \pm 1\%$

15.

	Resistance (Ω)	Conductance
1	47,000	21.277 μS
2	2,200,000	0.455 μS
3	33	30.3 mS
4	270	3.7 mS
5	4.7	212.7 mS
6	39,000	25.6 μS
7	100,000	10 μS
8	680	1.47 mS
9	10,000	100 μS
10	100	10 mS

17. 0.219

Chapter 3

1. $60\,V$
3. $15\,V$
5. $2\,A$
7. $0.35\,A$
9. (a) $13.5\,V$ (b) $8.75\,V$
11. (a) $0.4\,A$ (b) $0.21\,A$
13. $16.81\,V$

Chapter 4

1. $6\,A$
3. $6\,V$
5. $11.14\,\Omega$
7. $0.0125\,A$
9. $0.125\,\Omega$
11. $0.566\,A$
13. $342.86\,\Omega$
15. $12\,V$
17. $0.4\,A$
19. (a) $25{,}000\,\mu A$ (b) $25\,mA$
21. (a) $0.0004\,A$ (b) $0.4\,mA$
23. (a) $0.525\,mV$ (b) $0.000525\,V$
25. (a) $0.47\,k\Omega$ (b) $0.000\,47\,M\Omega$

27. (a) 0.215 A   (b) 215×10^3 μA
29. (a) 5.125 mA   (b) 0.005125 A
31. (a) 0.033 MΩ (b) 33,000 Ω
33. (a) 68×10^6 μS
    (b) 680×10^3 mS

Chapter 5

1. 0.78 A
3. 0.278 A
5. 500 Ω
7. 1.19 W
9. 7.07 V
11. 280 Ω
13. 2.2 V
15. 3.125 W
17. 8.214 kW
19. 102.273 Ω
21. (a) 40 V   (b) 1810 Ω
    (c) 0.022 A   (d) 5.967 V
    (e) 8.619 V   (f) 10.387 V
    (g) 15.028 V
23. $E_T = 30$ V; $R_T = 329$ Ω;
    $I_T = 91.18$ mA; $P_T = 2.735$ W;
    $V_1 = 4.28$ V; $V_2 = 3.0$ V;
    $V_3 = 6.2$ V; $V_4 = 10.94$ V;
    $V_5 = 3.55$ V; $V_6 = 2.0$ V;
    $P_1 = 0.39$ W; $P_2 = 0.273$ W;
    $P_3 = 0.565$ W; $P_4 = 0.997$ W;
    $P_5 = 0.323$ W; $P_6 = 0.182$ W

Chapter 6

1. 179.5 Ω
3. 18.02 Ω
5. (a) 9 W   (b) 27 W
7. 63.25 V
9. 109.09 Ω
11. (a) 7.61 A   (b) 4.39 A
13. 18.8 mA
15. 14.59 Ω
17. (a) 36.7 Ω (b) 27.2 mS
19. $R_3 = 6$ kΩ; $R_T = 1465$ Ω;
    $I_T = 16.4$ mA
21. increase to infinity (theoretically)
23. $R_T = 226$ Ω; $P_1 = 0.9$ W, $P_2 = 0.9$
    W, $P_3 = 0.2727$ W, $P_4 = 1.915$
    W, $P_T = 3.988$ W

25. $R_T = 16.24$ kΩ; $I_T = 1.48$ mA;
    $I_2 = 0.727$ mA

Chapter 7

1. (a) 6.076 V   (b) 5.924 V
3. (a) 0.632 A   (b) 0.253 A
    (c) 0.126 A
5. 0.774 A
7. (a) 1 A   (b) 0.67 A
9. (a) 7.483 V   (b) 3.495 mA
11. $I_B = 15$ μA; $V_{R_B} = 14.38$ V;
    $R_B = 0.959$ MΩ
13. 160.84 V
15. 610.5 Ω
17. $V_1 = 7.16$ V, $V_3 = 2.0$ V,
    $V_6 = 10.46$ V; $I_2 = 6$ mA,
    $I_4 = 15.24$ mA
19. $V_5 = 5.45$ V
21. $V_5 = 11.42$ V; $I_7 = 11.42$ mA
23. $V_{AB} = 17.9$ V; terminal A is +

Chapter 8

1. 0.69 A
3. 1.07 A
5. 1.58 A
7. 1.3 A
9. 0.12 A
11. $I_{R1} = 0.95$ A, $I_{R2} = 0.568$ A,
    $I_{R3} = 1.518$ A
13. $22.8 - 6.816 \cong 16$ V
15. $V_{R_1} = 12.72$ V, $V_{R_2} = 11.28$ V,
    $V_{R_3} = 6.56$ V
17. $V_{RL} = 9.6$ V; $I_{RL} = 0.96$ A
19. $I_{RL} = 0.182$ A; $V_{RL} = 4.36$ V
21. $I_{RL} = 0.035$ A; $V_{RL} = 1.765$ V
23. 0.77 A
25. (a) 87.15 A   (b) 1.12 A
27. 38.14 V
29. (a) $R_{\text{TH}} = 60$ Ω; $V_{\text{TH}} = 10$ V
    (b) $R_{\text{TH}} = 60$ Ω; $V_{\text{TH}} = 15$ V
    (c) $R_{\text{TH}} = 70$ Ω; $V_{\text{TH}} = 5$ V
    (d) $R_{\text{TH}} = 40$ Ω; $V_{\text{TH}} = 20$ V
31. (a) $R_N = 20$ Ω; $I_N = 0.5$ A
    (b) $R_N = 40$ Ω; $I_N = 0.57$ A
    (c) $R_N = 25$ Ω; $I_N = 0.5$ A
    (d) $R_N = 45$ Ω; $I_N = 0.5$ A

Chapter 9
1. 1024
3. 25.3 mils
5. 2.43 Ω
7. 107.29 ft
9. 0.566 Ω
11. 0.4760 Ω
13. 657 ft.
15. 0.567 Ω
17. (a) 97.6 V   (b) 105 V
19. Use a longer length of nichrome wire. For example, if 12 ft was used instead of 6 ft, the resistance would be $3.97 \times 2 = 7.94$ Ω, which would permit approximately 14 A.

Chapter 10
1.
3. (d)
5. (a)
7. (b)
9. (d)
11. (c)
13. 1.6 mA
15. (a)
17. A, 1.6 V; B, 600 kΩ; C, 0.05 V; D, 0.21 V; E, 90 kΩ; A, 2 Ω;·B, 12 V; C, 1 kΩ; D, 660 V
19. (a) 2 kΩ   (b) 105.26 Ω
    (c) 10.05 Ω
21. 450 μA
23. (c), (i), (h), (d), (a), (e), (b), (g), (f)
25. graticule
27. electron gun, deflection plates, fluorescent screen
29. (b)
31. 277.778 Hz
33. amplitude, period, shape
35. 8 mV_{pp}
37. 3.2 ms; 312.5 Hz
39. 12 μs; 83.33 kHz

Chapter 11
1. 1.76×10^{-3} J
3. 1.5×10^{-3}

5. 2300 μF
7. 360
9. 63.2 pF
11. 292.4
13. 197.55 pF
15. 614.2 nF
17. 0.0016 μF
19. $V_{C_1} = 1.25 \times 10^2$ V, $V_{C_2} = 7.5 \times 10^1$ V
21. 10,000 pF ± 5%
23. 39,000 pF ± 2%

Chapter 12
1. 12 A-t
3. 427 μWb
5. 31.6×10^{-6} Wb
7. 1.01×10^{-2} T
9. 2.743×10^{-3} Wb
11. 84.35 A-t/m
13. 3.746×10^3
15. 2.63×10^1
17. 4.23×10^{-7}

Chapter 13
1. 833.3 V
3. 236.8
5. 2.819 H
7. 2.8 A/s
9. 47.25 mH
11. 313.5
13. 3.16 H
15. 0.367
17. 4.3 H
19. 0.0434 H
21. 290 μH
23. 0.21
25. 2.13 A
27. 0.02 V
29. 17.926 μH

Chapter 14
1. 183.6 μs
3. 2.2 μF
5. 10.5 H
7. $1\tau \cong 29.79$ V , $2\tau \cong 29.46$ V

$3\tau \cong 28.52$ V, $4\tau \cong 25.95$ V
$5\tau \cong 18.9\,6$ V

9. approx. 1.8τ
11. 0.152 s
13. 166 V
15. 0.0403 A
17. 9.533 V
19. approx. 55 V
21. 12.32 s

Chapter 15

1. (a) 0.5236 (b) 0.7854
   (c) 1.5708
3. 188.4
5. 93.46 ns
7. 121.9 Hz
9. 4.918×10^2 m
11. (a) 20.174 V (b) 61.479 V
    (c) 36.564 V
13. 349 V
15. 54.15 V
17. 45.53 mA
19. 2.192 A
21. 110.65 mV
23. (a) 1.90 mW (b) 0.95 mW
25. 10.823 V
27. (a) 60 Hz (b) 120 Hz
    (c) 50 kHz (d) 3.33 MHz
    (e) 101 MHz
    (a) 5×10^6 (b) 545.45
    (c) 11.11
    (d) 3.058 (e) 8.57×10^{-2}
31. (a) 247.45 V (b) 118.776 V
    (c) 127.26 V
    (d) 21.58 mV (e) 183.82 μV

Chapter 16

1. (a) 753.98 Ω (b) 1507.96 Ω
   (c) 12.566 kΩ
3. 199 μH
5. 6.369×10^5 Hz
7. 1.062 mH
9. (a) 1.106×10^3 Ω
   (b) 4.423×10^2 Ω
11. 2.632×10^{-6} μF
13. (a) 500 Ω (b) 100 Ω

15. (a)   (b) 1.2 kΩ

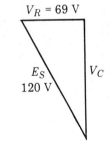

(c) 75 mA
(d) $V_1 = 15$ V; $V_2 = 30$ V;
    $V_3 = 45$ V
17. 1.59 μF
19. (a) 1.445×10^5 Hz
    (b) 4.676×10^2 Hz
    (c) 7.95 Hz
21. 7.667 kHz
23. 261 pF

Chapter 17

1. 45 V
3. 21.4 V
5. 54.9°

V_R = 69 V

E_S
120 V

V_C

7. 142.64 Ω
9.

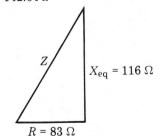

Z

X_{eq} = 116 Ω

R = 83 Ω
142.64 $\angle$ 54.4° Ω
11. (a) 0.281 A (b) 37.385 V
13. (a) 1.83 kΩ (b) 56.3° leading
15. 10.6 kΩ; 11.05° leading

17. 3.6 nF

19. 30.25 mW

21. $69 + j\,16\,\Omega$

23. $500\,\Omega; +j\,10\,\Omega$

25. 0.707, or 70.7%

27. 600 W; 498 W; 334.66 W

29. $447 + j\,400\,\Omega; 600\;\angle\,41.82°\,\Omega$

Chapter 18

1. $33\;\angle\,25.21°\,A$

3. $531 + j\,122\,\Omega$

5. $49.04\;\angle\,16.59°\,mA$

7. (a) $2.84\;\angle\,81°\,\Omega$
   (b) $4.55\;\angle\,-47°\,\Omega$

9. (a) $8.64 - j\,13.46\,V$
   (b) $469.95 + j\,356.71\,V$
   (c) $80.42 + j\,27.53\,V$

11. $51.18\;\angle\,60.79°\,\Omega$

13. (a) $72.62 + j\,48.26\,\Omega$
    (b) $87.19\;\angle\,33.61°\,\Omega$

15. $50.57\;\angle\,33.8°\,\Omega$

17. $102.2\;\angle\,188.58°\,\Omega$

19. $R = 145.18\,\Omega$ in series with
    $C = 0.78\,\mu F$

21. $Y = 0.00512\;\angle\,12.41°\,S$

23. (a) $59.2\;\angle\,-7.1°\,\Omega$
    (b) $0.84\;\angle\,7.1°\,A$

25. $11.9\;\angle\,29.82°\,V$

27. $39\;\angle\,5.69°\,\Omega; 38.8 = j\,3.87\,\Omega$

29. See Figure A–1.

Chapter 19

1. $1.684 \times 10^6\,Hz$

3. 49.5 pF

5. $1.71 \times 10^{-4}\,H$

7. $4.21\,\Omega$

9. (a) $90\,k\Omega$ (b) $8.88 \times 10^{-7}\,A$
   (c) $0°$

11. (a) $25\,mV$ (b) $3.75\,V$

13. 14.46–21.78 pF

15. 2.474 MHz

17. 52.08 kHz

19. $f_1 = 995\,kHz; f_2 = 1005\,kHz$

21. $L = 10.61\,\mu H; C = 10.61\,pF$

23. (a) $Q = 60$ (b) $Z_T = 8\,\Omega; I_T = 3\,A$

25. (a) $C = 244.4\,pF$ (b) $Z = 68.16\,k\Omega$

27. 21.22 kHz

29. $f_{\text{low}} = 9.93\,MHz; f_{\text{high}} = 10.067$
    MHz

Chapter 20

1. $7.96\,\mu F$

3. 36.2

5. (a) 22.387:1 (b) 0.159 V

7. $0.39\;\angle\,87.78°\,V$

9. $1.306\;\angle\,-14.93°\,V$

11. 0.579 mV

13. $k \cong 1 \times 10^8; k \cong 1 \times 10^8$

15. $V_{RL} = 0.80\;\angle\,122.4°\,V$

17. $25\,\mu V$

19. (a) $V_o = 4\,V$ (b) $dB = 46$
    (c) $dB = 46$

21. $L = 238.7\,mH; dB = -15$

23. $f = 1.59\,kHz; f_{co} = 796\,Hz$

25. $80.64°$

Chapter 21

1. 480 V

3. 19.05: 1, step-down

5. 0.417 A

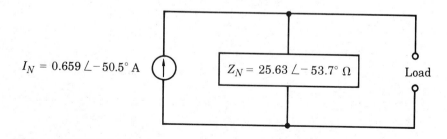

$I_N = 0.659\;\angle\,-50.5°\,A$ $Z_N = 25.63\;\angle\,-53.7°\,\Omega$ Load

Figure A–1

7. 0.083 A

9. 5.56 A

11. (a) 84.38 W   (b) 112.5 V
    (c) 0.938:1, step-down
    (d) 328.13

13. (a) $220 \angle 0° $ V
    (b) $3.38 \angle 32°$ A
    (c) $1.69 \angle 32°$A   (d) 632.93 W

15. 72 Ω

17. 4 kΩ

19. 2057.14 Ω

21. 2 A

23. 17.99 V

25. (a) 6.5 W   (b) 0.054 A
    (c) 12:1, step-down

27. (a) 60, 45, 15   (b) 150 V

29. (a) 1:4, step-up
    (b) $R_L = 17.14 \angle -8°$ Ω

(c) $I_p = 6.4 \angle 8°$ A

(d) $P_T = 38.01$ W

Chapter 22

1. $R_D = 0.225$ kΩ

3. $R_D = 21.7$ Ω

5. $r_{ac} = 4.69$ Ω

7. $R_D = 44.117$ MΩ

9. $R_s = 1.8$ kΩ, $I_{RL} = 7.5$ mA,
   $I_z = 3.333$ mA

11. $E_{avg} = 12.46$ V

13. $E_{avg} = 21.6$ V

15. $223.4 = V_p$

17. $R_s = 51.4$ Ω, $R_1 = 677$ Ω,
    $R_{Bldr} = 1.509$ kΩ

19. $R_s = 409$ Ω

21. % regulation = 0.63

23. $E_f = 115.94$ V

Index